Lecture Notes in Computer Science 6388

Commenced Publication in 1973
Founding and Former Series Editors:
Gerhard Goos, Juris Hartmanis, and Jan van Leeuwen

Devrim Ünay   Zehra Çataltepe
Selim Aksoy (Eds.)

Recognizing Patterns in Signals, Speech, Images, and Videos

ICPR 2010 Contests
Istanbul, Turkey, August 23-26, 2010
Contest Reports

 Springer

Volume Editors

Devrim Ünay
Bahçeşehir University
Electrical and Electronics Engineering Department
Çırağan Caddesi, Beşiktaş, 34353 Istanbul, Turkey
E-mail: devrim.unay@bahcesehir.edu.tr

Zehra Çataltepe
Istanbul Technical University
Electrical and Electronics Engineering Department
Maslak, Sariyer, 34469 Istanbul, Turkey
E-mail: cataltepe@itu.edu.tr

Selim Aksoy
Bilkent University
Department of Computer Engineering
06800 Ankara, Turkey
E-mail: saksoy@cs.bilkent.edu.tr

Library of Congress Control Number: 2010940503

CR Subject Classification (1998): I.5, I.2.10, I.2, H.3, I.4, J.3

LNCS Sublibrary: SL 6 – Image Processing, Computer Vision, Pattern Recognition, and Graphics

ISSN	0302-9743
ISBN-10	3-642-17710-7 Springer Berlin Heidelberg New York
ISBN-13	978-3-642-17710-1 Springer Berlin Heidelberg New York

springer.com

© Springer-Verlag Berlin Heidelberg 2010
Printed in Germany

Typesetting: Camera-ready by author, data conversion by Scientific Publishing Services, Chennai, India
Printed on acid-free paper   06/3180

Preface

The 20th ICPR (International Conference on Pattern Recognition) Conference took place in Istanbul, Turkey, during August 23–26, 2010. For the first time in the ICPR history, several scientific contests (http://www.icpr2010.org/contests.php) were organized in parallel to the conference main tracks. The purpose of these contests was to provide a setting where participants would have the opportunity to evaluate their algorithms using publicly available datasets and standard performance assessment methodologies, disseminate their results, and discuss technical topics in an atmosphere that fosters active exchange of ideas. Members from all segments of the pattern recognition community were invited to submit contest proposals for review.

In response to the Call-for-Papers, 21 substantial proposals were submitted, and the review process was highly selective leading to a striking acceptance rate of 38%. Accordingly, there were eight scientific contests organized under ICPR 2010, as listed below:

1. BiHTR: Bi-modal Handwritten Text Recognition
2. CAMCOM 2010: Verification of Video Source Camera Competition
3. CDC: Classifier Domains of Competence
4. GEPR: Graph Embedding for Pattern Recognition
5. ImageCLEF@ICPR: The CLEF Cross Language Image Retrieval Track
6. MOBIO: Mobile Biometry Face and Speaker Verification Evaluation
7. PR in HIMA: Pattern Recognition in Histopathological Images
8. SDHA 2010: Semantic Description of Human Activities

This volume includes the selected papers presented at the above-listed ICPR 2010 Contests. Each paper underwent a meticulous revision process guided by the referees listed in these proceedings.

We thank all contest organizers and the external referees for their excellent work, especially given the demanding time constraints. Without the willingness, strength, and organizational skills of the organizers, the ICPR 2010 Contests would not have been a spectacular success. Furthermore, we thank the Conference Chair Aytül Erçil and all the Organizing Committee members. It has been a wonderful experience to work with all of them.

Finally we thank all local organizers, including Rahmi Fıçıcı, Gülbin Akgün, members of TeamCon, and the volunteers, who worked tirelessly to create and maintain the main website of the contests and to pull off the logistical arrangements of the contests. It was their hard work that made the ICPR 2010 Contests possible and enjoyable.

August 2010

Devrim Ünay
Zehra Çataltepe
Selim Aksoy

Organization

The ICPR 2010 Contests were organized as part of the 20th International Conference on Pattern Recognition (ICPR), which was held during August 23–26, 2010 in Istanbul, Turkey.

Contest Chairs

Devrim Ünay   Bahcesehir University, Istanbul, Turkey
Zehra Çataltepe   Istanbul Technical University, Istanbul, Turkey
Selim Aksoy   Bilkent University, Ankara, Turkey

Contest Organizers

BiHTR
Moisés Pastor   Universidad Politecnica de Valencia, Spain
Roberto Paredes   Universidad Politecnica de Valencia, Spain
Thomas Deselaers   ETH Zurich, Switzerland
Luisa Mico   Universidad Politecnica de Valencia, Spain
Enrique Vidal   Universidad Politecnica de Valencia, Spain

CAMCOM 2010
Wiger van Houten   Netherlands Forensic Institute,
                           The Netherlands
Zeno J.M.H. Geradts   Netherlands Forensic Institute,
                           The Netherlands
Katrin Y. Franke   Gjøvik University College, Norway
Cor J. Veenman   Netherlands Forensic Institute,
                           The Netherlands

CDC
Tin Kam Ho   Bell Laboratories Alcatel-Lucent, USA
Núria Macià   Universitat Ramon Llull, Spain

GEPR
Mario Vento   University of Salerno, Italy
Pasquale Foggia   University of Salerno, Italy

ImageCLEF@ICPR

Barbara Caputo	IDIAP Research Institute, Switzerland
Stefanie Nowak	Fraunhofer IDMT, Germany
Jana Kludas	University of Geneva, Switzerland
Andrzej Pronobis	Royal Institute of Technology, Sweden
Henning Müller	University of Geneva, Switzerland
Jayashree Kalpathy-Cramer	Oregon Health and Science University, USA

MOBIO

Sébastien Marcel	Idiap Research Institute, Switzerland
Chris McCool	Idiap Research Institute, Switzerland
Timo Ahonen	University of Oulu, Finland
Jan Černocky	Brno University of Technology, Czech Republic

PR in HIMA

Metin N. Gürcan	Ohio State University, USA
Nasir Rajpoot	University of Warwick, UK
Anant Madabhushi	Rutgers The State University of New Jersey, USA

SDHA 2010

Michael S. Ryoo	Electronics and Telecommunications Research Institute (ETRI), Korea
J.K. Aggarwal	University of Texas at Austin, USA
Amit K. Roy-Chowdhury	University of California Riverside, USA

Referees

T. Ahonen	M. Gurcan	S. Nowak
S. Aksoy	T.K. Ho	R. Paredes
B. Caputo	J. Kalpathy-Cramer	M. Pastor
Z. Çataltepe	J.T. Lee	N. Rajpoot
H. Cernocky	N. Macia	B. Tamersoy
C.-C. Chen	A. Madabhushi	D. Ünay
P. Foggia	S. Marcel	W. van Houten
K.Y. Franke	C. McCool	C.J. Veenman
Z.J.M.H. Geradts	H. Müller	M. Vento

Table of Contents

ImageCLEF@ICPR – Information Fusion Task

ImageCLEF@ICPR – Visual Concept Detection Task

ImageCLEF@ICPR – Robot Vision Task

MOBIO – Mobile Biometry Face and Speaker Verification Evaluation

PR in HIMA – Pattern Recognition in Histopathological Images

SDHA 2010 – Semantic Description of Human Activities

Bi-modal Handwritten Text Recognition (BiHTR) ICPR 2010 Contest Report

Moisés Pastor and Roberto Paredes

Pattern Recognition and Human Language Technologies group
Department of Information Systems and Computation,
Technical University of Valencia
Cami de Vera s/n. 46022 Valencia, Spain
{mpastorg,rparedes}@dsic.upv.es
http://www.icpr2010.org/contests.php

Abstract. Handwritten text is generally captured through two main modalities: *off-line* and *on-line*. Each modality has advantages and disadvantages, but it seems clear that smart approaches to handwritten text recognition (HTR) should make use of both modalities in order to take advantage of the positive aspects of each one. A particularly interesting case where the need of this bi-modal processing arises is when an off-line text, written by some writer, is considered along with the on-line modality of the same text written by another writer. This happens, for example, in computer-assisted transcription of old documents, where on-line text can be used to interactively correct errors made by a main off-line HTR system.

In order to develop adequate techniques to deal with this challenging bi-modal HTR recognition task, a suitable corpus is needed. We have collected such a corpus using data (word segments) from the publicly available off-line and on-line IAM data sets.

In order to provide the Community with an useful corpus to make easy tests, and to establish baseline performance figures, we have proposed this handwritten bi-modal contest.

Here is reported the results of the contest with two participants, one of them achieved a 0% classification error rate, whilst the other participant achieved an interesting 1.5%.

1   Introduction

Handwritten text is one of the most natural communication channels currently available to most human beings. Moreover, huge amounts of historical handwritten information exist in the form of (paper) manuscript documents or digital images of these documents.

When considering handwritten text communication nowadays, two main modalities can be used: *off-line* and *on-line*. The off-line modality (the only one possible for historical documents) consists of digital images of the considered text. The on-line modality, on the other hand, is useful when an immediate communication is needed. Typically, some sort of electronic pen is used which

D. Ünay, Z. Çataltepe, and S. Aksoy (Eds.): ICPR 2010, LNCS 6388, pp. 1–13, 2010.

provides a sequence of x-y coordinates of the trajectory described by the pen tip, along with some info about the distance of the pen to the board or paper and/or the pressure exerted while drawing the text.

The difficulty of handwritten text recognition (HTR) varies widely depending of the modality adopted. Thanks to the timing information embedded in on-line data, on-line HTR generally allows for much higher accuracy than off-line HTR.

Given an on-line sample, it is straightforward to obtain an off-line image with *identical* shape as that of the original sample. Such an image is often referred to as *"electronic ink"* (of *e-ink*). Of course, the e-ink image lacks the on-line timing information and it is therefore much harder to recognize than the original on-line sample. Conversely, trying to produce the on-line trajectory that a writer may have produced when writing a given text image, is an ill-defined problem for which no commonly accepted solutions exist nowadays.

Given an on-line text to be recognized, several authors have studied the possibility of using *both* the on-line trajectory *and* a corresponding off-line version of this trajectory (its e-ink). This *multi-modal* recognition process has been reported to yield some accuracy improvements over using only the original on-line data [13,5].

Similar ideas are behind the data collected in [12], referred to as the IRONOFF corpus. In this data set, on-line and off-line sample pairs were captured simultaneously. A real pen with real ink was used to write text on paper while the pen tip position was tracked by an on-line device. Then, the paper-written text was scanned, providing an off-line image. Therefore, as in the e-ink case, for each written text sample, both the on-line and off-line shapes are *identical*. However, this is quite different from the thing we propose in this work where on-line and off-line sample pairs can be produced in different times and by different writers.

Another, more interesting scenario where a more challenging bi-modal (on/off-line) fusion problem arises is Computer Assisted Transcription of Text Images, called "CATTI" in [10]. In this scenario, errors made by an off-line HTR system are immediately fixed by the user, thereby allowing the system to use the validated transcription as additional information to increase the accuracy of the following predictions. Recently, a version of CATTI (called multi-modal CATTI or MM-CATTI) has been developed where user corrective feedback is provided by means of on-line pen strokes or text written on a tablet or touch-screen [11]. Clearly, most of these corrective pen-strokes are in fact on-line text aimed to fix corresponding off-line words that have been miss-recognized by the off-line HTR system. This allows taking advantage of both modalities to improve the feedback decoding accuracy and the overall multi-modal interaction performance. In this scenario, on-line HTR can be much more accurate than in conventional situations, since we can make use of several information derived from the interaction process. So far, we have only focused on contextual info derived from the available transcription validated by the user. But, now, we are trying to improve accuracy even further by using info from the off-line image segments which are supposed to contain the very same text as the one entered on-line by the user as feedback.

As compared with the use of e-ink or IRONOFF-style data to improve the plain on-line HTR accuracy, in the MM-CATTI scenario the on-line and off-line text shapes (for the same word) may differ considerably. Typically, they are even written by different writers and the on-line text tends to be more accurately written in those parts (characters) where the off-line text image is blurred or otherwise degraded. This offers great opportunities for significant improvements by taking advantage of the best parts of each shape to produce the recognition hypothesis.

In order to ease the development of adequate techniques for such a challenging bi-modal HTR recognition task, a suitable corpus is needed. It must be simple enough so that experiments are easy to run and results are not affected by alien factors (such as language model estimation issues, etc.). On the other hand, it should still entail the essential challenges of the considered bi-modal fusion problem. Also, considering the MM-CATTI scenario where a word is corrected at a time, we decided to compile this bi-modal *isolated word* corpus.

This corpus have compiled using data (word segments) from the publicly available off-line and on-line IAM corpora [6,4].

In order to provide the Community with an useful corpus to make easy tests, and to establish *baseline* performance figures, we have proposed this handwritten bi-modal contest.

The rest of the article is organized as follows. The next section presents the "biMod-IAM-PRHLT" corpus along with some statistics of this data set. Contest results are reported in Section 3. Finally, some conclusions are presented in the last section.

2   The biMod-IAM-PRHLT Corpus

In order to test the above outlined bi-modal decoding approaches on the interactive multi-modal HTR framework introduced in section 1, an adequate bi-modal corpus is needed. This corpus should be simple, while still entailing the essential challenges of the considered bi-modal HTR problems. To this end, a simple classification task has been defined with a relatively large number of classes (about 500): Given a bi-modal (on/off-line) sample, the class-label (word) it corresponds to must be hypothesized. Following these general criteria, a corpus, called *"biMod-IAM-PRHLT"*, has been compiled.

Obviously, the chosen words constituting the set of class-labels ("vocabulary") are not equiprobable in the natural (English) language. However, in order to encourage experiments that explicitly focus on the essential multi-modal fusion problems, we have pretended uniform priors by setting standard *test* sets with identical number of samples of each word. Nevertheless, the number of *training* samples available per word is variable, approximately reflecting the prior word probabilities in natural English.

The samples of the biMod-IAM-PRHLT corpus are word-size segments from the publicly available off-line and on-line IAM corpora (called IAMDB) [6,4], which contain handwritten sentences copied from the electronic-text LOB corpus [9]. The off-line word images were semiautomatically segmented at the IAM (FKI) from the original page- and line-level images. These word-level images are included in the off-line IAMDB. On the other hand, the on-line IAMDB was only available at the line-segment level. Therefore we have segmented and extracted the adequate word-level on-line samples ourselves, as discussed in section 2.1.

In order to select the (approximately 500) vocabulary words, several criteria were taken into account. First, only the words available in the word-segmented part of the off-line IAMDB were considered. From these words, only those which appear in at least one (line) sample of the on-line IAMDB were selected. To keep data management simple, all the words whose UTF-8 representation contained diacritics, punctuation marks, etc., were discarded (therefore all the remaining class-labels or words are plain ASCII strings). Finally, to provide a minimum amount of training (and test) data per class, only those words having at least 5 samples in each of the on-line and off-line modalities were retained. This yielded a vocabulary of 519 words, with approximately 10k on-line and 15k off-line word samples, which were further submitted to the data checking and cleaning procedures.

The resulting corpus is publicly available for academic research from the data repository of the PRHLT group(http://prhlt.iti.upv.es/iamdb-prhlt.html).

It is partitioned into training and test sub-corpora for benchmarking purposes. In addition to the test data included in the current version (referred to as *validation*), more on-line and off-line word-segmented test samples have been produced, but they are currently held-out so they will be used for benchmarking purposes. Basic results on this test set, referred to as *hidden test*, will be reported in this paper in order to establish the homogeneity of the held-out data with respect to the currently available (*validation*) data.

Figure 1 shows some examples of on/of-line word pairs contained in this corpus and some statistics will be shown later in section 2.2.

The correctness of all the off-line word segments included in the on-line IAMDB were manually checked at the IAM (FKI). For the on-line data, only the test samples have been checked manually; the quality of samples in the the much larger training set have been checked semiautomatically, as discussed in section 2.1.

2.1   On-Line Word Segmentation and Checking

As previously commented, the *off-line IAM Handwriting database* already contained data adequately segmented at the word level. The word segmentation of the *IAM On-Line Handwriting corpus* has been carried out semiautomatically.

Morphological HMMs were estimated from the whole on-line corpus. Then, for each text line image, a "canonical" language model was built which accounted only for the sequence of words that appear in the transcription of this line. Finally, the line image was submitted for decoding using the Viterbi algorithm. As a byproduct of this *"forced recognition"* process, a most probable horizontal segmentation of the line image into its constituent word-segments was obtained.

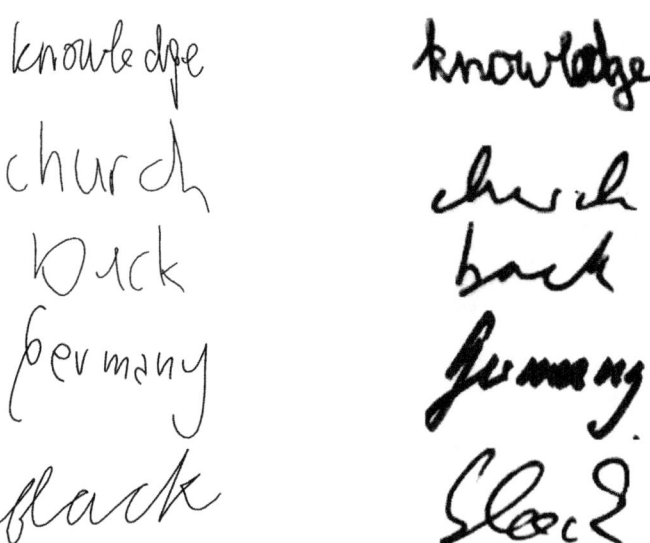

Fig. 1. Examples of (corresponding) off-line (left) and on-line (right) handwritten words

The correctness of this segmentation has been fully checked by hand only for the *validation* and *test* data. For the much larger *training* set, the quality has been checked semiautomatically. By random sampling, the number of segmentation errors of this on-line training partition was initially estimated at about 10%, but most of these errors have probably been fixed by the following procedure. An complete HTR system was trained using the just segmented on-line training word samples; then, the same samples were submitted to recognition by the trained system. The observed errors were considered as candidates to have word segmentation errors. These errors were manually checked and those samples which were found to be incorrectly segmented were either discarded or manually corrected. In the end, the amount of segmentation errors detected in this process was about 10%. Therefore, while the exact degree of correctness of the on-line training data is unknown, it can confidently expected to be close to 100%.

2.2   Corpus Statistics

Main figures of the *biMod-IAM-PRHLT* corpus are shown in Table 1. Other more detailed statistics are shown in the figures 2–4 below.

Figure 2 shows the amount of on-line and off-line samples available for each word class, in decreasing order of number of samples ("rank"). As previously mentioned, these counts approximately follow the real frequency of the selected vocabulary words in natural English.

Table 1. Basic statistics of the BIMOD-IAM-PRHLT corpus and their standard partitions. The *hidden test* set is currently held-out and Will be released for benchmarking in the future.

	Modality sub-corpus	
	on-line	off-line
Word classes (vocabulary)	519	519
Running words:		
training	8 342	14 409
validation	519	519
hidden test	519	519
total	9 380	15 447

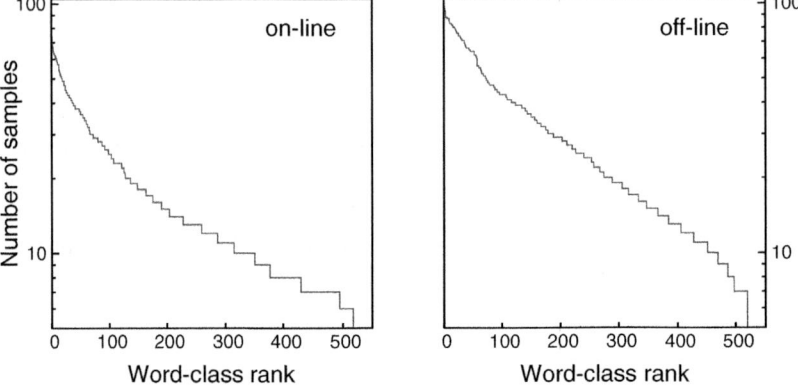

Fig. 2. Available samples per word class. The vertical axis shows the number of samples for the word classes shown in the horizontal axis in rank order.

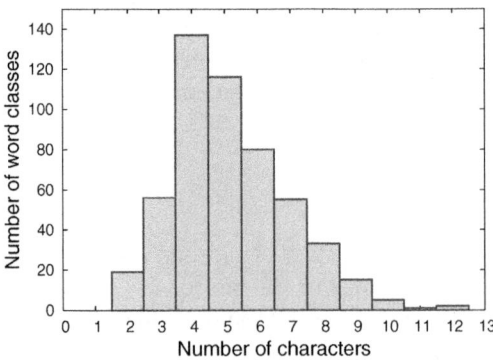

Fig. 3. Word length histograms. Number of word classes for each class-label length.

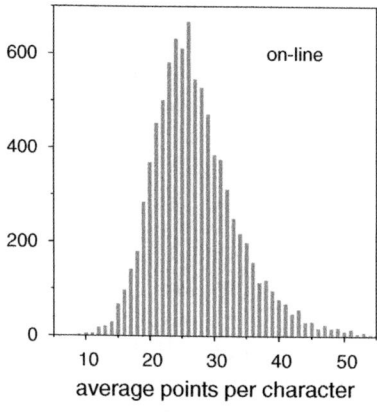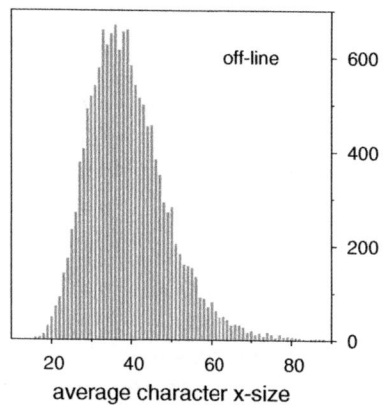

Fig. 4. Character size histograms. The vertical axis shows the number of word samples for each average character size.

Figure 3 shows the class-label (word) length distribution; By construction, all the words have at least two characters and the observed most frequent word length is 4 characters.

Finally, the histograms of figure 4 show the distribution of average sizes of characters in the the on-line trajectories and off-line images. Let s be a sample and N the number of characters of the word label of s. For an on-line sample, the average character size is measured as the number of points in the pen trajectory divided by N. For an off-line sample, on the other hand, the average character size is measured as the horizontal size (number of pixels) of the image divided N. Average character sizes are somewhat more variable in the off-line data. If character-based morphological HMMs are used as word models, these histograms can be useful to establish adequate initial HMM topologies for these models.

3 Participants

Three teams were participated at the present contest. The teams, the algorithms used, and the results obtained, will be explained in the next subsections.

Previous baseline results will be published by organizers of this contest in the proceedings of the ICPR 2010 [7]. In this work, basic experiments were carried out to establish baseline accuracy figures for the biMod-IAM-PRHLT corpus. In these experiments, fairly standard preprocessing and feature extraction procedures and character-based HMM word models have been used. The best results obtained for the validation partition were 27.6% off-line and 6.6% on-line classification error. Of course, no results were provided for the hidden test used in this contest. A weighted-log version of naive Bayes, assuming uniform priors, were used to balance the relative reliability of the on-line and off-line models. The best bi-modal classification score obtained was 4.0%.

3.1   UPV-April Team

This entry was submitted by the team composed of María José Castro-Bleda[1], Salvador España-Boquera[2], and Jorge Gorbe-Moya[3], from the *Departamento de Sistemas Informáticos y Computación* of the *Universidad Politécnica de Valencia*, (Spain), and Francisco Zamora-Martínez[4] from the *Departamento de Ciencias Físicas, Matemáticas y de la Computación, Universidad CEU-Cardenal Herrera* (Spain).

Off-line. Their off-line recognition system is based on hybrid HMM/ANN models, as fully described in [2]. Hidden Markov models (HMM) with a left-to-right without skips topology have been used to model the graphemes, and a single multilayer perceptron (MLP) is used to estimate all the HMM emission probabilities. Several preprocessing steps are applied to the input word image: slope and slant correction, and size normalization [3]. The best result obtained by the team was 12.7% classification error for validation and 12.7% for the hidden test.

On-line. Their on-line classifier is also based on hybrid HMM/ANN models using also a left-to-right without skips HMM topology. The preprocessing stage comprises uniform slope and slant correction, size normalization, and resampling and smoothing of the sequence of points. In order to detect the text baselines for these preprocessing steps, each sample is converted to an image which goes through the same process as the off-line images. Then, a set of 8 features is extracted for every point from the sequence: y coordinate, first and second derivative of the position, curvature, velocity, and a boolean value which marks the end of each stroke. The best result obtained by the team was 2.9% classification error for validation and 3.7% for the hidden test.

Combination of Off-line and On-line Recognizers. The scores of the 100 most probable word hypothesis are generated for the off-line sample. The same process is applied to the on-line sample. The final score for each sample is computed from these lists by means of a log-linear combination of the scores computed by both the off-line and on-line HMM/ANN classifiers. The best result obtained by the team was 1.9% classification error for validation and 1.5% for the hidden test (see table 3.4 for a summary of the results).

3.2   PRHLT Team

This submission is by Enrique Vidal[5], Francisco Casacuberta[6] and Alejandro H. Tosselli[7] from the PRHLT group (http://prhlt.iti.es) at the Instituto Tecnológico de Informática of the Technical University of Valencia (Spain).

---

[1] mcastro@dsic.upv.es
[2] sespana@dsic.upv.es
[3] jgorbe@dsic.upv.es
[4] fzamora@dsic.upv.es
[5] evidal@dsic.upv.es
[6] fcn@dsic.upv.es
[7] ahector@dsic.upv.es

Off-line. Their off-line recognition system is based on the classical HMM-Viterbi speech technology. Left-to-right, continuous density, hidden Markov models without skips have been used to model the graphemes. The HMM models were trained from the standard Off-line training data of the contest (i.e., no additional data from other sources were used), by means of the HTK toolkit [14].

Several standard preprocessing steps are applied to the input word image, which consists on median filter noise removal, slope and slant correction, and size normalization. The feature extraction process transforms a preprocessed text line image into a sequence of 60-dimensional feature vectors, each vector representing grey-level and gradient values of an image column or "frame" [1]. In addition, each of these vectors was extended by stacking 4 frames from its left context and 4 form its right context. Finally, Principal Component Analysis (PCA) was used to reduce these 180-dimensional vectors to 20 dimensions.

The best Off-line classification error rates obtained by the team were 18.9% for validation set and 18.9% for the hidden test set.

On-line. Their on-line classifier is also based on continuous density HMM models using also a left-to-right HMM topology without skips. As in the Off-line case, the HMM models were trained from the standard On-line training data of the contest, also without any additional data from other sources.

Each the e-pen trajectory of each sample was processed through only three simple steps: pen-up points elimination, repeated points elimination, and noise reduction (by simple low pass filtering). Each preprocessed trajectory was transformed into a new temporal sequence of 6-dimensional real-valued feature vectors [8], composed of normalized vertical position, normalized first and second time *derivatives* and curvature.

The best results obtained by the team were 4.8% classification error for the validation set and 5.2% for the hidden test set.

Combination of Off-line and On-line Recognizers. To obtain bi-modal results, the team used a weighted-log version of naive Bayes classifier (assuming uniform class priors):

$$\hat{c} = \operatorname*{argmax}_{1 \leq c \leq C} P(c \mid x, y) = \operatorname*{argmax}_{1 \leq c \leq C} \log P(x, y \mid c) \qquad (1)$$

where

$$\log P(x, y \mid c) = (1 - \alpha) \cdot \log P(x \mid c) + \alpha \cdot \log P(y \mid c) \qquad (2)$$

The weight factor α aims at balancing the relative reliability of the on-line (x) and off-line (y) models. To perform these experiments, all the log-probability values were previously shifted so that both modalities have a fixed, identical maximum log-probability value. Then, to reduce the impact of low, noisy probabilities, only the union of the K-best hypothesis of each modality are considered in the argmax of (1). Therefore, the accuracy of this classifiers depends on two parameters, α and K. Since these dependencies are not quite smooth, it is not straightforward to optimize these parameters only from results obtained on the validation data. Therefore, a Bayesian approach has been followed to smooth these dependencies. Namely,

$$P(x, y \mid c) \sim \sum_k P(K) \int P(x, y \mid c, K, \alpha) P(\alpha) \, d(\alpha) \qquad (3)$$

To simplify matters, the parameter prior distributions $P(K)$ and $P(\alpha)$ were assumed to be *uniform* in suitably wide intervals ($4 \le K \le 20$ and $0.3 \le \alpha \le 0.45$), empirically determined from results on the validation set. This allows to easily compute (3) by trivial numerical integration on α.

With this approach, the validation set error rate was 1.9%, while for the hidden test set, a 1.3% error rate was achieved.

3.3   GRFIA Team

This submission is by Jose Oncina[8] from the GRFIA group (http://grfia.dlsi.ua.es) of the University of Alicante.

Off-line and On-line. This group used the On-line and Off-line HMM decoding outputs (i.e., $P(x \mid c), P(y \mid c), 1 \le c \le C$) provided by the PRHLT group, therefore identical On-line and Off-line results were obtained.

Combination of Off-line and On-line Recognizers. Similarly, this group used the PRHLT classifier given by Eq. (3). However, in this case a rather unconventional classification strategy was adopted, which finally led to the best results for this contest.

The idea was to capitalize on the contest specification which established that a strictly uniform distribution of classes had been (rather artificially) set for both the validation and the test sets. More specifically, as stated in section 2, both the validation and the test sets do have "identical number of samples of each word (class)". Since each set has 519 samples and there are 519 classes, this implies that each set has exactly one sample per class. This restriction can be exploited by considering the classification of each (validation or test) set as a whole and avoiding the classifier to yield the same class-label for two different samples. While optimally solving this problem does entail combinatory computational complexities, a quite effective greedy strategy was easily implemented as follows.

First, the best score for each sample ($\max_c \log P(x, y \mid c)$) was computed and all the samples are sorted according to these scores. Then, following the resulting order, the classifier corresponding to Eq. (3) was applied to each sample and each resulting class label was removed from the list of class candidates (i.e., from the the argmax of (1)) for classification of the remaining samples.

With this approach, the validation set error rate was 1%, while for the hidden test set, a 0% error rate was achieved.

---

[8] oncina@dlsi.ua.es

3.4   Summary of Results

Finally in Table 2 a summary of the main *validation* and *hidden test* is presented. These *hidden test* results were obtained using the on- and off-line HMM parameters, as well as the parameters needed for the combination of modalities, determined using the *validation* set only.

Table 2. Summary of the best results (classification error rate %) using on-line and off-line classifiers alone and the bi-modal classifier. The relative improvement over the on-line-only accuracy is also reported.

Participant	data	uni-modal on-line	off-line	bi-modal	improvement
Baseline	*validation*	6.6	27.6	4.0	39%
UPV-April	*validation*	2.9	12.7	1.9	35%
	test	3.7	12.7	1.5	60%
PRHLT	*validation*	4.8	18.9	1.9	60%
	test	5.2	18.9	1.3	75%
PRHLT +	*validation*	4.8	18.9	1.0	79%
GRFIA	*test*	5.2	18.9	0.0	100%

4   Conclusions

The main conclusion from these results is that a simple use of both modalities does help improving the accuracy of the best modality alone. Morever, there is room for further improvements using more sophisticated multi-modal classifiers.

There are many pattern recognition problems where different streams represent the same sequence of events. This multi-modal representation offers a good opportunities to exploit the best characteristics of each modality to get improved classification rates.

We have introduced a controlled bi-modal corpus of isolated handwriting words in order to ease the experimentation with different models that deal with multi-modality. Baseline results are reported that include uni-modal results, bi-modal results and lower bounds by taking the best modality for each input pattern.

It can be seen that the UPV-April team achieve better results in uni-modal tests, but the PRHT team achieve better profit of the bi-modality, obtaining 60% and 75% of relative improvement on validation and hidden test respectively. With the a priory information that in the test there is only one sample per class, the GRFIA team impose the restriction of do not repeat any hypothesis, that is, every hypothesis is sorted by its reliability, then each hypothesis is classified on the most probably class not produced before. This way, the team classify every sample correctly.

From these results, we can conclude that, in this corpus, multi-modal classification can help to improve the results obtained from the best uni-modal classification. In future works, more sophisticated techniques would be applied to this corpus, and it is also planned to increase the samples of the corpus.

Acknowledgments

Work supported by the EC (FEDER/FSE), the Spanish Government (MEC, MICINN, MITyC, MAEC, "Plan E", under grants MIPRCV "Consolider Ingenio 2010" CSD2007-00018, MITTRAL TIN2009-14633-C03-01, erudito.com TSI-020110-2009-439, FPU AP2005-1840, AECID 2009/10), the Generalitat Valenciana (grant Prometeo/2009/014, and V/2010/067) and the Univ. Politécnica de Valencia (grant 20091027).

References

1. Toselli, A.H., Juan, A., Vidal, E.: Spontaneous handwriting recognition and classification. In: International Conference on Pattern Recognition, pp. 433–436 (August 2004)
2. España-Boquera, S., Castro-Bleda, M., Gorbe-Moya, J., Zamora-Martínez, F.: Improving Offline Handwritten Text Recognition with Hybrid HMM/ANN Models. IEEE Trans. Pattern Anal. Mach. Intell. (accepted for publication, 2010)
3. Gorbe-Moya, J., España-Boquera, S., Zamora-Martínez, F., Castro-Bleda, M.J.: Handwritten Text Normalization by using Local Extrema Classification. In: Proc. 8th International Workshop on Pattern Recognition in Information Systems, Barcelona, Spain, pp. 164–172 (2008)
4. Liwicki, M., Bunke, H.: Iam-ondb - an on-line english sentence database acquired from handwritten text on a whiteboard. In: 8th Intl. Conf. on Document Analysis and Recognition, vol. 2, pp. 956–961 (2005)
5. Marcus Liwicki, M., Bunke, H.: Combining on-line and off-line bidirectional long short-term memory networks for handwritten text line recognition. In: Proceedings of the 11th Int. Conference on Frontiers in Handwriting Recognition, pp. 31–36 (2008)
6. Marti, U., Bunke, H.: A full english sentence database for off-line handwriting recognition. In: Proc. of the 5th Int. Conf. on Document Analysis and Recognition, pp. 705–708 (1999)
7. Pastor, M., Toselli, A.H., Casacuberta, F., Vidal, E.: A bi-modal handwritten text corpus: baseline results. In: Ünay, D., Çataltepe, Z., Aksoy, S. (eds.) ICPR 2010. LNCS, vol. 6388, pp. 1–13. Springer, Heidelberg (2010)
8. Pastor, M., Toselli, A.H., Vidal, E.: Writing Speed Normalization for On-Line Handwritten Text Recognition. In: Proc. of the Eighth International Conference on Document Analysis and Recognition (ICDAR 2005), Seoul, Korea, pp. 1131–1135 (August 2005)
9. Johansson, G.L.S., Goodluck, H.: Manual of information to accompany the lancaster-oslo/bergen corpus of british english, for use with digital computers (1978)

10. Toselli, A.H., Romero, V., Rodríguez, L., Vidal, E.: Computer Assisted Transcription of Handwritten Text. In: International Conference on Document Analysis and Recognition, pp. 944–948 (2007)
11. Toselli, A.H., Romero, V., Vidal, E.: Computer assisted transcription of text images and multimodal interaction. In: Popescu-Belis, A., Stiefelhagen, R. (eds.) MLMI 2008. LNCS, vol. 5237, pp. 296–308. Springer, Heidelberg (2008)
12. Viard-Gaudin, C., Lallican, P., Knerr, S., Binter, P.: The ireste on/off (ironoff) dual handwriting database. In: International Conference on Document Analysis and Recognition, pp. 455–458 (1999)
13. Vinciarelli, A., Perrone, M.: Combining online and offline handwriting recognition. In: International Conference on Document Analysis and Recognition, p. 844 (2003)
14. Young, S., Odell, J., Ollason, D., Valtchev, V., Woodland, P.: The HTK Book: Hidden Markov Models Toolkit V2.1. Cambridge Research Laboratory Ltd. (March 1997)

Hybrid HMM/ANN Models for Bimodal Online and Offline Cursive Word Recognition

S. España-Boquera[1], J. Gorbe-Moya[1],
F. Zamora-Martínez[2], and M.J. Castro-Bleda[1]

[1] Departamento de Sistemas Informáticos y Computación
Universidad Politécnica de Valencia
46022 Valencia, Spain
[2] Departamento de Ciencias Físicas, Matemáticas y de la Computación
Universidad CEU-Cardenal Herrera
46115 Alfara del Patriarca (Valencia), Spain
{sespana,jgorbe,fzamora,mcastro}@dsic.upv.es

Abstract. The recognition performance of current automatic offline handwriting transcription systems is far from being perfect. This is the reason why there is a growing interest in assisted transcription systems, which are more efficient than correcting by hand an automatic transcription. A recent approach to interactive transcription involves multimodal recognition, where the user can supply an online transcription of some of the words. In this paper, a description of the bimodal engine, which entered the "Bi-modal Handwritten Text Recognition" contest organized during the 2010 ICPR, is presented. The proposed recognition system uses Hidden Markov Models hybridized with neural networks (HMM/ANN) for both offline and online input. The N-best word hypothesis scores for both the offline and the online samples are combined using a log-linear combination, achieving very satisfying results.

1 Introduction

Handwriting recognition can be divided into two main areas: offline and online handwriting recognition. An offline handwriting recognition system extracts the information from previously scanned text images whereas online systems receive information captured while the text is being written, usually employing a stylus and sensitive tablets.

Offline systems are applicable to a wider range of tasks, given that online recognition require the data acquisition to be made with specific equipment at the time of writing. Online systems are more reliable due to the additional information available, such as the order, direction and velocity of the strokes.

However, in all cases, achieving a perfect transcription requires human intervention. The aim of assisted handwriting transcription systems is minimizing the correction effort. An example of this kind of systems is the STATE system [1] which integrates both modalities of text recognition. The expert user may introduce text by writing it directly on the screen where the error is seen with

D. Ünay, Z. Çataltepe, and S. Aksoy (Eds.): ICPR 2010, LNCS 6388, pp. 14–21, 2010.
© Springer-Verlag Berlin Heidelberg 2010

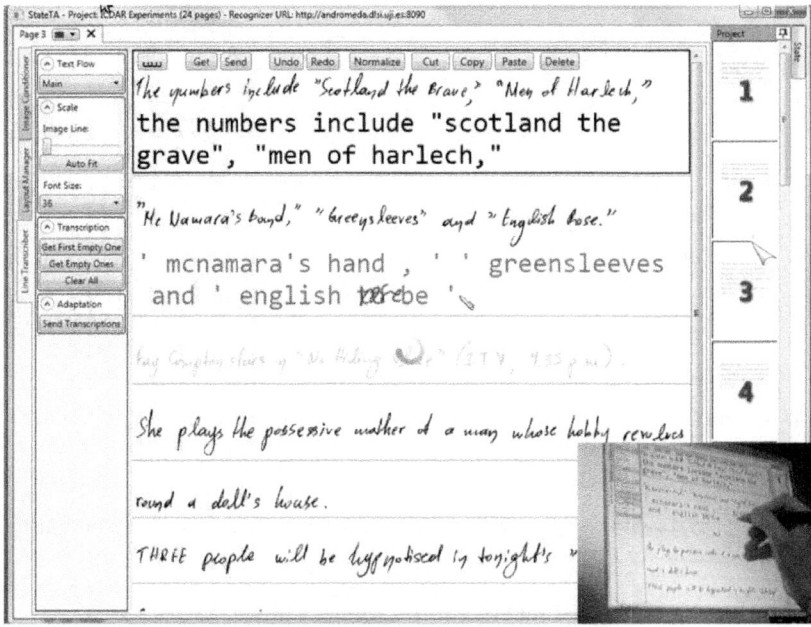

Fig. 1. User interacting with the STATE assisted transcription system

the help of an stylus sensitive device. The online handwritten text and gestures input can feed an interactive loop between the handwritten system and it can be exploited in a multimodal, combined recognition system (see a picture of the system in Figure 1).

This paper presents a bimodal recognition engine which combines our previous offline recognition system and the new developed online recognition system. The proposed engine entered the "Bi-modal Handwritten Text Recognition" contest organized during the 2010 ICPR.

The next section describes the bimodal task [2], extracted from the offline and online IAM database [3,4]. Section 3 describes the preprocessing steps applied to offline and online word samples. The bimodal handwriting recognition system is presented in Section 4, along with its performance. The work concludes with an analysis of the obtained results.

2 The Bimodal Corpus

The biMod-IAM-PRHLT corpus is a bimodal dataset of online and offline isolated handwritten words [2], publicly available for academic research. It is composed of 519 handwritten word classes with several online and offline word instances extracted from the publicly available IAM corpora [3,4]. Figure 2 shows some examples of bimodal samples from the corpus.

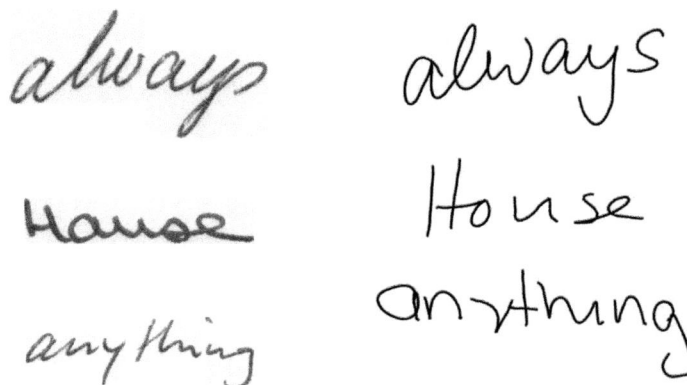

Fig. 2. Examples of word samples from the bimodal corpus: left, offline images, and right, online samples

Table 1. Basic statistics of the biMod-IAM-PRHLT corpus and their standard partitions. The vocabulary is composed of 519 word classes.

Running words	Online	Offline
Training set	8 342	14 409
Validation set	519	519
(Hidden) test set	519	519
Total	9 380	15 447

The writers of the online and offline samples are generally different. The offline samples are presented as grey-level images, and the online samples are sequences of coordinates describing the trajectory of an electronic pen.

The corpus is partitioned into training and validation for common benchmarking purposes. A test partition was held-out and it was used for the "Bi-modal Handwritten Text Recognition" contest organized during the 2010 ICPR. Both validation and test partitions are composed of a bimodal sample (a pair of an online and an offline instance) for each word in the vocabulary. Some basic statistics of this corpus are shown in Table 1.

3   Handwriting Preprocessing

3.1   Offline Preprocessing

The individual word images have been preprocessed using the method described in [5,6]. Neural networks have been used to estimate the slope and slant angles, and also to find the main body area of the text line in order to perform size normalization.

In order to correct the slope, a set of local extrema from the text contours of the image are extracted and then processed by a multilayer perceptron (MLP) which receives, for a given local extremum, a window of the image around that point. The task of this MLP consists in selecting which points are used to estimate the lower baseline by means of least squares fitting.

Following the slope correction, non-uniform slant correction based also on a MLP is performed. Instead of trying to estimate the slant angle, this MLP only detects whether a fragment of an image is slanted or not. For each angle in a given range, the input image is slanted, and the MLP gives a score to each pair (angle, column) of the original image. Then, a dynamic programming algorithm is used to find the maximum-score path across all the columns of the image which satisfies a smoothness constraint, to avoid abrupt slant changes. This path is an estimation of the local slant angle for each column of the image.

Finally, local extrema from the deslanted image are computed again and another MLP classifies each point into 5 classes: 'ascender', 'descender', 'lower baseline', 'upper baseline' and 'other'. Using this information, the sizes of ascenders, descenders and main body areas are normalized to a fixed height. Figure 3 (left) shows an example of all the steps of this process.

The feature extraction method is the same described in [7]. A grid of square cells is applied over the image and three values are extracted from each cell: normalized gray level, horizontal derivative of the gray level and vertical derivative of the gray level. Gray levels are weighted with a Gaussian which gives more importance to the pixels in the center of the cell and tends to zero in the borders. In all of the experiments a grid with 20 rows has been used.

3.2   Online Preprocessing

The online input consists of sequences of point coordinates with no stylus pressure information corresponding to writing strokes. The preprocessing stage comprises trace segmentation and smoothing of the sequence of points, uniform slope and slant correction, and size normalization. Each smoothed sample is converted to an image which goes through the same process as the offline images to obtain estimates of the slant angle and the text baselines. With this information, the online sequence is desloped and deslanted, and then scaled and translated so that the lower and upper baselines are located in $y = 0$ and $y = 1$ respectively (see Figure 3, right). Finally, a set of 8 features is extracted for every point from the sequence: y coordinate, first and second derivative of the position, curvature, velocity, and a boolean value which marks the end of each stroke.

4   Bimodal Handwriting Recognition Engine

4.1   HMM/ANN Optical Models

The recognition system is based on character-based hybrid HMM/ANN models, as fully described in [6]. Hidden Markov models (HMM) with a left-to-right

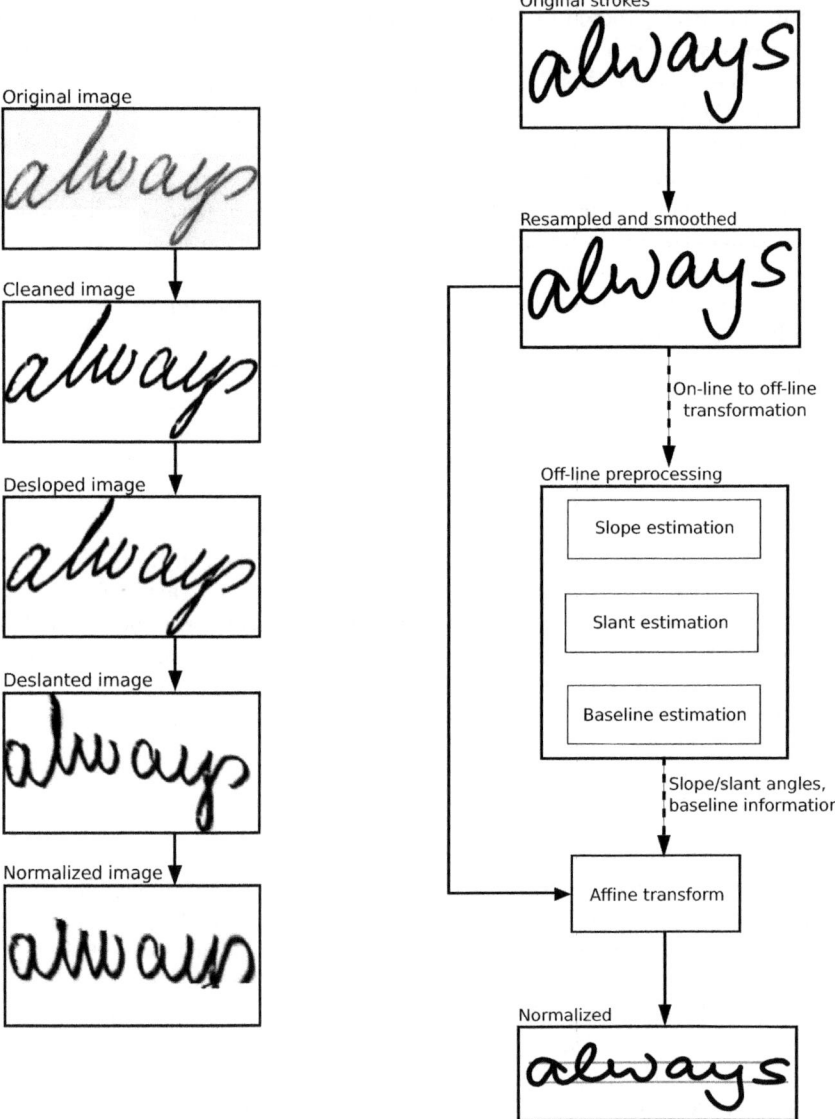

Fig. 3. Preprocessing example of an offline sample (left), and an online sample (right)

without skips topology have been used to model the graphemes, and a single multilayer perceptron is used to estimate all the HMM emission probabilities. Thus, the output unit $y_q(x_t)$ represents, after dividing by the priors, the probablity distribution of state q given the input at time t. Softmax ensures that this probability distribution is valid, i.e. $y_q(x_t) > 0$ for any HMM state q and $\sum_i y_i(x_t) = 1$.

At each MLP training step, error is computed according to cross entropy criterion and weights are updated with the standard backpropagation algorithm with momentum, using our own software [8].

The advantages of this approach are that the training criteria is discriminative (each class is trained with all the training samples, those which belong to the class and those which do not), and the fact that it is not necessary to assume an *a priori* distribution for the data.

Another advantage of this kind of hybrid HMM/ANN models is the lower computational cost compared to Gaussian mixtures, since a single forward evaluation of a unique MLP generates the values for all HMM states whereas a different Gaussian mixture is needed for each type of HMM state.

The recognition system presented in this work is based on HMM/ANN optical models for both offline and online recognition, using the following configurations:

- **Off-line**: We have used the same hybrid HMM/ANN models from [6]: a 7-state HMM/ANN using a MLP with two hidden layers of sizes 192 and 128. The input to the MLP was composed of the current frame plus a context of 4 frames at each side. It is worth noting that these models were trained with the training partition of the IAM-DB [3].
- **On-line**: Similarly, online models were trained with more data than that given in the bimodal corpus: the IAM-online training partition [4] was also used. Topologies of HMMs and MLP were just the same as the offline HMM/ANN models, but with a wider context at the input layer of the MLP: 12 feature frames at both sides of the actual input point.

4.2   Bimodal System

The scores of the 100 most probable word hypothesis were generated for the offline sample using the offline preprocessing and HMM/ANN optical models. The same process is applied to the online sample. The final score for each bimodal sample is computed from these lists by means of a log-linear combination of the scores computed by both the offline and online HMM/ANN classifiers:

$$\hat{c} = \operatorname*{argmax}_{1 \leq c \leq C}((1 - \alpha) \log P(x_{\text{off-line}}|c) + \alpha \log P(x_{\text{on-line}}|c)),$$

being C the number of word classes. The combination coefficients were estimated by exhaustive scanning of values over the biMod-IAM-PRHLT validation set. Table 2 shows a summary of the whole recognition engine.

Table 2. Bimodal recognition engine summary, including the combination coefficients

Bimodal system	offline	online
# input features per frame	60	8
# HMM states	7	7
MLP hidden layers' size	192-128	192-128
MLP input context (left-current-right)	4-1-4	12-1-12
Combination coefficient	$(1-\alpha)=0.55$	$\alpha=0.45$

5   Experimental Results

The ICPR contest organizers published baseline validation results for a bimodal recognition system described in [2]. The error rates were 27.6% for the offline modality and 6.6% for the online modality. A Naive Bayes combination of both unimodal systems resulted in a error rate of 4.0% for the bimodal task.

Our proposed system achieved a 12.7% error rate for the offline validation set, and a 2.9% for the online validation set. Combining both sources resulted in a 1.9% validation error rate. The error rates for the test dataset were 12.7% for the offline data, 3.7% for the online data, and 1.5% for the bimodal task.

Table 3 shows these results. Baseline system performance is also shown for comparison purposes with the validation data. Unimodal results, both for offline and online samples, are also shown. The last column illustrates the relative improvement of the bimodal system over the online (best unimodal) system. As can be observed, close to 60% of improvement is achieved with the bimodal system when compared to using only the online system for the test set.

Table 3. Performance of the bimodal recognition engine

		Unimodal		Bimodal	
System		Offline	Online	Combination	Relative improv.
Validation	Baseline [2]	27.6	6.6	4.0	39%
	HMM/ANN	12.7	2.9	1.9	34%
(Hidden) Test	HMM/ANN	12.7	3.7	1.5	59%

6   Conclusions

A perfect transcription for most handwriting tasks cannot achieved and human intervention is needed to correct it. Assisted transcription systems aim to minimize this human correction effort. An integration of online input into the offline transcription system can help in this process, as it is done in the STATE system [1]. For this end, we have developed a bimodal recognition engine which combines our previous offline recognition system and a new developed online recognition system. This work presents this system and tests it in the "Bi-modal Handwritten Text Recognition" contest organized during the 2010 ICPR.

Hybrid HMM/ANN optical models perform very well for both offline and online data, and their naive combination is able to greatly outperform each system. Nevertheless, more exhaustive experimentation is needed, with a larger corpus, in order to obtain more representative conclusions. As a future work we plan to fully integrate it in the assisted transcription system.

Acknowledgements

This work has been partially supported by the Spanish Ministerio de Educación y Ciencia (TIN2008-06856-C05-02) and the BFPI 06/250 scholarship from the Conselleria d'Empresa, Universitat i Ciencia, Generalitat Valenciana. The authors wish to thank the organizers for this contest.

References

1. Vilar, J.M., Castro-Bleda, M.J., Zamora-Martínez, F., España-Boquera, S., Gordo, A., Llorens, D., Marzal, A., Prat, F., Gorbe, J.: A Flexible System for Document Processing and Text Transcription. In: Meseguer, P. (ed.) CAEPIA 2009. LNCS (LNAI), vol. 5988, pp. 291–300. Springer, Heidelberg (2010)
2. Pastor, M., Vidal, E., Casacuberta, F.: A bi-modal handwritten text corpus. Technical report, Instituto Tecnológico de Informática, Universidad Politécnica de Valencia, Spain (2009)
3. Marti, U.V., Bunke, H.: The IAM-database: an English sentence database for offline handwriting recognition. International Journal of Document Analysis and Recognition 5, 39–46 (2002)
4. Liwicki, M., Bunke, H.: IAM-OnDB - an on-line English sentence database acquired from handwritten text on a whiteboard. In: International Conference on Document Analysis and Recognition, pp. 956–961 (2005)
5. Gorbe-Moya, J., España-Boquera, S., Zamora-Martínez, F.: Castro-Bleda, M.J.: Handwritten Text Normalization by using Local Extrema Classification. In: Proc. 8th International Workshop on Pattern Recognition in Information Systems, Barcelona, Spain, pp. 164–172. Insticc Press (June 2008)
6. España-Boquera, S., Castro-Bleda, M.J., Gorbe-Moya, J., Zamora-Martínez, F.: Improving Offline Handwritten Text Recognition with Hybrid HMM/ANN Models. IEEE Trans. Pattern Anal. Mach. Intell. (August 2010) (preprints), doi: 10.1109/TPAMI.2010.141
7. Toselli, A.H., Juan, A., González, J., Salvador, I., Vidal, E., Casacuberta, F., Keysers, D., Ney, H.: Integrated Handwriting Recognition and Interpretation using Finite-State Models. International Journal of Pattern Recognition and Artificial Intelligence 18(4), 519–539 (2004)
8. España-Boquera, S., Zamora-Martínez, F., Castro-Bleda, M.J., Gorbe-Moya, J.: Efficient BP Algorithms for General Feedforward Neural Networks. In: Mira, J., Álvarez, J.R. (eds.) IWINAC 2007. LNCS, vol. 4527, pp. 327–336. Springer, Heidelberg (2007)

Verification of Video Source Camera Competition (CAMCOM 2010)

Wiger van Houten[1], Zeno Geradts[1], Katrin Franke[2], and Cor Veenman[1]

[1] Netherlands Forensic Institute
Laan van Ypenburg 6
2497 GB The Hague, The Netherlands
{w.van.houten,z.geradts,c.veenman}@nfi.minjus.nl
[2] Gjøvik University
PO box 191 Teknologiveien 22
N-2815 Gjøvik, Norway
kyfranke@ieee.org

Abstract. Digital cameras are being integrated in a large number of mobile devices. These devices may be used to record illegal activities, or the recordings themselves may be illegal. Due to the tight integration of these mobile devices with the internet, these recordings may quickly find their way to internet video-sharing sites such as YouTube. In criminal casework it is advantageous to reliably establish the source of the video. Although this was shown to be doable for relatively high quality video, it is unknown how these systems perform for low quality transcoded videos. The CAMCOM2010 contest is organized to create a benchmark for source video identification, where the videos originate from YouTube. Despite the number of participants was satisfactory initially, only two participants submitted results, mostly due to a lack of time. Judging by the performance of the contestants, this is certainly not a trivial problem.

Keywords: YouTube, source video identification, camera identification.

1 Introduction

Digital imaging presents a lot of obvious advantages with respect to their analog counterparts. There is no need to chemically develop the images, virtually eliminating processing time. Results can be viewed immediately, and results can easily be edited to obtain a visually more pleasing image. With the internet, images and videos can be shared in seconds, and this may obfuscate the origin of the (original) image or video when there are no identifying characteristics (e.g. metadata) available.

For forensic purposes, however, we are interested in the origin of a certain image or video. In other words, we would like to know from which camera the file originates.

There are various reasons why this is of interest. First, it may link a witness or a suspect to a crime scene, and hence be used to verify statements. In particular,

D. Ünay, Z. Çataltepe, and S. Aksoy (Eds.): ICPR 2010, LNCS 6388, pp. 22–28, 2010.

a camera stolen in a burglary or during a homicide may be traced back to its original owner if some reference images are available. Another possibility is to trace back which images or videos were made by the same camera if there is no camera available. This can be of interest in child pornography cases, or when a film is recorded in a cinema by a digital camera and subsequently released to the internet. As such this problem is a verification problem: the question to be answered is whether a certain characteristic is present in both the digital camera itself as the natural image or video.

Although there has been a wide interest in the forensics community to solve this problem [1–3], this is mostly confined to (photo) cameras of relatively high quality. In [4] it was shown to be possible to identify the source camera from low resolution and heavily compressed JPEG images. All these methods rely on the presence of a perceptually invisible noise pattern, a 'fingerprint', in the video or image originating from the sensor. Hence, source identification is a matter of extracting and comparing these noise patterns, called Photo Response Non-Uniformity (PRNU) patterns, as visualized in Fig. 1. These patterns originate due to the fact that each pixel has a slightly different response to the same amount of light. Extracting the pattern essentially comes down to subtracting a low-pass filtered (denoised) version of the image from the original. The low-pass filtered version of the image represents the image in the absence of the noise-like PRNU pattern. Subtracting the filtered image from the original shows the characteristic pattern of deviations. The accuracy depends on the filter used to denoise the image; a simple low-pass filtered image is suboptimal as small details are removed. Hence, different filters result in different performance. Furthermore, different filters are suitable for different situations.

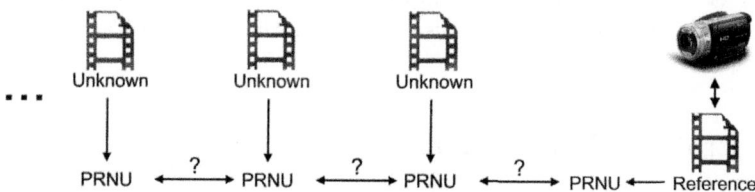

Fig. 1. Schematic overview of the problem. In the first scenario, the goal is to find the origin of an unknown video. This can be accomplished by extracting a characteristic (PRNU) from the unknown video, as well as from a reference video with known origin. When these PRNU patterns are similar, this may indicate that the unknown video has the same origin as the reference video. In the second scenario, no such reference video is available, and the goal is to establish which videos have a common origin.

The deviations are multiplicative with respect to the amount of light that falls on each pixel, and are easiest to extract from well illuminated images that contain smooth gradients. Hence, images that are well illuminated and have low frequency content allow for a more clear and accurate extraction of the sensor noise. On the other hand, high frequency textures or dark areas result in more uncertainty in the extracted pattern. When we need to answer whether a certain image or video originates from a certain camera, we make so-called reference or 'flatfield' images or videos that are well illuminated and contain no discernible details. In practice, a homogeneous grey area is recorded; these flatfield images or videos are used as reference. On the other hand, 'natural' videos are videos with normal content, i.e. they portray normal scenes.

With the advent of high speed mobile internet and the integration of cameras in mobile phones and laptops, the challenge shifts to low quality images and videos. To conserve bandwidth and processing power, these videos are often strongly compressed in a far from optimal video codec. The ever increasing popularity of video sharing sites as YouTube presents an extra challenge, as these low quality videos will be transcoded to a standard codec. These codecs remove or suppress high frequency noise, such as PRNU. Although small scale tests proved successful [5], it is hard to obtain data for large scale testing. For photographs, Flickr has been used in [6]. Images on Flickr generally contain EXIF metadata listing the camera model and type, and it is reasonable to assume that multiple images (with the same EXIF metadata for camera brand/model) from a single user have the same origin. However, such an online 'database' does not exist for videos; in particular, videos do not have a general metadata description such as EXIF for photographs.

The normal camera identification scheme is very successful, and we believe the challenge has shifted to heavily compressed videos. There is a certain point at which the compression becomes too strong and virtually removes the PRNU. Alternatively, the length of the video may be too short to get a reliable estimate of the PRNU pattern. Hence, it is desirable to know the limits of the PRNU technique. This necessitates sensitive methods to extract the PRNU. However, the lack of a video database (comparable to Flickr for photographs), makes it difficult to find these limits. These reasons motivated us to start the CAMCOM challenge.

2   Problem Description

As in the introduction, the problem with doing source video camera identification with videos from YouTube is the generally strong compression present, and this may result in visible compression artefacts. However, there is another problem with videos obtained from these cameras. Sensors in webcams, digital cameras or mobile phones are often used to make photographs as well as to record videos. These videos generally have a lower resolution than photographs made with the same sensor. Alternatively, video can be recorded in a wide range of resolutions and aspect ratios. This means that either a smaller subsection of the sensor is used, or that some form of scaling is used before the signal is output. These possibilities need to be taken into account when we do the source camera identification.

For example, when a sensor has a native resolution of 800x600, we can record video in a lower resolution such as 640x480. The signal may be downsampled by software, or a smaller section of the sensor can be used. Often it is also possible to record video in a higher resolution, depending on the driver software. Finally, the aspect ratio may change when the software driver allows this. All situations may occur at the same time.

3   Data

For the creation of the data set, we used approximately 100 low cost webcams with native resolutions of 640x480 or 352x288. A few screenshots can be found in Fig. 2. However, there is no need to record the videos in these resolutions as explained in the previous section. For example, a large amount of webcams we used had a native resolution of 352x288 (11:9 aspect ratio). The driver allowed to record video in 640x480 (4:3) resolution. When we analyzed the pattern, we found that a 320x240 subsection from the center was used which was subsequently upscaled to 640x480.

The natural videos contain footage of the surroundings of an office; they contain both high and low detailed scenes, and saturation occurred frequently. The contest was split up in two optional scenarios. Each scenario will be explained separately in the next sections.

Table 1. Overview of the dataset used in both scenarios. In this table, the number of videos and length (in seconds) of the videos used in each scenario is presented. The videos in both scenarios were not necessarily recorded in their native resolutions. In the first scenario, reference videos are available, and the natural videos are recorded in 3 different resolutions as indicated. In the second scenario, all videos were recorded in 640x480 resolution, but no reference video were available.

	Length (s)	Amount	Reference available?	Resolutions
Scenario I	10	148	Yes	320x240, 352x288, 640x480
	35	148	Yes	320x240, 352x288, 640x480
Scenario II	20	158	No	640x480
	60	100	No	640x480

3.1   Scenario I

The goal of the first scenario is to link videos of unknown origin with a set of 50 reference videos; these reference videos of known origin were supplied in their native resolutions. Ideally, the 50 reference cameras would be distributed to each contestant to make reference videos, but this is not feasible. Instead, we decided to record the reference videos ourselves, and make them available to the contestants. These videos were not uploaded to YouTube so that no additional transcoding occurred, and the PRNU has not been attenuated significantly. The videos are minimally compressed and have a length of approximately 30 seconds containing well-illuminated homogeneous areas. We also recorded 148 natural videos of approximately 10 seconds, and 148 natural videos of approximately 35 seconds. These natural videos are either recorded in their native format, or in some other resolution that the camera driver makes possible. Hence, each video may originate from a low or high resolution camera, independent of its resolution.

The contestant was asked to identify which of the videos had a corresponding reference video (if any). Multiple videos from each camera may be present. Each combination of natural video and reference video had to be scored. In this way, two contingency tables of 148x50 are obtained. This scenario mimics the situation where a forensic investigator needs to say whether a certain video originates from a certain camera that is available to the investigator.

3.2   Scenario II

The goal of the second scenario is to link natural videos; no reference videos are supplied. Or, in the verification framework, one natural video acts as a reference while the other videos need to be scored against this reference video. As in the previous scenario, two different video lengths were available: 158 videos of approximately 20 seconds, and 100 videos of approximately 60 seconds. Each set of a certain length needed to be assessed separately, multiple 'partners' may be present, and not all videos may have a corresponding 'partner'. To ease the process of matching, all videos are recorded in 640x480 resolution, and originate from low resolution cameras (352x288) and the higher resolution cameras (640x480). In this way two tables are obtained, of size 158x158 and 100x100. The difference with the first scenario is that there is no reference video with a known origin. This scenario is more difficult as the PRNU extraction from natural videos is more difficult.

Fig. 2. Sample screenshots from the videos

4   Results

Due to the limited response we only report some anonymized results. In Table 2 we report the Area Under Curve (AUC) and the Equal Error Rate (EER) of the systems. The ROC curves for scenario I and II are shown in Figs. 3 and 4, respectively. These measures do not tell the complete story, as an investigator may refrain from giving an answer about the source of the video. Hence, each contestant was also asked to submit a separate decision matrix in which the final decision was presented about the source of each test video (if any). It has to be mentioned that such a decision can be unrealistic in practice; a likelihood ratio

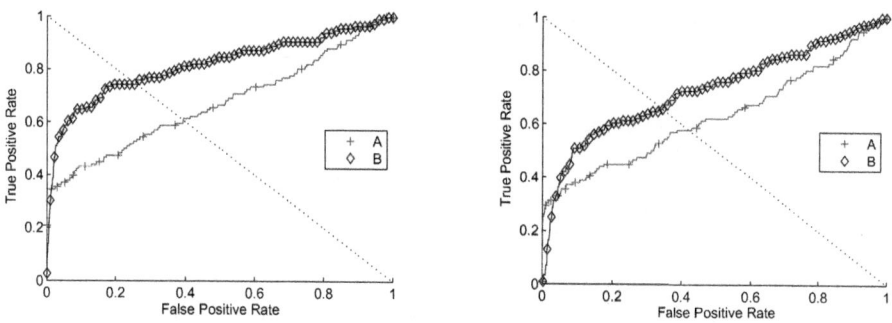

Fig. 3. ROC curves for Scenario I for both long (left) and short (right) videos

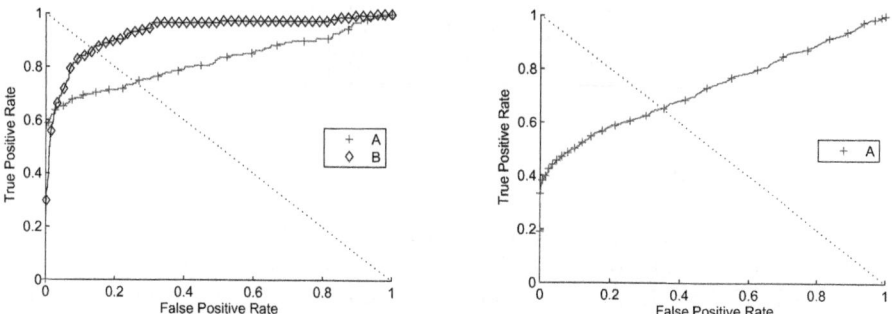

Fig. 4. ROC curves for Scenario II for both long (left) and short (right) videos. For the short videos, no valid contingency table was submitted by contestant A for the determination of the ROC curve.

or verbal scale of certainty can be more suitable in these comparisons. For the videos that had a corresponding partner, we use the decision matrix to calculate the Precision and Recall. Precision is defined as the number of True Positives divided by the sum of True Positives and False Positives. Recall is defined as the number of True Positives divided by the sum of True Positives and False Negatives. These numbers are shown in Table 3.

Even though the performance (based on the EER) of participant A is much lower than B, its precision is much higher. In both cases the Recall rate is low, suggesting conservative systems. The performance for the long videos of the second scenario is adequate, but for the short videos (10 seconds in scenario I, 20 seconds in scenario II) the performance is low. This low performance may be an indication that the limit of the PRNU has been reached, but this is hard to conclude from the limited amount of results available.

Table 2. Area Under Curve (AUC) and Equal Error Rate (EER) for both participants
A and B. The dagger (†) denotes that no valid results were submitted.

	AUC(A)	AUC(B)	EER(A)	EER(B)
ScI:Long	0.67	0.82	0.61	0.74
ScI:Short	0.63	0.73	0.58	0.66
ScII:Long	0.82	0.93	0.74	0.86
ScII:Short	0.73	†	0.65	†

Table 3. Precision P and Recall rate R for both participants. The dagger (†) denotes
that no valid results were submitted.

	P(A)	P(B)	R(A)	R(B)
ScI:Long	0.95	0.72	0.33	0.18
ScI:Short	0.79	0.52	0.13	0.12
ScII:Long	0.98	†	0.31	†
ScII:Short	1.00	†	0.17	†

5   Conclusion

Although the initial number of contestants was satisfactory, a large number of
participants eventually did not submit any results. The main reason mentioned
was a lack of time. Hence, it is difficult to draw conclusions from the limited
amount of data obtained. However, judging by the performance, it can be seen
there is certainly room for improvement. The dataset can be used in the future,
either for testing or as a benchmark as in this contest.

References

1. Geradts, Z., Bijhold, J., Kieft, M., Kurosawa, K., Kuroki, K., Saitoh, N.: Methods for
   identification of images acquired with digital cameras. In: Proc. of SPIE: Enabling
   Technologies for Law Enforcement and Security, vol. 4232, pp. 505–512 (2001)
2. Lukáš, J., Fridrich, J., Goljan, M.: Determining digital image origin using sensor
   imperfections. In: Proc. of SPIE, vol. 5685, p. 249 (2005)
3. Chen, M., Fridrich, J., Goljan, M., Lukáš, J.: Determining Image Origin and In-
   tegrity Using Sensor Noise. IEEE Trans. on Information Forensics and Security 3(1),
   74–90 (2008)
4. Alles, E.J., Geradts, Z.J.M.H., Veenman, C.J.: Source camera identification for
   heavily JPEG compressed low resolution still images. Journal of Forensic Sciences 54(3), 628–638 (2009)
5. van Houten, W., Geradts, Z.: Source video camera identification for multiply compressed videos originating from YouTube. Digital Investigation 6(1-2), 48–60 (2009)
6. Goljan, M., Fridrich, J., Filler, T.: Large Scale Test of Sensor Fingerprint Camera
   Identification. In: Proc. of SPIE, vol. 7254, pp. 72540I–72540I-12 (2009)

The Landscape Contest at ICPR 2010

Núria Macià[1], Tin Kam Ho[2], Albert Orriols-Puig[1], and Ester Bernadó-Mansilla[1]

[1] Grup de Recerca en Sistemes Intel·ligents
La Salle - Universitat Ramon Llull
C/ Quatre Camins, 2 08022 Barcelona, Spain
{nmacia,aorriols,esterb}@salle.url.edu
[2] Statistics and Learning Research Department
Bell Laboratories, Alcatel-Lucent
Murray Hill, NJ 07974-0636 USA
tkh@research.bell-labs.com
http://www.salle.url.edu/ICPR10Contest/

Abstract. *The landscape contest* provides a new and configurable framework to evaluate the robustness of supervised classification techniques and detect their limitations. By means of an evolutionary multiobjective optimization approach, artificial data sets are generated to cover reachable regions in different dimensions of data complexity space. Systematic comparison of a diverse set of classifiers highlights their merits as a function of data complexity. Detailed analysis of their comparative behavior in different regions of the space gives guidance to potential improvements of their performance. In this paper we describe the process of data generation and discuss performances of several well-known classifiers as well as the contestants' classifiers over the obtained data sets.

1 Introduction

In many applications researchers pursue perfection of classification techniques since there are obvious benefits to obtain the maximum accuracy, e.g., in performing medical diagnosis. However, there have been relatively few systematic studies on whether perfect classification is possible in a specific problem. Most attention has been devoted to fine tuning the techniques instead.

Over the last two decades the competitiveness of classification techniques[1], typically developed for a general purpose, has been claimed over a small and repetitive set of problems. Although a common test bed is useful and necessary to make fair comparisons among algorithms, it can lead to incomplete conclusions about the quality of the learning algorithms if we do not have control over its characteristics, such as similarity between the data sets. The study of how real-world problems distribute in data complexity dimensions is aimed at providing a remedy, so that the classifier's behavior can be understood in the context of the problems' characteristics. Also, stability of the classifier's performance over problems with similar characteristics can be assessed.

---

[1] Our discussions will be restricted to *supervised* classification techniques, also referred to as classifiers, learning algorithms, or learners.

D. Ünay, Z. Çataltepe, and S. Aksoy (Eds.): ICPR 2010, LNCS 6388, pp. 29–45, 2010.
© Springer-Verlag Berlin Heidelberg 2010

Problems that provide a good coverage of the data complexity space would be necessary to perform this kind of study. To this end, we design a contest, to be held before the 20th International Conference on Pattern Recognition in 2010, that uses a collection of problems selected on the basis of their complexity characteristics. Evaluation of the participating algorithms with this collection of problems would provide the landscape featuring the domains of competence of each algorithm in the data complexity space.

The purpose of this paper is to describe the contest and report some preliminary experiments which pursue (1) to support the belief that rather than a unique and globally superior classifier, there exist local winners, (2) to highlight the critical role of the test framework, and (3) to envisage how this space may help to understand the limitations of classifiers and offer guidance on the design of improvements that can push the border of their domains of competence.

The remainder of this paper is organized as follows. Section 2 describes the process that generates the data collection for the contest. Then, Section 3 summarizes the experiments and presents some preliminary results. Finally, Sections 4 and 5 elaborate on some observations and conclusions.

2   The Landscape Contest

The analysis of the performance of learning algorithms should rely on a known, controlled testing framework due to the observed dependence between the capabilities of learners and the intrinsic complexity of data. To develop such a framework, lately some studies have been done on the characteristics of real-world data to evaluate the learner performance, in particular, various notions of data complexity [3].

This section revises an earlier work, on using the characterization of the difficulty of classification problems to identify classifier domains of competence, which provides the basis of the current contest. Then, it describes the process to cover the complexity space with synthetic data sets.

2.1   Background

In [11], Ho and Basu presented a set of twelve measures that estimate different aspects of classification difficulty. Examples are the discriminative power of features, class separability, class geometry and topology, and sampling density w.r.t. feature dimensionality.

They aimed at providing a characterization of classification problems, represented by a training data set, to analyze the domains of competence of classifiers and obtain guidelines to select the most suitable technique for each problem. Observing that many of the proposed complexity measures are correlated, they looked for uncorrelated factors by projecting the problem distributions on the first four principal components extracted from the set of twelve complexity dimensions. These projections revealed a near continuum formed by real-world problems as well as problems synthesized with controlled difficulty. The continuum spreads across extreme corners of the space. Nevertheless, the projection of problems did not cover the entire space. A mystery remains on whether the empty regions were due to constraints of the selection of problems or to the nature of the complexity measures.

2.2 Synthesizing a New Landscape with Evolutionary Optimization

To achieve better coverage of the complexity space, a larger collection of problems is necessary. However, real-world problems with truth labels are expensive to obtain, and difficult to control for quality. Synthetic data offers an interesting alternative. They can be created with the least cost and time, and can be controlled to reach a good coverage of reachable regions in the complexity space. Though, to employ this alternative, it is important to ensure that the generated data have sufficient resemblance to real-world problems, and at the same time contain sufficient variety to represent different aspects of data complexity.

With this aim in mind, we proposed an approach [12] that starts with a sample of real-world problems where the class concepts are described by the data collected in the experiments or from synthetic data following tailored distributions. We then introduced perturbations to the sample problems to extend their coverage in the complexity space.

We formulated the problem as a search for useful problems: *given an original data set, search the best selection of instances that satisfies a desired complexity*. In fact, the desired complexity is specified as a multiobjective optimization problem because it involves the minimization or maximization of a set of complexity measures and the satisfaction of a set of internal constraints.

In the following, we present the formulation of the optimization problem, list the constraints implemented, and detail the process organization of the evolutionary multi-objective search.

Problem Formulation. First of all, we consider a set of n labeled instances (also referred to as examples) $\{i_1, i_2, ..., i_n\}$ coming from a real-world problem or a synthetic problem generated with reference to a physical process. The optimization problem consists in searching the selection of instances $I = \{i_1, i_2, ..., i_l\}$, $l \leq 1$, that minimizes or maximizes the set of objectives $O = \{o_1, o_2, ..., o_p\}$ where each o_i corresponds to a complexity measure. For instance, if we need to obtain a data set with as many points located on the class boundary as possible, we should maximize the complexity measure that evaluates the *fraction of points on the class boundary*, also called N1.

Constraints. The consistency of the resulting data sets also depends on the extrinsic complexities, such as the number of instances, the number of attributes, or a measure of balance between the classes. These external characteristics have been taken into account by including them into the approach as constraints which are responsible for (1) maintaining a minimum number of instances, (2) conserving a specific class balance, (3) ensuring that the calculation of the complexity measures is feasible, and (4) avoiding the duplicity of instances.

Organization of the Search Process. To create data sets whose complexity is bounded by several indicators of data complexity designed in [11], we used an evolutionary multiobjective strategy [5], more specifically a method based on NSGA-II [6].

The system requires defining (1) the meta-information, (2) the genetic representation of the solution of the problem, and (3) the fitness function to evaluate each candidate solution. These are described as follows.

Meta-information. The input of the system is a data set, whatever its distribution is— real-world distribution or synthesized distribution—, described by n learning instances, where each instance is defined by m continuous or nominal attributes. This data set will be altered through an instance selection process until reaching the desired complexity.

Knowledge representation. The NSGA-II system evolves a population composed of N individuals, where each individual is a candidate solution, i.e., an artificial data set characterize by a selection of instances from the input data set. The representation of the individual is an array of size $k \leq n$, where k is the number of instances selected by the current individual and n is the number of instances of the original data set. The array contains the indexes of the selected instances, e.g., if an individual contains $\{2,5,14\}$, it means that instances 2, 5, and 14 of the original data set are selected. The evolutionary multiobjective optimization (EMO) searches the best selection of instances that satisfies the required complexity, i.e., minimizes or maximizes the set of specified complexity measures. Then, the final data set is obtained by moving from the genotypic representation to the phenotypic representation, i.e., by grouping the instances coded in the individual in a file and adding the corresponding header to provide a data set in Weka format [16].

Objective functions. Each objective refers to a complexity measure. To calculate the fitness of the individual, we simply translate the genotype into the phenotype according to the information encoded in the individual and compute the value of the complexity measures.

The EMO approach follows the classic procedure of a genetic algorithm (GA) [9] but includes two sorting concepts: (1) the fast non-dominated sorting and (2) the crowding distance assignment, which contribute to optimize different objectives in a single simulation run. The fast non-dominated sorting procedure allocates the population into different fronts according to how well they satisfy the multiple objectives and, the crowding distance assignment estimates the density of the solutions surrounding a particular solution in the population. It tries to measure the diversity of the population and maintain it.

This process drives the population to the Pareto-optimal line (see the pseudo code in Algorithm 1).

At the beginning, the individuals of the initial populations P_0 and Q_0 are randomly initialized and evaluated. Then, the GA bases the search on the interaction of two specific and three primary genetic operators that are iteratively applied: *fast non-dominated sorting*, *crowding distance assignment*, *selection*, *crossover*, and *mutation*. The process is organized as follows.

First, populations P_t and Q_t are combined in a population R_t which is sorted according to the *non-dominated procedure*. Then, starting from the first front, the solutions of each front i are introduced into the new population P_{t+1} on the condition that there is enough room to allocate all the solutions of the given front. Otherwise, only the solutions with the highest *crowding distance* of front i are included into P_{t+1}, until filling all the population.

Thereafter, *selection* is applied to choose the parent population of the next generation, resulting in the offspring population Q_{t+1}. Pairs of these parents are selected without replacement, and they undergo *crossover* and *mutation* operators with probabilities χ and μ respectively. If neither crossover nor mutation are applied, the offspring are exact copies of the parents.

Algorithm 1. NSGA-II

Input: P_0 and Q_0 initial populations, N (Population size), t_{max} (Maximum number of generations)

$t = 0$

while $(t < t_{max})$ **do**

    Create $R_t = P_t \bigcup Q_t$

    Perform a non-dominated sorting to R_t and identify different fronts: F_i, $i = 1, 2, ..., n$ by using a ranking algorithm

    $P_{t+1} = P_t$

    $i = 1$

    while $(|P_{t+1}| + |F_i| < N)$ **do**

        $P_{t+1} = P_{t+1} \bigcup F_i$

        $i = i + 1$

    end while

    Include in P_{t+1} the most widely spread $(N - |P_{t+1}|)$ solutions by using the crowding distance

    Create Q_{t+1} from P_{t+1} by using crowded tournament selection, crossover, and mutation operators

    $t = t + 1$

end while

Output: Front F_1 from set $P_{t_{max}} \bigcup Q_{t_{max}}$

The whole process is repeated until the stop criterion is reached. In this case, the stop criterion is a maximum number of generations.

Regarding the genetic operators, we use the following strategies: (i) *Tournament selection.* Tournaments of s randomly chosen parents are held, and the parent with the best fitness, according to the crowded-comparison operator, is selected for recombination. (ii) *Two-point crossover.* Two cut-points along the individuals are randomly selected and the information in between of both parents are shuffled. The range of the cut-points corresponds to `[0,min(size individual1, size individual2)]` to maintain the consistency of the offspring. (iii) *Mutation.* Addition or deletion of instances brings some diversity into the population. First, we randomly assign a Boolean value to each instance of the input data set. Then, each instance whose value is *true* is searched in the individual to mutate. If the instance exists, we remove it. Otherwise, we add it at the end of the structure.

To sum up, this evolutionary technique allows us to stretch different dimensions of complexity at the same time. However, correlations among variables (i.e., measures) or a high number of independent variables affect the optimization performance. This suggests that we should perform generation in steps by combining the complexity measures three by three.

2.3 The Idyllic Landscape

The idyllic landscape refers to a landscape providing a complete coverage of the characteristic space – a space small enough in which we are able to map all the problems and have enough resolution in the complexity values. This framework can help us to

(1) better understand the domain of competence of different learners and (2) compare different learners on the complexity space. The creation of such a space is not trivial since there are many ways to characterize a problem and its linkage to the properties of the learners is not a direct function. As we mentioned previously, our proposal consists in characterizing the data sets geometrically by means of twelve complexity measures. Nevertheless, the generation of problems to cover this 12-dimensional space involves a complex parameterization. The resulting combinatorial problem entails an unattainable computational cost. This is investigated in a preliminary study that precedes the landscape contest. As a result we settle on what can be done with evolutionary multiobjective optimization.

2.4   The Real Landscape Obtained from Evolutionary Multiobjective Optimization

In order to prepare the data for the contest, four data set collections are created and named S1, S2, S3, and S4. We aim at testing (1) the learner behavior over the problem space (S1, S2, and S3) and (2) the learner's local behavior in their domain of competence (S4). The four data set collections are summarized as follows:

S1. Collection of data sets covering the reachable complexity space, designed for training the learner. All the instances are duly labeled.

S2. Collection of data sets covering the reachable complexity space, designed for testing the learner. No class labeling is provided.

S3. Collection of data sets with no class labeling, like S2, that will be used in a live test that will be run for a limited period of time (i.e., one hour).

S4. Collection of data sets with no class labeling, which covers specific regions of the complexity space where each learner dominates. Its design is aimed at determining the stability of dominance over the local neighborhood.

To this end, we generated 80,000 data sets running the EMO approach over five seed problems: Checkerboard, Spiral, Wave Boundary, Yin Yang, and Pima. The four first are data sets with a known concept (see Figure 1). *Checkerboard.* Classic non-linear problem with heavily interleaved classes following a checkerboard layout. *Spiral.* Problem with a non-linear class boundary following a spiral layout. *Wave Boundary.* Linearly separable problem defined by a sinusoidal function. *Yin Yang.* Linearly problem with small disjuncts. The last one is a real-world problem which belongs to the UCI repository (Pima Indians Diabetes) [2]. These five seed data sets were evolved for different objective configurations, plus all the combinations of the optimization of three complexity measures at each time. The selection of three complexity measures results in eight experiments which consist in maximizing (1) or minimizing (0) each dimension, (i.e., 000, 001, ..., 111). The entire generation process used eleven of the twelve complexity measures proposed in [10], with the omission of the maximum Fisher's discriminant ratio (F1).

Following the analysis performed in [11], we calculated the singular value decomposition (SVD) over all the complexity measures and built a space with the first two principal components (see Figure 2(a)). To limit the size of the contest, we decided to

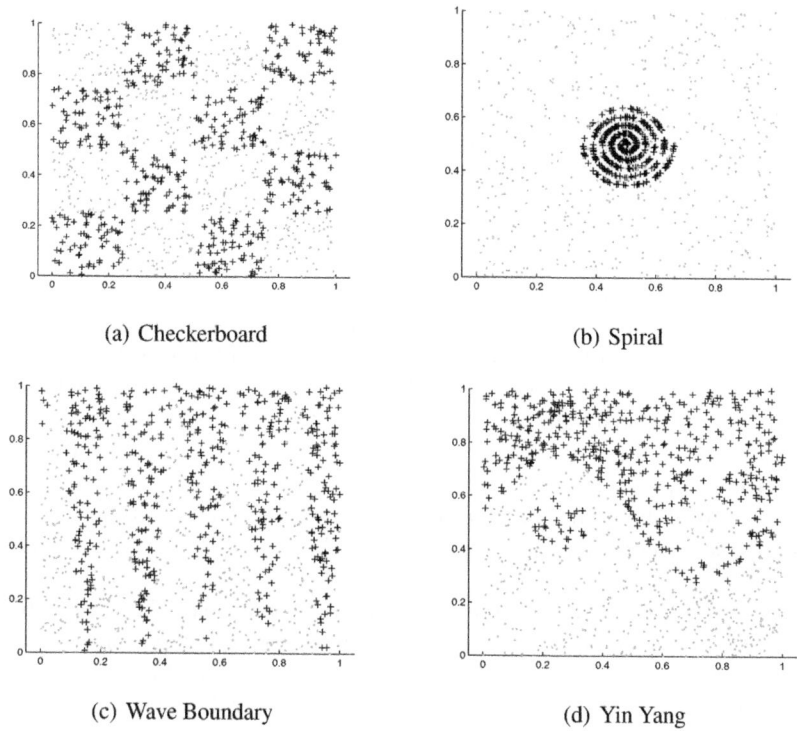

(a) Checkerboard

(b) Spiral

(c) Wave Boundary

(d) Yin Yang

Fig. 1. Seed data sets with a known structure

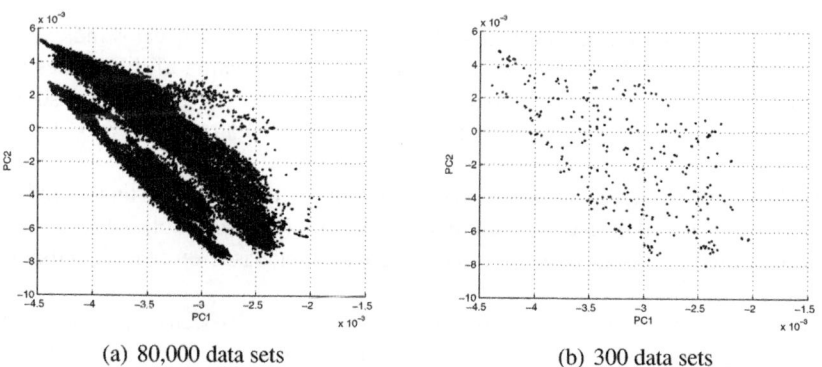

(a) 80,000 data sets

(b) 300 data sets

Fig. 2. Projection of the problems on the first and second principal components extracted from the complexity measurement: (a) the entire collection composed of 80,000 data sets and (b) 300 cherry-picked training data sets for use in the contest

select a sample from the collection. We divided the space into 100 cells and picked five data sets at random from each cell. Figure 2(b) plots the distribution of the 300-data set sample from the generated collection. These are the 300 data sets used in the contest.

3   Contest Execution and Results of Trial Runs

The landscape contest involves the running and evaluation of classifier systems over synthetic data sets. The contest is divided into two phases: (1) offline test and (2) live test. This section describes the experimental environment and presents some results of both phases.

3.1   Contest Description

For the offline test, participants ran their algorithms over two sets of problems, S1 and S2, and reported their results. In particular, we assessed (1) the predictive accuracy (i.e., test rate of correctly classified instances) applying a ten-fold cross-validation using S1 and (2) the class labeling of the test collection S2. Abilities such as robustness and scalability are evaluated indirectly, since they are implicit in the data complexity, and, thence reflected in the predictive accuracy. Learner's efficiency is shown in the second round of the contest (the live contest) where the results are to be submitted within a limited period of time. Finally, interpretability is not taken into consideration despite of its importance in certain knowledge domains, because it is out of the scope of this paper.

A live test was planned to take place during the conference. There, collections S3 and S4 were presented. S3 covers the data complexity space comprehensively, like S2, whereas S4 was generated according to the preliminary results submitted by the participants in order to determine the relative merits in each algorithm's respective domain of competence.

3.2   Contestants

The contest data set was released on March 31, 2010. Initially ten teams indicated their interest in participating in the contest. However, a combination of difficulties caused most teams to drop out over the next few months. At the end, three teams submitted their final results for the entire collection S2 by the June 1st due date, and all of them participated in the live test.

Table 1 summarizes the information of each team. The approaches they used include (1) a classifier based on a co-evolutionary algorithm, (2) a set of classifiers defined in a feature-based dissimilarity space, and (3) a classifier based on real-valued negative selection.

Interestingly, in addition to submitting results of their advocated approach, Team 2 applied nineteen classifiers in total to the training data S1 in order to compare performances of different families of learners. They documented a discovery that each classifier found a most favorable data set among the collection S1, and the nineteen favored sets were all distinct.

3.3   Results with Six Widely-Used Classifiers

Before the contestants submitted their results, we performed a test run of the contest using six widely-used classifiers belonging to different learning paradigms: C4.5 [14], IBk [1], Naïve Bayes (NB), PART [8], Random Tree (RT) [4], and SMO [13].

Table 1. Contestant description

Team 1 *Contestants:* Joaquín Derrac, Salvador García, and Francisco Herrera
        Affiliation: Universidad de Granada and Universidad de Jaén
        Contribution: IFS-CoCo in the landscape contest: Description and results

Team 2 *Contestants:* Robert P.W. Duin, Marco Loog, Elżbieta Pekalska, and David M.J. Tax
        Affiliation: Delft University of Technology and University of Manchester
        Contribution: Feature-based dissimilarity space classification

Team 3 *Contestants:* Luiz Otávio Vilas Boas Oliveira and Isabela Neves Drummond
        Affiliation: Universidade Federal de Itajubá
        Contribution: Real-valued Negative Selection (RNS) for classification task

---

C4.5 creates a decision tree by splitting the data recursively using the information gain of each attribute. IBk is an implementation of the nearest neighbor algorithm; to classify a previously unseen instance, it searches the k nearest neighbors and returns the majority class among them. Naïve Bayes is a probabilistic classifier based on the Bayes' theorem and the assumption of feature independence. PART is a decision tree algorithm that combines strategies from two methods to generate partial trees. RT builds a decision tree which is generated with a group of randomly selected features at each node from the original training set; the output of the new examples is inferred by considering the most popular class among the tree. SMO is an implementation of the *sequential minimal optimization* for efficiently training *support vector machines* [15]. All these methods were run using the Weka package [16] with the following configurations: (1) $k = 3$ for IBk and (2) the rest of the parameters were set to their default value.

Ten-Fold Cross-Validation. The performance of each technique was evaluated with the test classification accuracy, estimated using stratified ten-fold cross-validation [7]. Figures 7 and 8 represent the test accuracy over the complexity measurement space projected to the first two principal components. The x-axis refers to the first principal component and the y-axis refers to the second principal component. The color bar shows the gradation of the test accuracy; the darker the color, the lower the accuracy. For clarity of the plots, the accuracies are shown with a truncated scale from 25% to 100%. In Figures 7 and 8, we can see the results obtained by C4.5, IB3, NB, PART, RT, and SMO over the data sets generated using the five different seeds. Each column refers to each seed problem, namely, Checkerboard, Pima, Spiral, Wave Boundary, and Yin Yang.

We observe that C4.5 achieves a good performance except for a small group of problems located in the upper left corner, whereas IB3 behaves correctly for only half of the collection. According to the gradation of the accuracy, we believe that this measurement space is able to distinguish to some extent between easy and difficult problems. As all the learners involved in the experimentation failed learning the concept of the data sets located in the upper left corner, we can conclude that those data sets refer to difficult

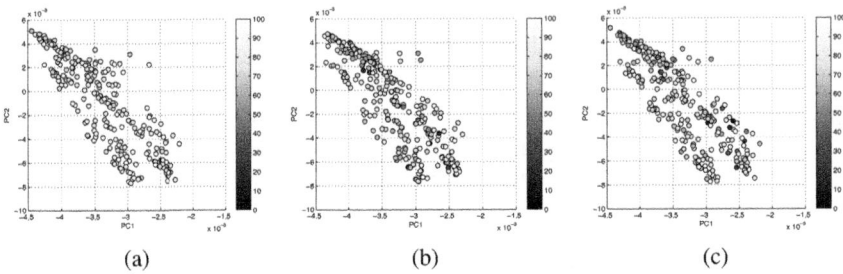

Fig. 3. C4.5 accuracies over the (a) training collection S1, (b) test collection S2, and (c) test collection S3

problems. On the other hand, regarding the algorithms based on decision trees, C4.5 and RT behave similarly whereas this pair differs from the PART results. This may indicate that we should relate the data complexity with the knowledge representation used by the learners instead of the learning paradigms.

Results with the Reserved Test Sets. Figures 3(a), 3(b), and 3(c) show the training accuracy of C4.5 over S1 and the test accuracies of C4.5 over the collections S2 and S3. These two collections, S2 and S3, were generated based on the same partition of the space used to generate S1. We used a larger random selection of problems in order to match each problem contained in S1 with problems generated with the same seed problem and are of comparable data complexity. Thus, problems contained in S2 and S3 have structurally similar counterparts in S1.

In general, we observe that the accuracies obtained during training remain in the same range as the accuracies attained during testing. However, for problems with similar complexity, the comparative advantages between classifiers are sometimes reversed. The problems seemingly easy in S1 could result in low accuracies in S2 and S3. This means that, for apparently easy problems, the accuracies are less consistent across different sample problems of the same complexity. This leads to a note of caution that data complexity alone is not sufficient to ensure similar classifier performances if the training and testing data may differ structurally. Additional measures of the structural similarity between training and testing data are needed to project classification accuracy. An extreme example is as follows: data sets with either a vertical linear boundary or a horizontal linear boundary can have the same geometric complexity; but if the learner is trained with a data set containing a vertical boundary and tested on another data set containing a horizontal boundary, the classification accuracy will be low. For data generated from the same seed problem, such large differences in structure are unlikely, but not impossible at local scales, especially when the samples are sparse.

3.4 Results with the Contestants' Classifiers

In analyzing the contestants' results, we notice that the methods proposed by Teams 1 and 2 behave similarly. For the majority of the problems they score the same accuracies. In a win/loss comparison over the 300 data sets, Team 1 outperforms the others in 121 problems and Team 2 outperforms in 61 problems. For the other 118 problems, either all

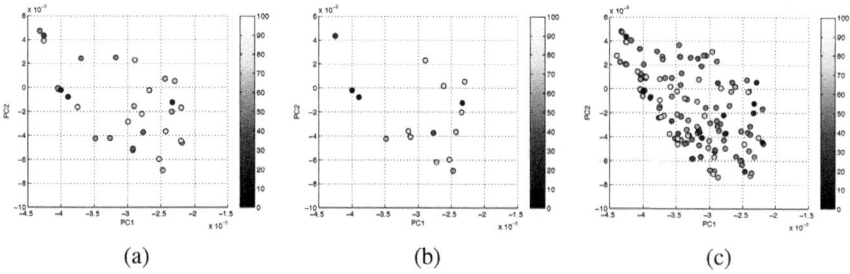

Fig. 4. Classifier accuracies over the test collection S2: (a) Team 1, (b) Team 2, and (c) Team 3

three techniques achieve the same score, or just Teams 1 and 2 come to a draw. A paired T-test shows that the difference between Teams 1 and 2 is not statistically significant, whereas their differences with Team 3 are. The accuracies of Team 3 are far below its rivals'; its average accuracy is 76% while the others' are about 92%.

Figure 4 plots the results with the reserved test set S2. For clarity, it plots only those data sets for which the learners achieved accuracies lower than 80%. In general, the number of correctly classified instances is high. Nonetheless, Figures 4(a) and 4(b) show some spots where the accuracies are extremely low and suggest performing an in-depth study with these specific data sets. Interestingly they are not located in the same region of the complexity space. For both learning paradigms, there is a common set of problems that cause degradations to their performances. These problems, despite belonging to different zones of the space, share the same underlying concept: a wave-shaped boundary (Wave Boundary). For Team 2, a checkerboard distribution poses some difficulties too. This points out the significant role of the "seeding" learning concept.

Regarding the domains of competence of classifiers, Team 2 performed a comparison with nineteen classifiers and observed that each of the nineteen classifiers they tried has a unique data set for which it is the best. This is a very interesting result, and can be a subject for further study. Figure 5 shows the location of these nineteen data sets in the PC1-PC2 space. This kind of study may help to open up a new methodology to test machine learning techniques by using prototypical data sets.

3.5 Announcement of Awards

The contest was composed of three different tests: (1) offline test, (2) real-time test, and (3) the neighborhood dominance test. (2) and (3) were held for a fixed allocation of time during the conference. The winner of each test was determined to be the followings.

Champion of the offline test: The winner is... Team 1
Their test results, obtained over the collection S2, were high with an average accuracy of 92.36%. Given that the difference to their rival's is not statistically significant, we resort to win/loss/tie counting, and the counts show that their system outperformed in 121 and behaved equally with their rival's in 118 out of 300 data sets.

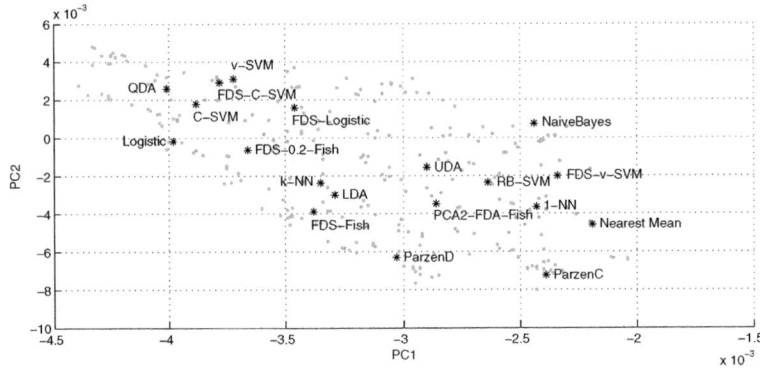

Fig. 5. Data sets within the collection S1 that are favored by the nineteen classifiers attempted by Team 2

Champion of the real-time test: The winner is... Team 1
Their test results, obtained over the collection S3, were high with an average accuracy of 81.73%, which again did not differ from their rival's with statistical significance. Yet, by win/loss/tie counting, their system outperformed in 167 and behaved equally with their rival's in 28 out of 300 data sets.

Champion of the neighborhood dominance test: The winner is... Team 2
Teams 1 and 2 were able to defend the dominance of their competence neighborhood, whereas Team 3 could not. In addition, Team 2 was unbeaten in Team 3's dominance region. While Teams 1 and 2 were again close rivals, Team 2 showed a notable dominance in its competence region, being superior to Team 1 and presenting small differences in Team 1's competence neighborhood.

We were pleasantly surprised by the unexpected exploration and discoveries the participants made using the training collection, and the perseverance the participants showed in overcoming the difficulties associated with the contest. Therefore we decided to give two more awards to the teams most accomplished in these dimensions:

Excellence in offline training: The winner is... Team 2
For their extra analysis of the behavior of nineteen classifiers, and their discovery of the nineteen "golden" data sets.

Perseverance in participation: The winner is... Team 3
For their perseverance in completing the competition despite the difficulties.

We sincerely thank all the participants for making this contest a success.

4 Discussion

The performed experiments have shown interesting results that point out the importance of the problem structure. In order to better understand the learners' behavior or to make some guidelines to choose the right learner any time, we have to address in detail three key points related to the construction of the testing framework: (1) complexity

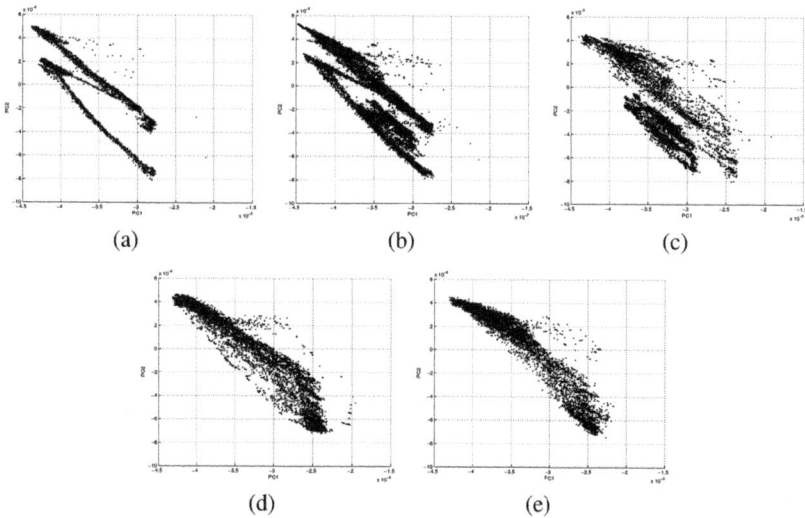

Fig. 6. Coverage derived from each seed data set: (a) Checkerboard, (b) Pima, (c) Spiral, (d) Wave Boundary, and (e) Yin Yang

measures, (2) structural dimension, and (3) completeness. In the following, each one of these aspects is elaborated.

Complexity Measures. The proposed landscape is built using eleven of the twelve complexity measures proposed in [10]. However, it is important to analyze the contribution of each measure and how the problem distribution may be modified depending on the insertion of more complexity measures or the deletion of some. Moreover, it would be interesting to determine whether this space suffices to provide some guidelines that link data characteristics to learner properties, or whether we have to carry out an individual study for each complexity measure.

Structural Limits from Seeding Data. The coverage of the proposed space is based on the difficulty of the problems originating from only five seed data sets each representing a different pair of class concepts. The nature of the seed distributions may influence the resulting testing framework. Figure 6 shows how each seed data set leads to the coverage of a different region of the space, with some overlapping. Further work should be planned to determine the effect of the seed data on the resulting coverage, and whether coverage originating from different seed data would have any significant difference.

Completeness. The two aforementioned aspects lead to the concern about whether and how the completeness of the space could be guaranteed. What is the minimum number of dimensions needed to fully represent the difficulty of a problem? Which of these dimensions are most suitable? What would be the proper seed data that have the farthest reach over the space? In our proposal, we are able to cover more than 50% of the volume of space (in the first two principal components). It is to be seen whether the remaining regions are inherently empty, or could be reached using some other seed data.

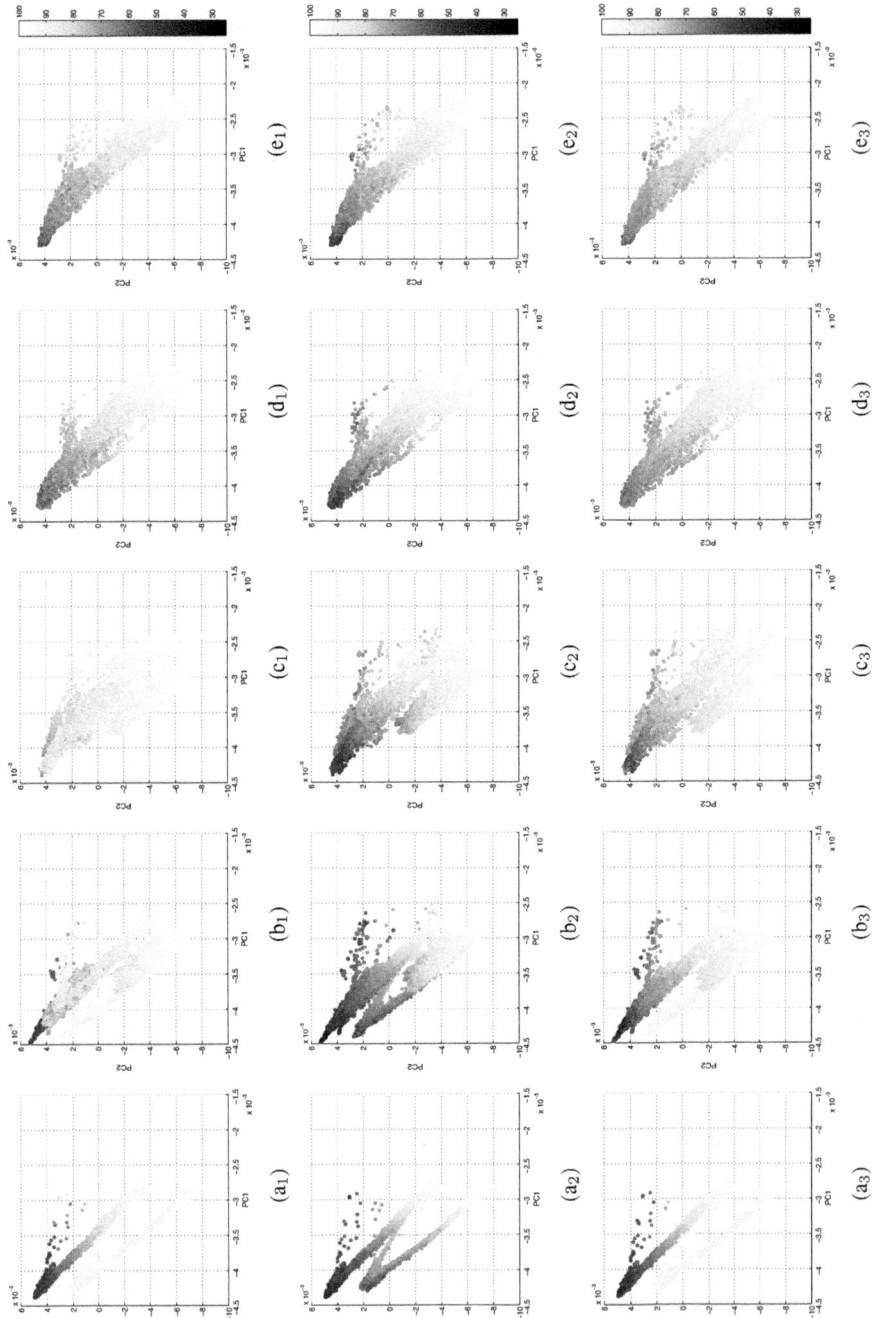

Fig. 7. C4.5, IB3, and NB test accuracy over ten-fold cross-validation (a_1, a_2, a_3) Checkerboard, (b_1, b_2, b_3) Pima, (c_1, c_2, c_3) Spiral, (d_1, d_2, d_3) Wave Boundary, and (e_1, e_2, e_3) Yin Yang respectively

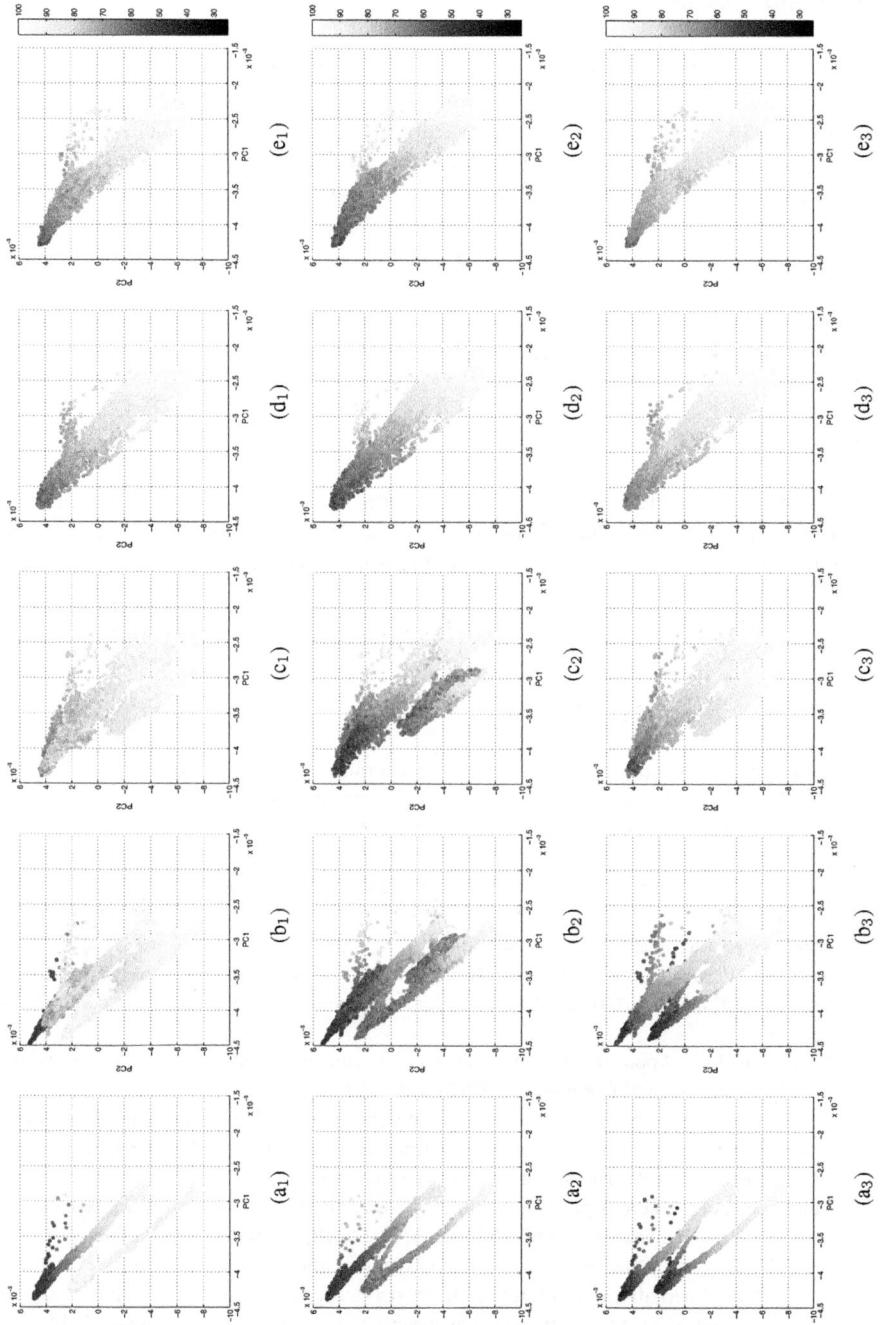

Fig. 8. PART, RT, and SMO test accuracy over ten-fold cross-validation (a_1, a_2, a_3) Checkerboard, (b_1, b_2, b_3) Pima, (c_1, c_2, c_3) Spiral, (d_1, d_2, d_3) Wave Boundary, and (e_1, e_2, e_3) Yin Yang respectively

5   Conclusions

The problems generated using our approach offer some potential to perfect classification techniques since it is possible to tailor collection of data characterized by specific complexities to evaluate and debug the learners. Nevertheless, we believe that this is only a small step towards offering guidelines to select the right learner according to the complexity of the data. The consolidation of the independent and relevant features for characterizing a problem is still a pending task. We have some early success in creating a set of problems that can reach a large range in data complexity. Analysis of classifier performances in the reached regions highlights their differences in a systematic way. We believe that we are on the way to understand the crucial role that data complexity plays in the analysis and evaluation of classification techniques.

Acknowledgements

The authors thank the *Ministerio de Educación y Ciencia* for its support under the project TIN2008-06681-C06-05, the *Generalitat de Catalunya* for the grant BE Ref. 2009BE-00, *Fundació Crèdit Andorrà*, and *Govern d'Andorra*.

References

1. Aha, D.W., Kibler, D.F., Albert, M.K.: Instance-based learning algorithms. Machine Learning 6(1), 37–66 (1991)
2. Asuncion, A., Newman, D.J.: UCI Machine Learning Repository, University of California (2007), http://www.ics.uci.edu/~mlearn/MLRepository.html
3. Basu, M., Ho, T.K. (eds.): Data Complexity in Pattern Recognition. Springer, Heidelberg (2006)
4. Breiman, L.: Random forests. Machine Learning 45(1), 5–32 (2001)
5. Coello, C.A., Lamont, G.B., Veldhuizen, D.A.V.: Evolutionary algorithms for solving multi-objective problems, 2nd edn. Springer, New York (2007)
6. Deb, K., Pratap, A., Agarwal, S., Meyarivan, T.: A fast and elitist multiobjective genetic algorithm: NSGA-II. IEEE Transactions on Evolutionary Computation 6(2), 182–197 (2002)
7. Dietterich, T.G.: Approximate statistical tests for comparing supervised classification learning algorithms. Neural Computation 10(7), 1895–1924 (1998)
8. Frank, E., Witten, I.H.: Generating accurate rule sets without global optimization. In: Proceedings of the Fifteenth International Conference on Machine Learning, pp. 144–151. Morgan Kaufmann Publishers Inc., San Francisco (1998)
9. Goldberg, D.E.: The design of innovation: Lessons from and for competent genetic algorithms, 1st edn. Kluwer Academic Publishers, Dordrecht (2002)
10. Ho, T.K., Basu, M.: Measuring the complexity of classification problems. In: Proceedings of the Fifteenth International Conference on Pattern Recognition, pp. 43–47 (2000)
11. Ho, T.K., Basu, M.: Complexity measures of supervised classification problems. IEEE Transactions on Pattern Analysis and Machine Intelligence 24(3), 289–300 (2002)

12. Macià, N., Orriols-Puig, A., Bernadó-Mansilla, E.: In search of targeted-complexity problems. In: Genetic and Evolutionary Computation Conference (2010)
13. Platt, J.: Fast training of support vector machines using sequential minimal optimization. In: Advances in Kernel Methods - Support Vector Learning. MIT Press, Cambridge (1998)
14. Quinlan, J.R.: C4.5: Programs for machine learning. Morgan Kaufmann Publishers, San Mateo (1995)
15. Vapnik, V.: The nature of statistical learning theory. Springer, New York (1995)
16. Witten, I.H., Frank, E.: Data mining: Practical machine learning tools and techniques, 2nd edn. Morgan Kaufmann, San Francisco (2005)

Feature-Based Dissimilarity Space Classification

Robert P.W. Duin[1], Marco Loog[1], Elżbieta Pękalska[2], and David M.J. Tax[1]

[1] Faculty of Electrical Engineering, Mathematics and Computer Sciences,
Delft University of Technology, The Netherlands
r.duin@ieee.org, m.loog@tudelft.nl, d.m.j.tax@tudelft.nl
[2] School of Computer Science, University of Manchester, United Kingdom
pekalska@cs.man.ac.uk

Abstract. General dissimilarity-based learning approaches have been proposed for dissimilarity data sets [1,2]. They often arise in problems in which direct comparisons of objects are made by computing pairwise distances between images, spectra, graphs or strings.

Dissimilarity-based classifiers can also be defined in vector spaces [3]. A large comparative study has not been undertaken so far. This paper compares dissimilarity-based classifiers with traditional feature-based classifiers, including linear and nonlinear SVMs, in the context of the ICPR 2010 Classifier Domains of Competence contest. It is concluded that the feature-based dissimilarity space classification performs similar or better than the linear and nonlinear SVMs, as averaged over all 301 datasets of the contest and in a large subset of its datasets. This indicates that these classifiers have their own domain of competence.

1 Introduction

Pairwise dissimilarities constitute a natural way to represent objects. They may even be judged as more fundamental than features [4]. Vector spaces defined by pairwise dissimilarities computed between objects like images, spectra and time signals offer an interesting way to bridge the gap between the structural and statistical approaches to pattern recognition [1,2]. Structural descriptions may be used by the domain experts to express their specific background knowledge [5,6]. Such descriptions often rely on graphs, strings, or normalized versions of the raw measurements, while maintaining the object connectivity in space, frequency or time. A well chosen dissimilarity measure is used to compare objects to a fixed set of representation objects. Such dissimilarity vectors construct a vector space, the so-called dissimilarity space. Traditional classifiers, designed for feature spaces, can be constructed in the dissimilarity space.

This dissimilarity approach may also be used on top of a feature representation [3]. It offers thereby an alternative to kernel approaches based on similarities. Dissimilarity measures are more general than kernels. The later have to obey the Mercer condition so that the implicit construction of classifiers, such as Support Vector Machine (SVM), can be possible in the related kernel spaces. The dissimilarity approach has the advantage that any measure can be used as well as any classifier that works in vector spaces.

It is the purpose of this paper to present a large scale comparison between traditional classifiers built in a feature vector space and some appropriate classifiers built in the

D. Ünay, Z. Çataltepe, and S. Aksoy (Eds.): ICPR 2010, LNCS 6388, pp. 46–55, 2010.

dissimilarity space defined over the original feature space. This dissimilarity space is built by the Euclidean distances to the set of chosen representation objects. The dimension of the space equals the size of the representation set. Various studies are available on the selection of this set out of the training set [7,8] and classification results depend on such a selection procedure. To simplify our experiments we will restrict ourselves to representation sets that are equal to the training set. It means that the number of training objects is identical to the dimension of the space. Consequently, we focus on classifiers in the dissimilarity space that can handle this situation.

2   The Dissimilarity Representation

The dissimilarity representation has extensively been discussed, e.g. in [1] or [9], so we will only focus here on some aspects that are essential for this paper.

Traditionally, dissimilarity measures were optimized for the performance of the nearest neighbor rule. In addition, they were also widely used in hierarchical cluster analysis. Later, the resulting dissimilarity matrices served for the construction of vector spaces and the computation of classifiers. Only more recently proximity measures have been designed for classifiers that are more general than the nearest neighbor rule. These are usually similarities and kernels (but not dissimilarities) used in combination with SVMs. So, research on the design of dissimilarity measures such that they fit to a wide range of classifiers is still in an early stage. In this paper we focus on the Euclidean distance derived in the original feature space. Most traditional feature-based classifiers use the Euclidean distance measure in one way or the other as well. It is our purpose to investigate for which datasets such classifiers can be improved by transforming the feature space into a dissimilarity space, both relying on the same Euclidean distance.

Given a set of pairwise dissimilarities between all training objects, the so-called dissimilarity matrix, we studied two ways of constructing a vector space [1]: the postulation of a dissimilarity space and a (pseudo-Euclidean) embedded space. Because the dissimilarity matrix we compute here is the Euclidean distance matrix in the feature space, the resulting embedded space is the original feature space. Therefore, we will just deal with the dissimilarity space, which is introduced now more formally.

2.1   Dissimilarity Space

Let $\mathcal{X} = \{o_1, \ldots, o_n\}$ be a training set of objects o_i, $i = 1, \ldots, n$. In general, these are not necessarily vectors but can also be real world objects or e.g. images or time signals. Given a dissimilarity function and/or dissimilarity data, we define a data-dependent mapping $D(\cdot, R) : \mathcal{X} \to \mathbb{R}^k$ from $\mathcal{X}$ to the so-called *dissimilarity space* [10,11,12]. The k-element set R consists of objects that are representative for the problem. This set, the representation or prototype set, may be a subset of $\mathcal{X}$. In the dissimilarity space each dimension $D(\cdot, p_i)$ describes a dissimilarity to a prototype p_i from R. Here, we will choose $R := \mathcal{X}$. As a result, every object is described by an n-dimensional dissimilarity vector $D(o, \mathcal{X}) = [d(o, o_1) \ \ldots \ d(o, o_n)]^T$, which is a row of the given dissimilarity matrix D. The resulting vector space is endowed with the traditional inner product and the Euclidean metric. Since we have n training objects in an n-dimensional space, a classifier such as SVM is needed to handle this situation. Other solutions such as dimension reduction by PCA or prototype selection are not considered here with one exception, i.e. the use of a representation set, randomly selected out of the training set and consisting of 20% of the training objects. We will then compare classifiers built in complete dissimilarity spaces with classifiers built in the reduced spaces (defined over smaller representation sets), yielding five times as many objects as dimensions.

Since the dissimilarity space is defined by the Euclidean distance between the objects in the feature space and, in addition, we also use Euclidean distance over the dissimilarity space, it can easily be shown that asymptotically (for growing representation sets and training sets) the nearest neighbors in the dissimilarity space are identical to the nearest neighbors in the feature space. This does not hold, however, for finite sets. This is an advantage in case of noisy features: nearest neighbors in the dissimilarity space are more reliable than in the feature space because noise is reduced by averaging in the process of computing distances.

2.2   Feature-Based Dissimilarity Space Classification

Feature-based Dissimilarity Space (FDS) classification is now defined as follows:

1. Determine all pairwise distances as an $n \times n$ dissimilarity matrix D between the n objects in the training set $\mathscr{X} = \{o_1, \ldots, o_n\}$. D_{ij} is the Euclidean distance between the i-th and j-th objects.
2. Define the dissimilarity space as a Euclidean vector space X by $X = D$. Hereby, an i-th object is represented by a dissimilarity vector of the D_{ij}-values, $j = 1, \ldots, n$.
3. Train classifiers on the n training objects represented in the n-dimensional dissimilarity space.
4. Test objects are represented in the dissimilarity space by their Euclidean distances to the objects in the training set and applied to the trained classifier in this space.

Traditional classifiers can now be used as FDS classifiers. We will study three classifiers here: the (pseudo-)Fisher linear discriminant, the logistic classifier and linear SVM. Concerning computational effort, the dimensionality is increased to the feature size n. In particular, the Fisher discriminant may suffer from this as it relies on the inverse of an $n \times n$ covariance matrix.

In order to illustrate some properties of FDS classification on a two-dimensional spiral dataset we compare two classifiers: an optimized radial basis SVM computed in the feature space (implicitly in the radial basis kernel space) and a Fisher discriminant in the dissimilarity space; see Figure 1. The later classifier is overtrained and results in a hyperplane having exactly the same distances to all objects in the training set. This works out such that in the feature space the distances to the most neighboring objects are still about equal on all positions of the spiral. This does not hold for SVM whose constant kernel is too narrow on the outside of the spiral and too large in the center.

3   Observations on the Datasets

We use the ICPR2010 Landscape contest for a systematic comparison with a set of other classifiers of the 301 contest datasets. All but one datasets are just two class problems.

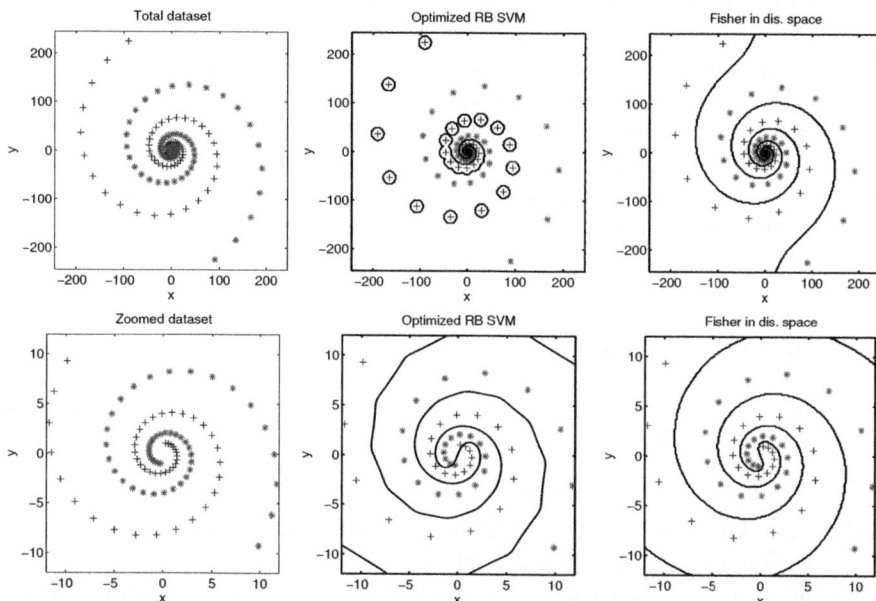

Fig. 1. A spiral example with 100 objects per class. Top row shows the complete data sets, while bottom row presents the zoom of the spiral center. 50 objects per class, systematically sampled, are used for training. The middle column shows the training set and the SVM with an optimized radial basis function; 17 out of 100 test objects are erroneously classified. The right column shows the Fisher linear discriminant (without regularization) computed in the dissimilarity space derived from the Euclidean distances. All test objects are correctly classified.

They are either 8-dimensional or 20-dimensional and the class sizes vary between 5 and 938. The largest dataset has 20 classes in 20 dimensions and has almost 10000 objects. We observe that a small amount of artificial noise has been added in a number of cases. A two-dimensional linear subspace obtained by 2-dimensional PCA is informative for a number of datasets. As we intend to compare a large set of classifiers we restrict ourselves by some choices.

1. No feature scaling is included. This would be profitable for about 100 datasets, but it would double the set of experiments to be run with and without feature scaling. For other datasets feature scaling would also emphasize the artificially added noise.
2. Since many classifiers cannot be directly applied to the set of 10000 objects, the following scheme has been followed:
   (a) The 65 datasets with more than 500 objects are split at random in equally sized subsets smaller than 500.
   (b) For each of these subsets a version of the particular classifier is trained obtaining either posterior probabilities or class confidences in the [0,1] interval.
   (c) During classification these base classifiers are combined by the sum rule.
3. Class sizes can be very skewed. It is assumed that the observed class frequencies are representative for the test sets. So no corrections are made for the skewness.

4. Due to a large set of datasets we skip very time consuming snd advanced classifiers such as adaboost, neural networks and SVMs with kernels optimized by grid search.

5. Instead of a grid search we use 'smart' choices for the regularization parameter of the SVMs and the choice of the kernels. This will not directly influence our conclusions as we use the same SVM procedures in the feature space as well as in the dissimilarity space.

6. We also include one classifier in the 2D PCA space as good results are achieved in 2D subspaces for some datasets. This would not be an obvious choice for general problems but it seems appropriate in the setting of artificially generated datasets.

4 Experiments

We train all classifiers by 10-fold cross-validation. The total execution time is about five days. The results of our experiments are presented in Table 1. The classifiers are:

1-NN, the 1-nearest neighbor rule.

k-NN, the k-nearest neighbor rule. k is optimized by LOO (Leave-One-Out) cross-validation over the training set.

ParzenC, densities estimated by a single Parzen kernel. Its width is optimized by LOO cross-validation over the training set.

ParzenD, densities estimated by different Parzen kernels per class, using an ML-estimator for the kernel width. The variances of the kernels are for every dimensions proportional to the corresponding class variances.

Nearest Mean, the nearest mean classifier.

UDA, uncorrelated discriminant analysis assuming normally distributed classes with different diagonal covariance matrices. This routine is similar to the so-called Gaussian Naive Bayes rule.

LDA, linear discriminant analysis assuming normally distributed classes with a common covariance matrix for all classes.

QDA, quadratic discriminant analysis assuming normally distributed classes with different covariance matrices.

Naive Bayes, using histogram density estimates with 10 bins per feature.

Logistic, linear logistic classifier.

FDS-0.2-Fish, feature-based dissimilarity space classifier, using randomly selected 20% of the training set for representation and the Fisher discriminant for classification.

FDS-Fish, FDSC using the (pseudo-)Fisher discriminant for classification. For a complete dissimilarity space whose dimension is equal to the size of the training set, first the null-space is removed and then the linear classifier is constructed that perfectly separates the classes in the training set. This classifier is overtrained in comparison to a linear SVM as it uses all objects as 'support' objects.

FDS-Logistic, FDSC using the linear logistic classifier. Similarly to the (pseudo-)Fisher rule, this classifier is overtrained in the complete dissimilarity space.

FDSC-C-SVM, FDSC using the C-SVM rule to compute a linear classifier. The value of C is set to 1 and is rather arbitrary.

Table 1. The averaged results per classifier: the mean classification error, the number of times the classifier is the best and the average classifier rank

	Mean error	# Best Scores	Mean rank
1-NN	0.204	6.0	13.7
k-NN	0.165	12.5	8.6
ParzenC	0.172	13.0	9.6
ParzenD	0.209	6.0	12.7
Nearest Mean	0.361	1.0	16.8
UDA	0.168	43.0	10.4
LDA	0.202	16.5	9.6
QDA	0.216	1.0	12.0
NaiveBayes	0.162	30.0	9.3
Logistic	0.204	13.5	10.4
FDS-0.2-Fish	0.191	7.5	12.1
FDS-Fish	0.162	9.0	8.5
FDS-Logistic	0.157	11.0	7.8
FDS-C-SVM	0.170	13.0	8.1
FDS-v-SVM	0.159	8.0	7.1
PCA2-FDS-Fish	0.143	70.5	7.7
C-SVM	0.195	12.0	8.5
v-SVM	0.208	12.5	9.8
RB-SVM	0.160	15.0	7.3

FDSC-v-SVM, FDSC using the v-SVM rule to compute a linear classifier. v is estimated from the class frequencies and the LOO 1-NN error. This error serves as an estimate of the number of support objects. It is corrected for the sometimes very skewed class frequencies.

PCA2-FDSC-Fish, the feature space is first reduced to two dimensions by PCA. This space is converted to a dissimilarity space, in which the (pseudo-)Fisher discriminant is computed.

C-SVM, C-SVM in a feature space with $C = 1$.

v-SVM, the v-SVM rule described above, now in a feature space.

RB-SVM, the radial basis SVM using an estimate for the radial basis function based on the Parzen kernel as found by ParzenC. As a 'smart' choice we use five times the width of the Parzen kernel as found by ParzenC and the v-SVM rule as described above.

All experiments are performed by PRTools [13]. The LIBSVM package is used for training SVM [14]. All classifiers in the dissimilarity space are linear, but they correspond to nonlinear classifiers in the original feature space thanks to the nonlinearity of Euclidean distance. All other classifiers are computed in the original feature space.

The best classifier, PCA2-FDSC-Fish, makes use of the analyst observation that a number of datasets is in fact just 2D. If we abstain from this classifier then still the dissimilarity-based classifiers perform very well, comparable or better than the radial basis SVM. A plausible explanation is that FDSC can be understood as a SVM with a variable kernel as illustrated in Section 2. It has, however, the disadvantage that the linear classifier in the dissimilarity space still depends on all objects and is not restricted to a set of support objects. It may thereby be outlier sensitive.

Table 2. Classification errors for most characteristic datasets. Best results per dataset are underlined.

	D242	D47	D200	D168	D116	D291	D180	D100	D298	D82	D183	D171	D292	D286	D97	D5	D29	D24	D218
1-NN	.120	.328	.083	.250	.350	.321	.049	.235	.611	.302	.074	.166	.049	.101	.164	.530	.526	.416	.060
k-NN	.160	.220	.175	.254	.173	.272	.046	.126	.425	.300	.056	.166	.049	.101	.128	.513	.379	.296	.047
ParzenC	.144	.254	.068	.284	.207	.289	.051	.162	.449	.296	.052	.366	.051	.126	.125	.533	.453	.322	.047
ParzenD	.132	.323	.135	.222	.357	.403	.060	.235	.618	.279	.069	.186	.051	.086	.145	.560	.440	.382	.056
Nearest Mean	.277	.427	.425	.401	.167	.580	.114	.490	.385	.378	.511	.419	.521	.373	.178	.537	.435	.330	.509
UDA	.179	.267	.099	.274	.213	.075	.034	.126	.425	.253	.078	.284	.087	.220	.092	.527	.366	.270	.073
LDA	.177	.263	.310	.252	.197	.557	.020	.129	.412	.250	.069	.304	.049	.207	.095	.517	.366	.279	.047
QDA	.170	.272	.120	.277	.280	.164	.071	.123	.495	.260	.069	.282	.067	.202	.132	.503	.362	.335	.056
NaiveBayes	.158	.293	.117	.240	.207	.170	.049	.123	.203	.265	.069	.294	.056	.188	.115	.507	.371	.305	.043
Logistic	.172	.263	.304	.254	.197	.554	.034	.129	.409	.236	.069	.301	.054	.220	.099	.513	.379	.279	.052
FDS-0.2-Fish	.219	.310	.175	.285	.277	.216	.034	.132	.498	.281	.048	.337	.067	.200	.102	.550	.440	.352	.047
FDS-Fish	.158	.289	.182	.307	.217	.167	.031	.126	.462	.289	.056	.133	.041	.086	.095	.533	.466	.365	.047
FDS-Logistic	.130	.289	.086	.240	.217	.167	.031	.123	.462	.277	.056	.137	.036	.081	.095	.543	.466	.365	.047
FDS-C-SVM	.170	.280	.188	.262	.187	.226	.031	.123	.422	.289	.065	.142	.054	.072	.089	.493	.457	.343	.043
FDS-v-SVM	.157	.246	.123	.270	.193	.236	.029	.126	.432	.289	.061	.210	.051	.136	.066	.487	.388	.288	.043
PCA2-FDS-Fish	.158	.319	.093	.302	.240	.197	.040	.123	.302	.352	.061	.161	.044	.099	.086	.120	.427	.403	.060
C-SVM	.200	.250	.286	.242	.187	.430	.029	.129	.412	.248	.061	.308	.056	.202	.086	.517	.332	.292	.043
v-SVM	.200	.254	.286	.267	.193	.528	.037	.179	.415	.272	.061	.335	.056	.244	.092	.490	.366	.249	.043
RB-SVM	.160	.254	.125	.270	.197	.174	.031	.129	.445	.279	.061	.419	.054	.136	.095	.503	.362	.288	.039

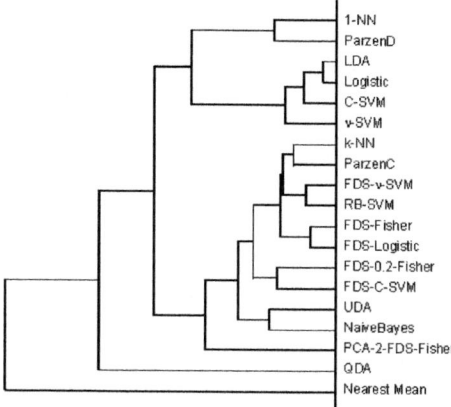

Fig. 2. Dendrogram

The classification of the test sets in the competition is based on the best classifier per dataset determined by the 10-fold cross-validation experiments. This classifier is trained by the entire training set. The performance of this classification rule can be estimated by the average of the minimum cross-validation error per dataset (of the best classifier for that dataset). This appears to be 0.106. This number is optimistically biased as it is based on the minima of 19 stochastic numbers. A better estimate would demand an additional 10-fold cross-validation loop over all classifiers. This would have increased the computation time to about 50 days. The meaning of such error estimates is of course very limited. It is an estimate of the expected error of the best classifier found by 10-fold cross-validation for a new dataset randomly selected out of the same distribution of datasets that generated the set used in contest.

We applied a cluster analysis to the 19×301 matrix of all classification errors in order to obtain a better view of the similarities of the various classifiers (as seen through the eyes of the set of contest datasets). In Figure 2 the dendrogram is shown resulting from a complete linkage cluster procedure. A number of obvious relations can be observed: k-NN and ParzenC, or the linear classifiers group (LDA, logistic and linear SVMs). The various FDS classifiers constitute a group with the RB-SVM, which make sense as all are nonlinear classifiers in the feature space.

5   Discussion

Our experiments are run on the set of contest datasets. The results show that the best linear classifiers in the dissimilarity space (FDSC-v-SVM) perform overall better than the linear as well as nonlinear SVM in the feature space. This result is of minor interest as it depends on the distribution of datasets in the contest. One should realize that for any classification rule, datasets can be either found or constructed for which this rule outperforms all other rules. As each classifier relies on assumptions, estimators or approximations, a dataset for which these are exactly fulfilled is an ideal dataset for that classifier.

Table 3. Rank correlations of the classification errors for the most characteristic datasets

	1-NN	k-NN	ParzenC	ParzenD	Nearest Mean	UDA	LDA	QDA	NaiveBayes	Logistic	FDS-0.2-Fish	FDS-Fish	FDS-Logistic	FDS-C-SVM	FDS-v-SVM	PCA2-FDS-Fish	C-SVM	v-SVM	RB-SVM
1-NN	1.0	-0.0	0.3	0.8	-0.4	-0.3	-0.5	-0.1	-0.0	-0.5	-0.3	0.3	0.7	-0.1	-0.2	0.3	-0.7	-0.7	-0.4
k-NN	-0.0	1.0	0.4	-0.0	0.2	-0.5	-0.1	-0.1	-0.2	-0.1	-0.1	-0.0	-0.1	0.1	0.2	-0.1	-0.1	0.1	-0.0
ParzenC	0.3	0.4	1.0	0.2	-0.1	-0.3	-0.5	-0.2	-0.2	-0.5	0.3	0.1	0.1	-0.3	0.2	-0.0	-0.3	-0.1	0.2
ParzenD	0.8	-0.0	0.2	1.0	-0.5	-0.4	-0.3	-0.0	0.0	-0.3	-0.1	0.2	0.7	-0.1	-0.5	-0.0	-0.4	-0.6	-0.4
Nearest Mean	-0.4	0.2	-0.1	-0.5	1.0	0.1	0.3	-0.2	0.2	0.5	-0.3	-0.5	-0.6	0.0	-0.1	-0.1	0.3	0.4	0.1
UDA	-0.3	-0.5	-0.3	-0.4	0.1	1.0	0.1	0.5	0.2	0.3	0.0	-0.3	-0.3	-0.3	-0.2	-0.1	0.2	0.1	0.1
LDA	-0.5	-0.1	-0.5	-0.3	0.3	0.1	1.0	-0.3	-0.1	0.8	-0.2	-0.4	-0.5	-0.2	0.0	-0.5	0.7	0.6	0.2
QDA	-0.1	-0.1	-0.2	-0.0	-0.2	0.5	-0.3	1.0	0.4	-0.1	0.0	-0.2	-0.0	-0.3	-0.3	0.1	-0.3	-0.2	0.0
NaiveBayes	-0.0	-0.2	-0.2	0.0	0.2	0.2	-0.1	0.4	1.0	0.1	-0.2	-0.5	-0.1	-0.0	-0.5	-0.0	-0.1	-0.1	-0.0
Logistic	-0.5	-0.1	-0.5	-0.3	0.5	0.3	0.8	-0.1	0.1	1.0	-0.4	-0.6	-0.6	-0.2	-0.1	-0.4	0.6	0.7	-0.0
FDS-0.2-Fish	-0.3	-0.1	0.3	-0.1	-0.3	0.0	-0.2	0.0	-0.2	-0.4	1.0	0.3	0.2	-0.1	-0.0	-0.2	0.1	-0.0	0.3
FDS-Fish	0.3	-0.0	0.1	0.2	-0.5	-0.3	-0.4	-0.2	-0.5	-0.6	0.3	1.0	0.7	0.3	0.1	0.4	-0.5	-0.6	-0.3
FDS-Logistic	0.7	-0.1	0.1	0.7	-0.6	-0.3	-0.5	-0.0	-0.1	-0.6	0.2	0.7	1.0	0.1	-0.2	0.3	-0.6	-0.9	-0.6
FDS-C-SVM	-0.1	0.1	-0.3	-0.1	0.0	-0.3	-0.2	-0.3	-0.0	-0.2	-0.1	0.3	0.1	1.0	0.3	0.3	-0.1	-0.1	-0.2
FDS-v-SVM	-0.2	0.2	0.2	-0.5	-0.1	-0.2	0.0	-0.3	-0.5	-0.1	-0.0	0.1	-0.2	0.3	1.0	0.1	0.1	0.3	0.3
PCA2-FDS-Fish	0.3	-0.1	-0.0	-0.0	-0.1	-0.1	-0.5	0.1	-0.0	-0.4	-0.2	0.4	0.3	0.3	0.1	1.0	-0.6	-0.4	-0.4
C-SVM	-0.7	-0.1	-0.3	-0.4	0.3	0.2	0.7	-0.3	-0.1	0.6	0.1	-0.5	-0.6	-0.1	0.1	-0.6	1.0	0.7	0.4
v-SVM	-0.7	0.1	-0.1	-0.6	0.4	0.1	0.6	-0.2	-0.1	0.7	-0.0	-0.6	-0.9	-0.1	0.3	-0.4	0.7	1.0	0.5
RB-SVM	-0.4	-0.0	0.2	-0.4	0.1	0.1	0.2	0.0	-0.0	-0.0	0.3	-0.3	-0.6	-0.2	0.3	-0.4	0.4	0.5	1.0

On the basis of the above it is of interest to observe that all classifiers that we studied here are the best ones for one or more datasets. This proves that the contest is sufficiently rich to show the variations in classification rules that we applied. Simple rules like Nearest Mean and 1-Nearest Neighbor are sometimes the best, as well as much more advanced rules like the Radial Basis SVM and the dissimilarity space classifiers.

To make the analysis less dependent on the accidental collection of problems in the contest, for every classifier we selected the dataset for which it is the best and for which the second best classifier is most different. This set of 19 datasets, one for each classifier, can be considered as a set of prototypical problems. Table 2 presents the 10-fold cross-validation errors for these prototypical datasets. Table 3 shows the rank correlations between the classifiers on the basis of the classification errors for these datasets.

In Table 2 the differences between the datasets can be clearly observed. A dataset that might be judged as very simple is D116, as Nearest Mean is the best. Dataset D291 is interesting as all linear classifiers fail and perform close to random, while UDA (Naive Gaussian) is very good. Dataset D29 shows an almost random performance for all classifiers and inspired us to include the two-dimensional subspace classifier PCA2-FDS-Fish. We were somewhat disappointed by our 'smart' choice for v in the v-SVM classifier as it turned out to be very often worse than the C-SVM with the rather arbitrary choice of $C = 1$. This holds both for feature spaces as well as dissimilarity spaces.

The similarities and dissimilarities between the classifiers can be better judged from the rank correlations between the performances on the 19 prototypical datasets; see Table 3. Strong positive correlations indicate similar classifiers, e.g. 1-NN and ParzenD, LDA and Logistic or the two linear SVMs. Strong negative correlations can be observed between the linear and nonlinear classifiers, e.g. the FDS classifiers in the dissimilarity space. It is interesting that there is no correlation between the 1-NN and k-NN rules.

6   Conclusions

We have presented a set of classifiers based on a dissimilarity representation built on top of a feature representation. Linear classifiers in the dissimilarity space correspond to nonlinear classifiers in the feature space. The nonlinearity has not to be set by some kernel but results naturally from the object distances in the training set as they are used for representation. Consequently, there are no parameters to be defined if classification rules like the Fisher discriminant or the logistic classifier are applied. The contest shows a large set of examples for which this classification scheme outperforms traditional classifiers including linear and nonlinear SVMs.

Acknowledgments. We acknowledge financial support from the FET programme within the EU FP7, under the SIMBAD project (contract 213250) as well as the Engineering and Physical Sciences Research Council in the UK.

References

1. Pękalska, E., Duin, R.: The Dissimilarity Representation for Pattern Recognition. Foundations and Applications. World Scientific, Singapore (2005)
2. Pękalska, E., Duin, R.: Beyond traditional kernels: Classification in two dissimilarity-based representation spaces. IEEE Transactions on Systems, Man, and Cybernetics, Part C: Applications and Reviews 38(6), 729–744 (2008)
3. Pękalska, E., Duin, R.P.W.: Dissimilarity-based classification for vectorial representations. In: ICPR, vol. (3), pp. 137–140 (2006)
4. Edelman, S.: Representation and Recognition in Vision. MIT Press, Cambridge (1999)
5. Riesen, K., Bunke, H.: Graph classification based on vector space embedding. IJPRAI 23(6) (2009)
6. Xiao, B., Hancock, E.R., Wilson, R.C.: Geometric characterization and clustering of graphs using heat kernel embeddings. Image Vision Comput. 28(6), 1003–1021 (2010)
7. Fischer, A., Riesen, K., Bunke, H.: An experimental study of graph classification using prototype selection. In: ICPR, pp. 1–4 (2008)
8. Pękalska, E., Duin, R., Paclík, P.: Prototype selection for dissimilarity-based classifiers. Pattern Recognition 39(2), 189–208 (2006)
9. Jacobs, D., Weinshall, D., Gdalyahu, Y.: Classification with Non-Metric Distances: Image Retrieval and Class Representation. IEEE TPAMI 22(6), 583–600 (2000)
10. Duin, R., de Ridder, D., Tax, D.: Experiments with object based discriminant functions; a featureless approach to pattern recognition. Pattern Recognition Letters 18(11-13), 1159–1166 (1997)
11. Graepel, T., Herbrich, R., Bollmann-Sdorra, P., Obermayer, K.: Classification on pairwise proximity data. In: Advances in Neural Information System Processing, vol. 11, pp. 438–444 (1999)
12. Pękalska, E., Paclík, P., Duin, R.: A Generalized Kernel Approach to Dissimilarity Based Classification. J. of Machine Learning Research 2(2), 175–211 (2002)
13. Duin, R., Juszczak, P., de Ridder, D., Paclík, P., Pękalska, E., Tax, D.: PR-Tools (2004), http://prtools.org
14. Chang, C.C., Lin, C.J.: LIBSVM: a library for support vector machines (2001), Software available at http://www.csie.ntu.edu.tw/~cjlin/libsvm

IFS-CoCo in the Landscape Contest: Description and Results

Joaquín Derrac[1], Salvador García[2], and Francisco Herrera[1]

[1] Dept. of Computer Science and Artificial Intelligence, CITIC-UGR
University of Granada. 18071, Granada, Spain
jderrac@decsai.ugr.es, herrera@decsai.ugr.es
[2] Dept. of Computer Science. University of Jaén. 23071, Jaén, Spain
sglopez@ujaen.es

Abstract. In this work, we describe the main features of IFS-CoCo, a coevolutionary method performing instance and feature selection for nearest neighbor classifiers. The coevolutionary model and several related background topics are revised, in order to present the method to the ICPR'10 contest "Classifier domains of competence: The Landscape contest". The results obtained show that our proposal is a very competitive approach in the domains considered, outperforming both the benchmark results of the contest and the nearest neighbor rule.

Keywords: Evolutionary Algorithms, Feature selection, Instance selection, Cooperative coevolution, Nearest neighbor.

1 Introduction

Data reduction [15] is one of the main process of data mining. In classification, it aims to reduce the size of the training set mainly to increase the efficiency of the training phase (by removing redundant instances) and even to reduce the classification error rate (by removing noisy instances).

The k-Nearest Neighbors classifier (NN) [3] is one of the most relevant algorithms in data mining [21]. It is a Lazy learning method [1], a classifier which does not build a model in its training phase. Instead of using a model, it is based on the instances contained in the training set. Thus, the effectiveness of the classification process relies on the quality of the training data. Also, it is important to note that its main drawback is its relative inefficiency as the size of the problem grows, regarding both the number of examples in the data set and the number of attributes which will be used in the computation of its similarity functions (distances) [2].

Instance Selection (IS) and Feature Selection (FS) are two of the most successful data reduction techniques in data mining. Both are very effective in reducing the size of the training set, filtrating and cleaning noisy data. In this way, they are able to enhance the effectiveness of classifiers (including NN), improving its accuracy and efficiency [11,12].

D. Ünay, Z. Çataltepe, and S. Aksoy (Eds.): ICPR 2010, LNCS 6388, pp. 56–65, 2010.

Evolutionary algorithms (EAs)[6] are general purpose search algorithms that use principles inspired by nature to evolve solutions to problems. In recent years, EAs have been successfully used in data mining problems[8,9], including IS and FS (defining them as combinatorial problems) [4,10].

Coevolution is a specialized trend of EAs. It tries to simultaneously manage two or more populations (also called species), to evolve them and to allow interactions among individuals of any population. The goal is to improve results achieved from each population separately. The Coevolution model has shown some interesting characteristics over the last few years [20], being applied mainly in the optimization field [19].

In this work, we show the application of IFS-CoCo (Instance and Feature Selection based on Cooperative Coevolution, already published in [5]) over the benchmark domains defined for the contest: Classifier Domains of Competence: The Landscape Contest. The performance of our approach will be tested throughout the S1 benchmark set of the contest, and compared with the 1-NN classifier and other reference results.

This work is organized as follows: Section 2 gives an overview of the background topics related to our approach. Section 3 describes the main features of IFS-CoCo. Section 4 presents our participation in the Landscape contest and the results achieved. Section 5 concludes the work.

2   Background

In this section, two main topics related with our proposal will be reviewed: Evolutionary Instance and Feature Selection (Section 2.1), and Coevolutionary Algorithms (Section 2.2). Definitions and several cases of application will be shown in order to provide a solid background to present our approach.

2.1   Evolutionary Instance and Feature Selection

In recent years, EAs have arisen as useful mechanisms for data reduction in data mining. They have been widely employed to tackle the FS and IS problems.

The FS problem can be defined as a search process of P features from an initial set of M variables, with $P <= M$. It aims to remove irrelevant and/or redundant features, with the aim of obtaining a simpler classification system, which also may improve the accuracy of the model in classification phase[12].

The IS problem can also be defined as a search process, where a reduced set S of instances is selected from the N examples of the training set, with $S <= N$. By choosing the most suitable points in the data set as instances for the reference data, the classification process can be greatly improved, concerning both efficiency and accuracy [11].

In [4], a complete study of the use of EAs in IS is done, highlighting four EAs to complete this task: Generational Genetic Algorithm (GGA), Steady-State Genetic Algorithm (SGA), CHC Adaptive Search Algorithm(CHC) [7] and Population-Based Incremental Learning (PBIL). They concluded that EAs outperform classical algorithms both in reduction rate and classification accuracy. They also concluded that CHC is the most appropriate EA to make this task, according to the algorithms they compared. Several researching efforts have been also applied to develop EA based FS methods. For example, [17] studies the capabilities of CHC applied to the FS problem.

Beyond these applications, it is important to point out that both techniques can be applied simultaneously. Despite the most natural way to combine these techniques is to use one first (e.g. IS), store its results and to apply them to the second technique (e.g. FS), some authors have already tried to get some profit from the combined use of both approaches [10].

2.2   Coevolutionary Algorithms

A Coevolutionary Algorithm (CA) is an EA which is able to manage two or more populations simultaneously. Coevolution, the field in which CAs can be classified, can be defined as the co-existence of some interacting populations, evolving simultaneously. In this manner, evolutionary biologist Price [14] defined coevolution as *reciprocally induced evolutionary change between two or more species or populations*. A wider discussion about the meaning of Coevolution in the field of EC can be found in the dissertation thesis of Wiegand [18].

The most important characteristic of Coevolution is the possibility of splitting a given problem into different parts, employing a population to handle each one separately. This allows the algorithm to employ a *divide-and-conquer* strategy, where each population can focus its efforts on solving a part of the problem. If the solutions obtained by each population are joined correctly, and the interaction between individuals is managed in a suitable way, the Coevolution model can show interesting benefits in its application.

Therefore, the interaction between individuals of different populations is key to the success of Coevolution techniques. In the literature, Coevolution is often divided into two classes, regarding the type of interaction employed:

Cooperative Coevolution: In this trend, each population evolves individuals representing a component of the final solution. Thus, a full candidate solution is obtained by joining an individual chosen from each population. In this way, increases in a collaborative fitness value are shared among individuals of all the populations of the algorithm [13].

Competitive Coevolution: In this trend, the individuals of each population compete with each other. This competition is usually represented by a decrease in the fitness value of an individual when the fitness value of its antagonist increases [16].

In this work, we will focus our interest on Cooperative Coevolution, since its scheme of collaboration offers several advantages for the development of approaches which integrate several techniques related, e.g. data reduction techniques.

3   IFS-CoCo: Instance and Feature Selection Based on Cooperative Coevolution

In this section we present IFS-CoCo, providing a description of its most important charasteristics. A full study concerning several advanced topics about its behavior (including optimization of its parameter, capabilities when applied to medium sized data sets, and more) can be found in [5].

Our approach performs several data reduction process (instance selection, feature selection, and both) in order to build a multiclassifier based on the well-known nearest neighbor rule (three 1-NN classifiers whose output is agregated by a majority rule). It defines a cooperative coevolutionary model composed of three populations, which evolve simultaneously:

 - An Instance Selection population (IS).
 - A Feature Selection population (FS).
 - A dual population, performing both Instance and Feature Selection (IFS).

Figure 1 depicts its organization (N denotes the number of instances in the training set, whereas M denotes the number of features). In isolation, the three populations can be seen as genetic-based search methods, where their respective chromosomes encode the features/instances currently selected. However, in contrast to existing evolutionary approaches for Instance Selection [4] or to wrapper approaches for Feature Selection [12], the evaluation of the quality of the chromosomes (i.e. the fitness function) is not performed in isolation.

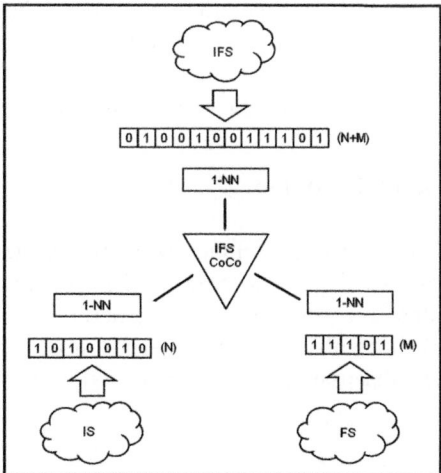

Fig. 1. Population scheme of IFS-CoCo: The three populations (IS, FS and IFS) define three 1-NN classifiers whose output are merged by a majority vote

Performing an evaluation of the fitness function of IFS-CoCo requires three chromosomes (one for each population). Once they have been gathered, the fitness value of a chromosome is computed as follows:

$$Fitness(J) = \alpha \cdot \beta \cdot clasRate(J)$$
$$+(1 - \alpha) \cdot ReductionIS(J)$$
$$+(1 - \beta) \cdot ReductionFS(J) \qquad (1)$$

– **clasRate(J):** Classification accuracy over the training set. In order to compute this accuracy, a multiclassifier is built based on three 1-NN classifiers. Each of them will employ as reference set only the subset defined by each chromosome selected.

  Thus, each chromosome defines a reduced version of the original training set, which may give a different output than the rest when classifying a training instance. In order to join these outputs, a majority voting process is performed by the 1-NN classifiers. Its result is taken as the final output of the multiclassifier.

  Finally, the *clasRate* is computed as the classification accuracy over the training set by the multiclassifier. This value is assigned to the three chromosomes employed to compute the fitness function.
– **ReductionIS(J):** Ratio of instances discarded from the original training set.
– **ReductionFS(J):** Ratio of features discarded from the original training set.

The search process of the three populations is conducted by the CHC algorithm [7]. The populations evolve sequentially, performing a generation in each step, before starting to evolve the next population. This process is carried out until the specified number of evaluations runs out. Then, the best chromosome of each population is gathered, in order to build a final multiclassifier, ready to classify test instances. This multiclassifier will work in the same manner as all the multiclassifiers employed in the coevolutionary process.

4   The Landscape Contest

In this section, we describe the experimental study performed. Since it is a part of the Landscape contest, data sets (Section 4.1) are fixed by the organizers. Comparison algorithms and configuration parameters employed are also considered (Section 4.2). Results obtained in the study are shown (Section 4.3) and analyzed (Section 4.4), discussing the strengths and limitations of our approach.

4.1   Problems

The data sets considered belong to the S1 benchmark provided by the organization[1]. It consists of 300 real-valued small data sets (with less than 1000 instances)

---

[1] http://www.salle.url.edu/ICPR10Contest/?page_id=21

and a large data set with roughly 10000 instances, being the majority of them two-class problems.

These data sets have been partitioned by using the ten fold cross-validation (10-fcv) procedure, and their values have been normalized in the interval $[0, 1]$ to equalize the influence of attributes with different ranges (a desirable charasteristic for NN-based classifiers, e.g. IFS-CoCo).

4.2   Comparison Algorithms and Configuration Parameters

In order to test the performance of IFS-CoCo, we have selected two additional methods as reference: 1-NN rule (the baseline classifier whose performance is enhanced by the preprocessing techniques of IFS-CoCo) and the benchmark results offered at the contest web page[2]. We will denote these results as *Benchmark* method throughout the study.

The configuration parameters of IFS-CoCo are the same that were used in its original presentation [5]:

- Number of evaluations: 10000
- Population size: 50 (for each population)
- α weighting factor: 0.6
- β weighting factor: 0.99

Given the scale of the experiment, we have not performed a fine-tuning process of the parameters for each data set. Instead, we have employed the same configuration in all runs, expecting a suitable behavior in all the cases (for a wider discussion about the tuning of α and β parameters, see Section 5.3 of [5]).

4.3   Results Achieved on Benchmark S1

Table 1 shows a summary of the results achieved in the S1 benchmark. Accuracy denotes the average accuracy obtained through a 10-folds cross validation procedure, whereas Wins denotes the number of data sets in which each method achieves the best result of the experiment.

Table 1. Summary of results

Method	Accuracy	Wins
IFS-CoCo	87.73	173
1-NN	74.17	10
Benchmark	81.84	117

Moreover, Figures 2 and 3 depict a comparison between IFS-CoCo and *Benchmark* or 1-NN, respectively. The dots symbolize the accuracy achieved in test

---

[2] http://www.salle.url.edu/ICPR10Contest/DataSets/TrainingAccuracy.txt

phase by the two classifiers in a concrete dataset (thus, 301 points are represented). A straight line splits the graphic, exactly at the points where the accuracy measure of both classifiers is equal. Therefore, those points below (right) of the line represent data sets where IFS-CoCo behaves better than the comparison algorithm, whereas those points above (left) of the line represent the opposite.

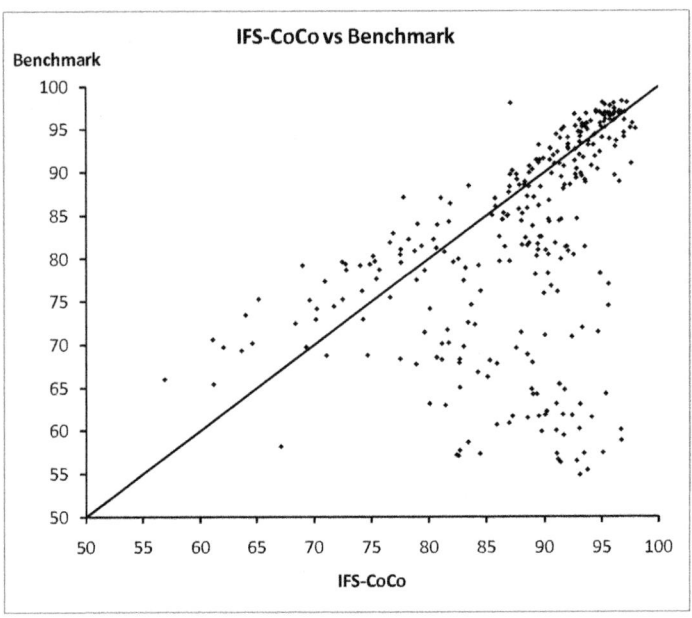

Fig. 2. Graphical comparison of IFS-CoCo vs Benchmark

These graphics emphasize the superiority of IFS-CoCo over the comparison algorithms. In comparison with *Benchmark*, our approach greatly improves its results in a large number of problems. The majority of the problems in which *Benchmark* improves IFS-CoCo are *easy* problems (those in which both classifiers achieved more than a 90% of accuracy), where there are no great differences. On the contrary, in harder problems, differences are much greater.

Furthermore, the differences between IFS-CoCo and 1-NN are even greater: There are only a few points in which 1-NN outperforms our approach, most of them depicting *easy* problems.

4.4   Strength and Limitations of Our Approach

As we have shown in the former subsection, our approach is able to improve the performance of the comparison methods. The simultaneous search for the best instances and features allows IFS-CoCo to dynamically adapt its behavior to different kinds of problems (i.e. giving more importance to features in some

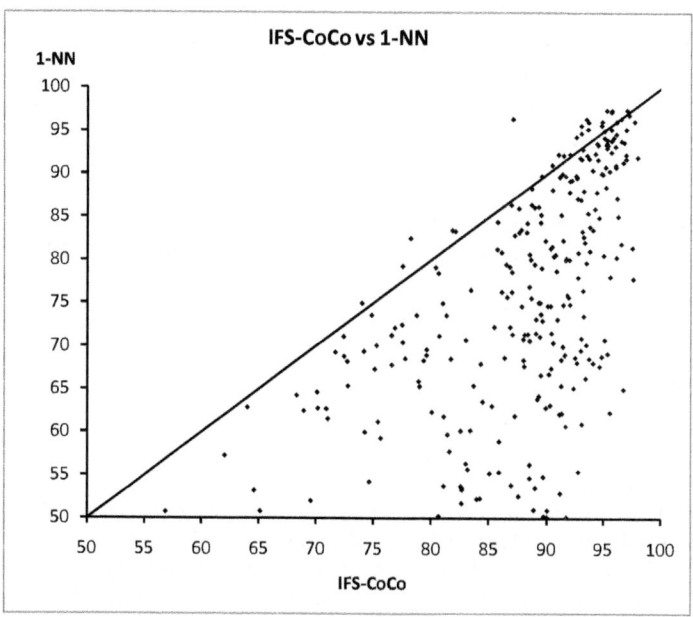

Fig. 3. Graphical comparison of IFS-CoCo vs 1-NN

problems and instances in the rest), showing a robust behavior in most of the domains considered (e.g. the greatest improvement of *Benchmark* over IFS-CoCo is in data set d60, 10.98%, whereas the greatest improvement of IFS-CoCo over benchmark is almost four times higher, 38.32%, in d17).

Moreover, the much reduced size of the subsets selected by IFS-CoCo to build the final classifier allows to classify quickly the test sets, being 1-NN often slower than the multiclassifier in test phase.

On the other hand, the main limitation of IFS-CoCo is the computation time in training phase. The cost of computing 10000 times the fitness function by means of three 1-NN classifiers is high, thus an important amount of time is required in order to let IFS-CoCo select the best possible subsets from the training data (in S1 phase, IFS-CoCo spent almost 4 days to finish the 10-folds cross validation procedure over the 301 data sets).

5   Conclusions

In this work, we have shown the preliminary results of IFS-CoCo in The Landscape contest. These results highlight the good performance of our approach in general classification domains, outperforming those achieved by the 1-NN rule. Moreover, it also outperforms the benchmark results offered by the organizers of the contest.

The main capability of our approach (the ability of working simultaneously both in the instances' and features' space) has been the key to the robust behavior shown in these problems, performing well in most of the domains considered. However, a future analysis of the characteristics of the data sets employed could give a new insight about the strength and limitations of our approach.

Acknowledgments. This work was supported by Project TIN2008-06681-C06-01. J. Derrac holds a research scholarship from the University of Granada.

References

1. Aha, D.W. (ed.): Lazy Learning. Springer, Heidelberg (1997)
2. Chen, Y., Garcia, E.K., Gupta, M.R., Rahimi, A., Cazzanti, L.: Similarity-based classification: Concepts and algorithms. Journal of Machine Learning Research 10, 747–776 (2009)
3. Cover, T.M., Hart, P.E.: Nearest neighbor pattern classification. IEEE Transactions on Information Theory 13, 21–27 (1967)
4. Cano, J.R., Herrera, F., Lozano, M.: Using evolutionary algorithms as instance selection for data reduction in KDD: An experimental study. IEEE Transactions on Evolutionary Computation 7(6), 561–575 (2003)
5. Derrac, J., García, S., Herrera, F.: IFS-CoCo: Instance and feature selection based on cooperative coevolution with nearest neighbor rule. Pattern Recognition 43(6), 2082–2105 (2010)
6. Eiben, A.E., Smith, J.E.: Introduction to Evolutionary Computing. Natural Computing. Springer, Heidelberg (2003)
7. Eshelman, L.J.: The CHC adaptative search algorithm: How to have safe search when engaging in nontraditional genetic recombination. In: Rawlins, G.J.E. (ed.) Foundations of Genetic Algorithms, pp. 265–283. Morgan Kaufmann, San Francisco (1991)
8. Freitas, A.A.: Data Mining and Knowledge Discovery with Evolutionary Algorithms. Springer, Heidelberg (2002)
9. Ghosh, A., Jain, L.C. (eds.): Evolutionary Computation in Data Mining. Springer, Heidelberg (2005)
10. Ho, S.Y., Liu, C.C., Liu, S.: Design of an optimal nearest neighbor classifier using an intelligent genetic algorithm. Pattern Recognition Letters 23(13), 1495–1503 (2002)
11. Liu, H., Motoda, H. (eds.): Instance Selection and Construction for Data Mining. The International Series in Engineering and Computer Science. Springer, Heidelberg (2001)
12. Liu, H., Motoda, H. (eds.): Computational Methods of Feature Selection. Chapman & Hall/Crc Data Mining and Knowledge Discovery Series. Chapman & Hall/CRC, Boca Raton (2007)
13. Potter, M.A., De Jong, K.A.: Cooperative coevolution: an architecture for evolving coadapted subcomponents. Evolutionary Computation 8, 1–29 (2000)
14. Price, P.W.: Biological Evolution. Saunders College Publishing, Philadelphia (1998)
15. Pyle, D.: Data Preparation for Data Mining. The Morgan Kaufmann Series in Data Management Systems. Morgan Kaufmann, San Francisco (1999)

16. Rosin, C.D., Belew, R.K.: New Methods for competitive coevolution. Evolutionary Computation 15, 1–29 (1997)
17. Whitley, D., Guerra-Salcedo, C.: Genetic search for feature subset selection: A comparison between CHC and GENESIS. In: Proceedings of the Third Annual Conference on Genetic Programming, Wisconsin, pp. 504–509 (1998)
18. Wiegand, R.P.: An Analysis of Cooperative Coevolutionary Algorithms. Ph. D. thesis, George Mason University, Fairfax, Virginia (2003)
19. Jansen, T., Wiegand, R.P.: The Cooperative Coevolutionary (1+1) EA. Evolutionary Computation 12, 405–434 (2004)
20. Wolpert, D., Macready, W.: Coevolutionary Free Lunches. IEEE Transactions on Evolutionary Computation 9, 721–735 (2005)
21. Wu, X., Kumar, V. (eds.): The Top Ten Algorithms in Data Mining. Data Mining and Knowledge Discovery. Chapman & Hall/CRC, Boca Raton (2009)

Real-Valued Negative Selection (RNS) for Classification Task

Luiz Otávio Vilas Boas Oliveira and Isabela Neves Drummond

Universidade Federal de Itajubá - UNIFEI
Av. BPS, 1303, Pinheirinho, Itajubá - MG, 37500 903, Brazil

Abstract. This work presents a classification technique based on artificial immune system (AIS). The method consists of a modification of the real-valued negative selection (RNS) algorithm for pattern recognition. Our approach considers a modification in two of the algorithm parameters: the detector radius and the number of detectors for each class. We present an illustrative example. Preliminary results obtained shows that our approach is promising. Our implementation is developed in Java using the Weka environment.

1 Introduction

Artificial Immune Systems (AIS) are computational methods inspired by biological immune systems to solve complex problems. We highlight their uses in the pattern recognition task, where the cells and molecules that do not belong to the body are recognized and eliminated by the immune system. The recognition task is identified in the process of negative selection and clonal selection, which are mechanisms present in the immune system.

The negative selection algorithm (NSA) [1] was developed for anomaly detection, with applications in computer protection, based on negative selection of T lymphocytes in the thymus. The algorithm is implemented in two phases: censoring and monitoring. In the censoring phase, we define the set of self strings (S) to be protected. Then, random strings are generated and they are match against the strings in S. If a string match any one in S, it is rejected. Otherwise, the string is stored in a set of detectors (R). In the monitoring, the state of self strings can be monitored by continually matching strings in S against strings in the defined collection R. If a detector is ever activated, it must be characterized as a non-self string.

In this work, we proceed to investigate how the detector radius and the number of the detectors for each class can be modified in order to improve the algorithm. Our initial proposal involves the calculation of the area covered by each detector and the possibility of finding the best number of detectors per class.

This paper is organized as follows. In Section 2 we briefly study the original NSA algorithm for real values. In Section 3, we formally present our modified RNS describing the calculation of the modified parameters. In Section 4 we show an application of the proposed approach, using an illustrative example. In

D. Ünay, Z. Çataltepe, and S. Aksoy (Eds.): ICPR 2010, LNCS 6388, pp. 66–74, 2010.

Section 5 we present the results obtained for the contest using the proposal data sets, and finally, Section 6 states the conclusion and points to future research directions.

2 Real-Valued Negative Selection Algorithm (RNS)

The time and space complexity of the NSA is exponential on the window size used for the detection (the number of bits that should be used when comparing two binary strings). This may represent a scalability limitation, since a large window size may be necessary [2].

Another important issue is the representation of the problem space. For some applications, the binary coding creates difficulty. A real-valued representation can be employed instead to characterize the self / non-self space and evolve another set of points (detectors) to cover the non-self space.

The Real-Valued Negative Selection (RNS) [2] uses as input the self samples, which are represented by n-dimensional points, and tries to evolve detectors (another set of points) to cover the non-self space. From an initial set of randomly generated detectors, an iterative process is performed to update the positions of the detectors driven by two goals: the first is to keep the detector away from the self points and the second is to maximize the coverage of non-self space by keeping the detectors separated.

The portion of the hyperspace covered by the detectors can be configured through the parameter r that specifies the radius of detection for each detector. Since it is not desirable that the detectors match self points, r is also the shortest allowed distance between a detector and a self point.

To determine if a detector d matches a self point, the algorithm identifies the k nearest neighbors of d in the self set. Then, we calculate the median distance of the k neighbors. If this distance is less than r, it is considered that the detector d matches the self point. This strategy makes the algorithm more robust to noise and outliers [2].

The function $\mu_d(x)$ is the membership function of the detector d. It is the membership degree of x to the detection space around d, and is defined by Equation 1:

$$\mu_d(x) = e^{-\frac{|d-x|^2}{2r^2}} \qquad (1)$$

For each detector we assign an age, which is increased every time the detector is inside the self set. If the detector reaches a certain age (indicated by the parameter t) without leaving the self space, it becomes "old", and it is replaced by a new randomly generated detector. The age of a detector that is not in self space is always zero.

If the distance between the detector d and the nearest self point is less than the radius r, i.e., if the detector d matches self points, we verify its age t to determine if it should be replaced. If the detector is not replaced, we calculate the motion direction of the detector d in the space, taking into account only the self points. The motion direction is represented by a set of points dir. If the detector does not match the self points, the direction is calculated taking into account only the detectors. This direction is used to determine the new position of the detector.

3   The RNS-Based Classifier

The original algorithm creates detectors to recognize non-self antigens, discarding the detectors that match the self set. For pattern recognition, in particular, the classification task, it is necessary that the detectors are able to separate possible classes in order to make a correct classification of the input element. Thus, for each class we consider the points belonging to the other classes as self points, and the ones belonging to the class as non-self.

We generate a set of detectors for each class, from the non-self points, which are moved away from the self region. In this approach we do not employ the parameter r, described in previous section. Instead of r given by the user, each detector has its own *radius*, calculated based on the higher value between two distances: the one obtained between the detector and the nearest self point, and the distance obtained between the detector and the nearest non-self point. Figure 1 shows these two distances as n_1 and n_2 in a simple example, and illustrates a radius for one detector. In this way, each detector has its own *radius*. They are different among themselves, and are dependent on the data set.

Our second modification is made to the number of detectors for each class, considering that not always when we have a data set with k elements and n classes, these classes are composed of k/n elements. In many real problems, the sizes of the classes are not proportional or equal to each other. In a previous test we used an overall percentage of the detectors generated by the method as an input parameter. Thus, this percentage is multiplied by the total number of instances in the data set presented as input. The number of detectors so obtained are divided equally among the classes.

We can easily verify that this is not the best way to find the number of the detectors we need for each class, because the number of instances per class is not the same, and in some cases can be very different. Therefore, our approach still has

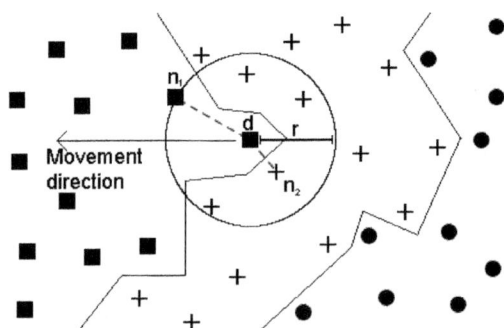

Fig. 1. Representation of a three classes problem: The radius of detector d based on distances n_1 and n_2

the percentage as a parameter of the algorithm, but we employ it to determine the number of detectors for each class. For example, if a given data set has 20 instances for class A and 80 for class B, and the user sets 10% as detectors percentage, it will generate 2 detectors for class A and 8 for class B, which is different from the initial approach that would use 5 detectors for each class.

Algorithm 1. *MODIFIED-RNS* $(\eta,\ t,\ k,\ c,\ \gamma)$
η: *adaptation rate, i.e., the rate at which the detectors will adapt on each step*
t: *once a detector reaches this age it will be considered mature*
k: *number of neighbors to take into account*
c: *number of steps*
γ: *detectors rate, i.e., percentage of detectors to be generated*
Auxiliary functions
Calculate-Radius(d)
d: detector whose radius will be calculated
nearestSelf = find the nearest self neighbor
nearestNonSelf = find the nearest non-self neighbor
$n_1 = distance(d,nearestSelf)$
$n_2 = distance(d,nearestSelf)$
$d.radius = max(n_1,n_2)$

    for $i = 1$ **to** c **do**
      for *each class of the input set* **do**
      (*Tested class is considered non-self and remainder is considered self*)
      $detectorsNumber = \gamma * numberofinstances/numberofclasses$
      Generates a random population with detectorsNumber detectors
      (*This population contains only non-self points*)
        for *each detector d* **do**
        NearCells =k-nearest neighbors of d in the self set
        NearCells is ordered with respect to the distance to d.
        $NearestSelf = median\ of\ NearCells$
        if $distance(d, NearestSelf) < d.radius$
         then   *Increase age of d*
         $dir = \frac{\sum_{c \in NearCells}(d-c)}{|NearCells|}$
        if *age of d* $> t$ *(*Detector is old*)*
         then *Replace d by a new random detector*
        else
        $d = d + \eta * dir$
        Calculate-Radius(d)
        end-if
        else
        age of d $= 0$
        $dir = \frac{\sum_{d' \in Detectors} \mu_d(d')(d-d')}{\sum_{d' \in Detectors} \mu_d(d')}$
        $d = d + \eta * dir$
        Calculate-Radius(d)
        end-if
        end-for
      end-for
    end-for

We are also investigating another approach. Considering that the maximum number of detectors for each class is the number of instances given as input, we propose a study of the percentage variation of the detectors number. There is an expectation that the greater the number of detectors is, the better the classifier we obtain, because in our method the detectors determine the classifier, i.e., a set of detectors defines a region for each class. However, since each detector has a radius trying to cover a non-self space, we know that this region depends on the data set and how their points span the class space. Thus, in order to check the best number of detectors for each class on the different classification problems, this approach is proposed.

After the detectors are generated, the classification of new elements is performed. Given the input, we calculate all the distances from the input to each detector. The shortest distance defines the element's class.

Besides the commented parameters, the method still work as in the original algorithm, with an adaptation rate that drives the evolution of the algorithm controlling the movement of the detector in the problem space; and the age of a detector at which the algorithm will substitute it by a new one, like it was detailed in Section 2. Algorithm 1 presents the schema of the method.

4   Illustrative Example

In the following we present an illustrative example using the Wine data set available at the UCI Machine Learning Repository [3]. The data are from a chemical analysis of wines grown in the same region but derived from three different cultivars. Each of the 178 instances has 13 attributes and the class sizes are 59, 71 and 48 respectively. In a classification context, it is not a very challenging data set, but a good one for first testing a classifier.

We perform our tests using the Weka environment. Our implementation is in Java language taking data in the Weka format. We define a 10-fold cross-validation test and 100 performances in Weka experimenter. The data sets were preprocessed to normalize their attribute values. For result analysis and comparison, we calculate the rate of correctly classified instances (classifier accuracy) and the *kappa* index.

For the modified RNS method considered in this work, we have also used the adaptation rate, detector age, and number of neighbors as parameters. The results showed in Table 1 include the 5 best classifiers' accuracies among over 100 runs, in order to illustrate the ranges of the parameters. In this case, the radius of each detector is a fixed value.

The proposed radius modification was performed in a second experiment, where we hold the best configuration obtained for the first test, taking as fixed values: adaptation rate, number of neighbors and detector age; and with one radius for each detector (calculated as showed in Figure 1), varying the number of detectors considered. These results are presented in Table 2 and we can observe that the best classifiers have as number of detectors a value corresponding to 80% of the number of instances for each class. Note that this rate defines the

Table 1. Classification results for Wine data set using fixed radius and 10% of detectors proportional to each class of the problem

#	adapt. rate	radius	age	n. neighbors	classif. acc	kappa
1	0.000150	0.5	1	6	92.23	0.88
2	0.000025	3.0	1	6	92.74	0.89
3	0.000025	0.5	1	6	92.18	0.88
4	0.000025	0.5	1	6	91.91	0.88
5	0.000025	0.5	1	11	92.29	0.88

Table 2. Classification results for Wine dataset varying the detectors percentage from 10% to 100% of each entry class

detectors rate	classifier accuracy
0.10	91.73
0.20	93.71
0.30	93.96
0.40	94.61
0.50	94.27
0.60	94.27
0.70	94.76
0.80	95.12
0.90	94.50
1.00	94.50

number of random detectors that will be generated for the classifier according to the size of the class.

In the current phase of our project we are testing the method for image classification. Some preliminary results show that the approach can be employed as an image classifier. Our first test was carried out using synthetic MR (Magnetic Resonance) brain images available at the Simulated Brain Database (SBD) [4]. Some preliminary results are presented in Table 3. The experiment was performed using the Weka environment in the same way we did for the Wine data set described above.

Table 3. Classification results for MRI data set using fixed radius and 10% of detectors proportional to each class of the problem

#	adapt. rate	radius	age	n. neighbors	classif. acc	kappa
1	0.000025	0.4	1	3	84.63	0.81
2	0.000025	0.4	1	3	84.63	0.81
3	0.000025	0.4	2	3	84.66	0.81
4	0.000025	0.4	1	3	84.65	0.81
5	0.000025	0.4	1	3	84.63	0.81

5   Experimental Results

In this section we present the results obtained for "The Landscape Contest" with brief comments. For the offline test we have run our algorithm over S1 and S2 problems. The S1 data set is composed of 301 subsets as training bases for the classifier, and we got the results using a 10-fold cross-validation. The analysis is done over the predictive accuracy for each subset in the S1 collection. The results are present in Tables 4, 5, 6 and 7. The S2 collection is given without class labeling, and is provided to test the classifier. Probably in a later work we will have the S2 data sets as a benchmark to evaluate a classifier.

We have observed a predictive accuracy of over 60% for 168 subsets, somewhat around half of the training bases in S1. For almost 100 subsets the classifier reaches accuracies of over 80%. The subsets in S1 are two-class problems, except for the last one that is a multiclass problem with 20 classes. A brief analysis shows that better results (correct rates of over 80%) were obtained from subsets with some similar features: they have 230 to 540 instances and 9 and 21 attributes. A more careful analysis should be made, in order to identify whether the results

Table 4. Classification results for S1 problem: sets D1 to D80

set	rate	set	rate	set	rate	set	rate	set	rate	set	rate	set	rate	set	rate
D1	50.16	D11	64.80	D21	51.55	D31	51.22	D41	68.13	D51	68.75	D61	84.49	D71	68.21
D2	53.67	D12	57.34	D22	61.80	D32	78.87	D42	58.54	D52	56.08	D62	80.27	D72	77.22
D3	51.09	D13	49.40	D23	55.51	D33	76.11	D43	50.35	D53	48.97	D63	81.55	D73	62.55
D4	50.16	D14	49.65	D24	55.36	D34	64.47	D44	70.06	D54	52.11	D64	75.71	D74	57.42
D5	53.00	D15	50.05	D25	49.14	D35	65.89	D45	73.00	D55	56.23	D65	73.20	D75	71.93
D6	63.00	D16	49.21	D26	56.42	D36	66.12	D46	50.87	D56	52.97	D66	76.19	D76	50.80
D7	62.93	D17	50.07	D27	52.49	D37	64.23	D47	52.15	D57	86.71	D67	71.39	D77	75.89
D8	67.84	D18	50.16	D28	47.98	D38	65.40	D48	61.88	D58	84.33	D68	50.62	D78	71.15
D9	66.77	D19	52.64	D29	49.56	D39	64.22	D49	68.83	D59	84.26	D69	64.33	D79	47.24
D10	66.22	D20	53.79	D30	51.94	D40	63.81	D50	56.74	D60	77.07	D70	66.45	D80	74.58

Table 5. Classification results for S1 problem: sets D81 to D150

set	rate	set	rate	set	rate	set	rate	set	rate	set	rate	set	rate
D81	48.06	D91	89.42	D101	87.20	D111	54.15	D121	60.53	D131	95.28	D141	54.81
D82	53.37	D92	90.36	D102	57.65	D112	80.74	D122	43.67	D132	95.32	D142	65.15
D83	68.08	D93	83.44	D103	82.66	D113	76.97	D123	58.44	D133	92.02	D143	66.13
D84	51.45	D94	83.44	D104	75.42	D114	81.33	D124	47.77	D134	92.32	D144	86.92
D85	55.86	D95	79.60	D105	78.11	D115	82.39	D125	53.67	D135	92.98	D145	90.33
D86	54.76	D96	84.22	D106	77.83	D116	52.00	D126	51.21	D136	91.76	D146	59.43
D87	50.94	D97	84.86	D107	52.77	D117	78.61	D127	48.76	D137	90.27	D147	62.64
D88	46.69	D98	84.26	D108	59.46	D118	79.17	D128	59.44	D138	90.29	D148	59.89
D89	46.69	D99	84.93	D109	59.21	D119	53.30	D129	52.76	D139	91.40	D149	64.26
D90	50.07	D100	87.08	D110	65.82	D120	76.13	D130	53.48	D140	91.56	D150	91.07

Table 6. Classification results for S1 problem: sets D151 to D228

set	rate	set	rate	set	rate	set	rate	set	rate	set	rate	set	rate
D151	61.71	D162	77.91	D173	51.07	D184	55.59	D195	92.84	D206	78.85	D218	95.68
D152	88.29	D163	68.60	D174	50.17	D185	68.09	D196	94.15	D207	65.54	D219	96.70
D153	88.32	D164	53.37	D175	53.44	D186	84.42	D197	92.02	D208	53.53	D220	96.83
D154	90.15	D165	50.43	D176	46.38	D187	91.81	D198	84.85	D209	49.70	D221	96.96
D155	73.13	D166	54.09	D177	96.91	D188	58.47	D199	86.93	D210	74.48	D222	90.43
D156	86.04	D167	50.71	D178	96.67	D189	94.69	D200	71.42	D211	40.38	D223	96.08
D157	85.25	D168	55.38	D179	92.38	D190	91.05	D201	90.98	D212	63.05	D224	95.66
D158	69.58	D169	50.99	D180	92.28	D191	86.75	D202	87.65	D214	53.21	D225	93.10
D159	84.61	D170	54.39	D181	95.92	D192	92.71	D203	85.66	D215	51.22	D226	96.08
D160	85.81	D171	51.08	D182	94.96	D193	92.32	D204	79.06	D216	96.96	D227	94.02
D161	49.90	D172	55.69	D183	88.74	D194	94.93	D205	76.17	D217	96.10	D228	83.69

Table 7. Classification results for S1 problem: sets D229 to D301

set	rate	set	rate	set	rate	set	rate	set	rate	set	rate	set	rate
D229	96.17	D240	92.68	D251	53.69	D262	96.52	D273	84.54	D284	49.50	D295	94.60
D230	93.47	D241	52.78	D252	96.85	D263	94.55	D274	87.36	D285	52.93	D296	52.99
D231	66.54	D242	79.65	D253	96.99	D264	61.43	D275	87.24	D286	44.19	D297	46.32
D232	92.62	D243	78.25	D254	98.10	D265	62.74	D276	87.22	D287	49.25	D298	53.48
D233	95.21	D244	71.36	D255	97.39	D266	94.00	D277	52.58	D288	52.98	D299	50.33
D234	94.63	D245	45.56	D256	95.70	D267	95.13	D278	72.16	D289	50.00	D300	56.62
D235	86.39	D246	79.65	D257	90.07	D268	54.22	D279	57.63	D290	52.13	D301	11.95
D236	92.33	D247	53.82	D258	97.83	D269	58.26	D280	50.00	D291	50.16		
D237	92.63	D248	56.52	D259	95.52	D270	67.06	D281	48.77	D292	94.10		
D238	91.11	D249	65.47	D260	98.01	D271	68.23	D282	47.76	D293	95.07		
D239	90.93	D250	49.21	D261	96.08	D272	84.51	D283	47.27	D294	93.86		

depend on some other features of the data distribution in the space, such as the number of instances per class or the dimensionality of the problem.

6 Conclusion

In this work, we have described a negative selection based algorithm. We have investigated the effects of detector radius and the number of detectors per class, changing them from the original version of the algorithm. To illustrate our approach we work on two different applications, a data base from UCI repository [3] and an image classification problem using brain magnetic resonance images. The implementation was applied to other data bases available at UCI, such as Iris Plant, Spambase and Optdigits. The initial experiments have led to promising results.

For the contest proposed, our results were reported, but the rates obtained should be analyzed along with the data complexity measures, so that the method can be better characterized. We are still working on the parameters of the RNS-based classifier described here. In the future, new ways for determining the detector radius will be studied and tested. We are also testing the method for classifying a protein data base and satellite images.

Acknowledgements

Isabela Neves Drummond acknowledges financial support from CNPq (process 478684/2009-6).

References

1. Forrest, S., Perelson, A.S., Allen, L., Cherukuri, R.: Self-nonself discrimination in a computer. In: IEEE Symposium on Research in Security and Privacy. IEEE Computer Society Press, Los Alamitos (1994)
2. Gonzalez, F., Dasgupta, D., Kozma, R.: Combining negative selection and classification techniques for anomaly detection. In: Congress on Evolutionary Computation, Hawaii, pp. 705–710. IEEE, Los Alamitos (2002)
3. Frank, A., Asuncion, A.: UCI machine learning repository (2010), http://archive.ics.uci.edu/ml/
4. BrainWeb: Simulated brain database (June 2006), http://www.bic.mni.mcgill.ca/brainweb/

Graph Embedding for Pattern Recognition

Pasquale Foggia and Mario Vento

Dip. di Ing. dell'Informazione e Ing. Elettrica
Università di Salerno
Via Ponte don Melillo, I-84084 Fisciano (SA), Italy
pfoggia@unisa.it, mvento@unisa.it
http://nerone.diiie.unisa.it/contest

Abstract. This is the report of the first contest on Graph Embedding for Pattern Recognition, hosted at the ICPR2010 conference in Istanbul. The aim is to define an effective algorithm to represent graph-based structures in terms of vector spaces, to enable the use of the methodologies and tools developed in the statistical Pattern Recognition field. For this contest, a large dataset of graphs derived from three available image databases has been constructed, and a quantitative performance measure has been defined. Using this measure, the algorithms submitted by the contest participants have been experimentally evaluated.

1 Introduction

In Pattern Recognition, statistical and structural methods have been traditionally considered as two rather separate worlds, although many researchers have attempted to reduce the distance between these two approaches. The goal inspiring these attempts was to find a way for exploiting the advantages of a structural representation (such as a graph) in terms of expressiveness, and at the same time preserving the ability to use the wealth of effective vector-based algorithms from Statistical Pattern Recognition.

A possible solution to this issue is the use of Graph Embedding, which is a methodology aimed at representing a whole graph (possibly with attributes attached to its nodes and edges) as a point in a suitable vectorial space. Of course there are countless ways for mapping a graph to a vector; but interesting embeddings are characterized by the fact that they preserve the similarity of the graphs: the more two graphs are similar, the closer the corresponding points are in the target vector space.

Graph embedding, in this sense, is a real bridge joining the two worlds: once the object at hand has been described in terms of graphs, and the latter represented in the vectorial space, all the problems of matching, learning and clustering can be performed using classical Statistical Pattern Recognition algorithms.

Possible approaches proposed for performing the graph embedding include spectral methods [8,14], based on the eigendecomposition of the adjacency matrix or of some matrix related to it. In [2], a statistical technique known as

D. Ünay, Z. Çataltepe, and S. Aksoy (Eds.): ICPR 2010, LNCS 6388, pp. 75–82, 2010.

Multi-Dimensional Scaling (MDS) is used to embed a graph, characterized by the matrix of the geodesic distances between nodes, into a manifold. In [13], an embedding algorithm for the special case of trees is proposed. The algorithm is based on the computation of the minimum common super-tree of a set of trees; then each tree is represented as a vector whose elements encode the nodes of the super-tree which are present in the tree. In [4] random walks are used to derive a graph embedding; in particular the embedding encodes the expected time needed for a random walk to travel between two nodes. In [11] the embedding is built by choosing at random a small set of graphs as prototypes, and representing a graph with the vector of the graph-edit distances from each prototype. Since graph-edit distance calculation is an NP-complete problem, an approximation of this measure is actually used.

While an explicit graph embedding enables the use of all the methodologies and techniques devised for vector spaces, for many of these techniques it would be sufficient to have an algorithm that just performs the scalar product between the vectors associated to two graphs. Recently there has been a growing interest about the so called *graph kernels*, that are functions applied on graphs that satisfy the main mathematical properties of a scalar product; see [7,3,10] for some examples. In a sense, such methods could be considered as an implicit kind of graph embedding, although such an embedding would not permit all the operations defined on a vector space. For the sake of generality, we have decided to open the contest also to implicit graph embedding techniques.

While the authors proposing explicit or implicit embedding methods have usually tested their algorithms on graphs derived from specific applications, it is difficult to compare them to each other for the lack of a common, standardized benchmark. Thus the main motivation for this contest is to provide a direct comparison between different methods. Of course we do not expect that a single method can outperform all the others on every application; it is more likely that changing the application, and thus the characteristics of the graphs, the algorithm with the best performance will be a different one.

Since many applications of interest for the Pattern Recognition community work on 2D images, we have chosen to test the algorithms on graphs derived by 2D images. Of course 2D images do not exhaust the possible variety of graphs characteristics; in future editions of the contest the comparison will be possibly extended to other kinds of input data.

We have chosen a method to encode each input image as a graph. Such an encoding is by no way unique; actually we have chosen, among the many possibilities proposed in the literature, a representation, the *Region Adjacency Graph*, that is relatively easy to compute, yields graphs which are not very large (say tens to hundreds of nodes for typical images), but still retains sufficient information about the image structure for many interesting applications such as classification, clustering and retrieval, as shown by its wide adoption in the past years.

Then we have defined a performance measure to evaluate the ability of the embedding to ensure that similar graphs are mapped to similar vectors. We have not based this measure on the graph-edit distance for two reasons:

- it would have been computationally impractical to compute the graph-edit distance between any to graphs, except for very small graphs;
- furthermore, since we are ultimately interested in verifying the effectiveness of graph embedding techniques to solve Pattern Recognition problems, it is not obvious that graph-edit distance is the definition of graph similarity more appropriate towards this goal.

Therefore, the chosen performance index is based on the use of a cluster validity index, and on the assumption that the classes of the image subjects are strongly related to the similarity concept we are trying to capture with the embedding.

In the rest of the paper we will describe in more detail the construction of the databases used for the contest and the definition of the performance index. Then a short description of the contest participants will be provided, followed by a presentation of the results of our experiments.

2   The Databases

For this contest we have chosen three large image databases publicly available: the Amsterdam Library of Object Images (ALOI), the Columbia Object Image Library (COIL) and the Object Databank by Carnegie-Mellon University (ODBK).

Fig. 1. Some examples of the images in the ALOI database

The Amsterdam Library of Object Images [5] is a collection of images of 1000 small objects. Each object has been acquired several times changing the orientation and the illumination, for a total of 110,250 images. See fig. 1 for some examples. From this database, we have selected 50 objects and 72 views for each object, for a total of 3600 images.

The Columbia Object Image Library [9] is a collection of images of 100 small objects, each acquired from 72 different orientations. See fig. 2 for some examples of the pictures. Also from this database we have selected 50 objects, for a total of 3600 images.

Fig. 2. Some examples of the images in the COIL database

Fig. 3. Some examples of the images in the ODBK database

The Object Databank [12] has been obtained from a collection of 209 3D object models, that have been rendered with photo-realistic quality using 14 different view points. See fig. 3 for some examples of the pictures. From this database we have used 208 objects and 12 of the 14 view points, for a total of 2496 images.

The images of each database have been divided into a first set that has been distributed to the contest participants before the contest evaluation, in order to allow them to tune their algorithms, and a second set, not shared with the participants, that has been used for the final performance evaluation.

Each image has been smoothed using a Gaussian filter, and then it has been segmented using a Pyramidal segmentation algorithm (provided by the OpenCV

Table 1. Statistics on the graphs in the datasets. The aloi-1, coil-1 and odbk-1 datasets have been provided to the contest participants, while the aloi-2, coil-2 and odbk-2 dataset have been used for the final performance evaluation.

	aloi-1	aloi-2	coil-1	coil-2	odbk-1	odbk-2
Max nodes	103	134	107	100	636	528
Avg nodes	29.24	18.37	33.31	34.88	65.18	56.91
Max edges	112	156	97	92	598	519
Avg edges	28.37	17.25	32.30	32.33	62.45	54.37
N. of graphs	1800	1800	1800	1800	1248	1248
N. of classes	25	25	25	25	104	104

library [1]). Finally, from the segmentation the Region Adjancency graph has been constructed as the image graph-based representation. The nodes of the graph have as attributes the relative size and the average color components of the corresponding regions, while the edges of the graph have no attributes.

In Table 1 some statistics about the databases are provided.

3   The Performance Index

In order to measure the performance of an embedding algorithm, we have considered the dataset divided in classes on the basis of the object represented in the original image, and then we have used a clustering validation index to evaluate the separation between the classes when represented by the vectors explicitly or implicitly produced by the algorithm.

More specifically, we compute first the distances d_{ij} between each pair of graphs g_i and g_j; for the explicit embedding, d_{ij} is the Euclidean distance, while for the implicit embedding the distance is computed as follows:

$$d_{ij} = \sqrt{p_{ii} + p_{jj} - 2p_{ij}}$$

where p_{ij} is the scalar product between g_i and g_j.

Given the distances, the *C index* by Hubert and Schultz [6] is computed. The C index is defined as follows: first we compute the set S_w of the distances d_{ij} such that g_i and g_j lie in the same class; M is the cardinality of S_w. Then, the sets S_{min} and S_{max} are computed taking respectively the M shortest distances and the M largest distances among all the possible values of d_{ij}. Finally, the index is computed as:

$$C = \frac{sum(S_w) - sum(S_{min})}{sum(S_{max}) - sum(S_{min})}$$

Notice that the smaller the value, the better is the separation of the classes; the index value is in the interval $[0, 1]$ reaching 0 in the ideal case in which all the inter-class distances are smaller than all the intra-class distances.

The C index has been chosen on the basis of the following considerations:

- it requires only the distances d_{ij}, and not the centroids of each class, that would not be trivial to define for the implicit embeddings;
- the value does not change if all the distances are multiplied by a same positive costant, so it is independent of the scale used for the vectors or the products;
- it provides an integral measure that is not significantly affected by outliers.

Since we have three datasets, a way to combine the three corresponding indices must be defined in order to obtain a single figure to evaluate an algorithm. We have decided to use the geometric mean of the indices for attributing a ranking to each algorithm.

Please remember that for the contest evaluation, three different sets of graphs (aloi-2, coil-2 and odbk-2) have been used from the ones provided in advance to the contest participants (aloi-1, coil-1 and odbk-1).

4   Contest Participants

The table 2 presents a synoptic view of the contest participants, for which a short description will be provided thereafter.

Table 2. The contest participants

Team	Institution	Type	Impl. language
Y. Osmanlıoğlu, F. Yılmaz, M. F. Demirci	Drexel University, TOBB University of Economics and Technology	implicit	C/C++
S. Jouili, S. Tabbone	Laboratoire Lorrain de Recherche en Informatique et ses Applications	explicit	Java
K. Riesen, H. Bunke	University of Bern	explicit	Java
M. M. Luqman, J. Lladós, J.-Y. Ramel, T. Brouard	Université François Rabelais de Tours, Universitat Autónoma de Barcelona	explicit	Matlab

4.1   The Algorithm by Osmanlıoğlu et al.

This algorithm is based on the embedding of the nodes of each graph in a vector space using the so called *caterpillar decomposition*, that builds a set of disjoint paths that cover the graph, and encodes each node with a vector whose components are associated to the paths needed to reach the node.

Once the nodes are embedded in a vector space, the scalar product between two graphs is computed using a point matching algorithm based on the *Earth Mover's distance* (EMD), which is defined as a linear programming problem.

4.2   The Algorithm by Jouili and Tabbone

The algorithm starts by computing a dissimilarity measure between each pair of graphs, using a graph matching technique. From the dissimilarity matrix a positive and semidefinite matrix is obtained; using the eigendecomposition of this matrix, the embedding is obtained by choosing the eigenvectors corresponding to the largest eigenvalues.

4.3   The Algorithm by Riesen and Bunke

This method assumes that a dissimilarity function between two graphs has been defined; in particular the implementation provided uses an approximation of graph edit distance, although the method can work with other dissimilarity measures.

Given a set of graphs, the method starts by choosing n prototypes that will be used as references for constructing the vectors. As the authors state, the choice of the prototype graphs, and also of their number n, is a critical issue. In the current implementation, the method attempts to select the prototypes that best possibly reflect the distribution of the set of graphs. Once the prototypes have been chosen, each graph is represented as a vector whose components correspond to the dissimilarity from each of the prototypes.

4.4   The Algorithm by Luqman et al.

This method is based on the encoding of meaningful information about the nodes and edges of the graph using fuzzy intervals. In particular, the first two components of the vector are the number of nodes and edges. They are followed by a an encoding of the histogram of node degrees and of the node attributes that uses fuzzy intervals obtained by means of a technique based on Akaike Information Criterion. The algorithm is able to learn these intervals using labeled training data, if available, for an improved performance.

5   Contest Results

The results of the performance evaluation on the test datasets are given in table 3.

Table 3. The performance index computed on the test datasets. The cells in boldface are the best results of each column. Note that lower index values correspond to better results.

Team	aloi-2	coil-2	odbk-2	geom. mean
Osmanlıoğlu et al.	0.088	**0.067**	**0.105**	**0.085**
Jouili and Tabbone	0.136	0.199	0.138	0.155
Riesen and Bunke	**0.048**	0.128	0.132	0.093
Luqman et al.	0.379	0.377	0.355	0.370

From the reported data it results that the algorithm by Osmanlıoğlu et al. has the best performance in two of the three datasets (coil-2 and odbk-2), while the algorithm by Riesen and Bunke is the best on the aloi-2 dataset. Considering the geometric mean of the index over the three datasets, these two algorithms are close, with a slight advantage for Osmanlıoğlu et al.

Note that the algorithm by Osmanlıoğlu et al. performs an implicit graph embedding; if we restrict our attention only to explicit graph embedding, the best algorithm is definitely the one by Riesen and Bunke.

As regards the computational cost of the algorithms, it would not be fair to make a comparison, since the implementations are in different programming languages. Just to give a rough idea, the program by Osmanlıoğlu et al., which has been the faster (since it is the only one written in C++), has required in the average about three hours and half to process each dataset on a 2.0GHz Xeon CPU.

6   Conclusions

In this paper we have presented the results of the first contest on Graph Embedding for Pattern Recognition, hosted by the International Conference on Pattern Recognition in Istanbul.

We have described the construction of the datasets used for the contest and the definition of the performance index. Then we have briefly introduced the contest participants, and we have presented and discussed the contest results.

This contest can be considered as a first step towards an assessment of the effectiveness of graph embedding algorithms when using graphs derived from typical Pattern Recognition applications. It is auspicable that this kind of quantitative comparison will be extended in the future to other embedding algorithms, and to graphs obtained from other application contexts.

References

1. The Open Source Computer Vision library,
   http://opencv.willowgarage.com/wiki/
2. Bai, X., Yu, H., Hancock, E.R.: Graph matching using spectral embedding and alignment. In: 17th Int. Conference on Pattern Recognition, pp. 398–401 (2004)
3. Borgwardt, K., Kriegel, H.P.: Shortest-path kernels on graphs. In: Proc. 5th Int. Conference on Data Mining, pp. 74–81 (2005)
4. Emms, D., Wilson, R., Hancock, E.R.: Graph embedding using quantum commute times. In: Escolano, F., Vento, M. (eds.) GbRPR 2007. LNCS, vol. 4538, pp. 371–382. Springer, Heidelberg (2007)
5. Geusebroek, J.M., Burghouts, G.J., Smeulders, A.W.M.: The Amsterdam library of object images. Int. Journal of Computer Vision 61(1), 103–112 (2005)
6. Hubert, L., Schultz, J.: Quadratic assignment as a general data-analysis strategy. British Journal of Mathematical and Statistical Psychology 29, 190–241 (1976)
7. Kashima, H., Inokuchi, A.: Marginalized kernels between labeled graphs. In: Proc. 20th Int. Conference on Machine Learning, pp. 321–328 (2003)
8. Luo, B., Wilson, R.C., Hancock, E.R.: Spectral embedding of graphs. Pattern Recognition 36(10), 2213–2230 (2003)
9. Nene, S.A., Nayar, S.K., Murase, H.: Columbia object image library (COIL-100). Tech. Rep. CUCS-006-96, Dep. of Computer Science, Columbia University (1996)
10. Neuhaus, M., Bunke, H.: Edit distance-based kernel functions for structural pattern classification. Pattern Recognition 39, 1852–1863 (2006)
11. Riesen, K., Neuhaus, M., Bunke, H.: Graph embedding in vector spaces by means of prototype selection. In: Escolano, F., Vento, M. (eds.) GbRPR 2007. LNCS, vol. 4538, pp. 383–393. Springer, Heidelberg (2007)
12. Tarr, M.J.: The object databank,
    http://www.cnbc.cmu.edu/tarrlab/stimuli/objects/index.html
13. Torsello, A., Hancock, E.R.: Graph embedding using tree edit-union. Pattern Recognition 40, 1393–1405 (2007)
14. Wilson, R.C., Hancock, E.R., Luo, B.: Pattern vectors from algebraic graph theory. IEEE Trans. on Pattern Analysis and Machine Intelligence 27(7), 1112–1124 (2005)

Graph Embedding Using Constant Shift Embedding*

Salim Jouili and Salvatore Tabbone

LORIA - INRIA Nancy Grand Est UMR 7503 - University of Nancy 2
BP 239, 54506 Vandoeuvre-lès-Nancy Cedex, France
{salim.jouili,tabbone}@loria.fr

Abstract. In the literature, although structural representations (e.g. graph) are more powerful than feature vectors in terms of representational abilities, many robust and efficient methods for classification (unsupervised and supervised) have been developed for feature vector representations. In this paper, we propose a graph embedding technique based on the constant shift embedding which transforms a graph to a real vector. This technique gives the abilities to perform the graph classification tasks by procedures based on feature vectors. Through a set of experiments we show that the proposed technique outperforms the classification in the original graph domain and the other graph embedding techniques.

Keywords: Structural pattern recognition, graph embedding, graph classification.

1 Introduction

In pattern recognition, object representations can be broadly divided into statistical and structural methods [1]. In the former, the object is represented by a feature vector, and in the latter, a data structure (e.g. graphs or trees) is used to describe the components and their relationships into the object. In the literature, many robust and efficient methods for classification (unsupervised and supervised), retrieval and other related tasks are henceforth available [5]. Most of these approaches are often limited to work with a statistical representation. Indeed, the use of feature vectors has numerous helpful properties. That is, since recognition is to be carried out based on a number of measurements of the objects, and each feature vector can be regarded as a point in an n-dimensional space, the measurement operations can be performed in simple way such that the Euclidean distance as the distance between objects. When a numerical feature vector is used to represent the object, all structural information is discarded. However a structural representation (e.g. graph) is more powerful than feature

---

* This work is partially supported by the French National Research Agency project NAVIDOMASS referenced under ANR-06-MCDA-012 and Lorraine region. For more details and resources see http://navidomass.univ-lr.fr

D. Ünay, Z. Çataltepe, and S. Aksoy (Eds.): ICPR 2010, LNCS 6388, pp. 83–92, 2010.
© Springer-Verlag Berlin Heidelberg 2010

vector in terms of representational abilities. The graph structure provides a flexible representation such that there is no fixed dimensionality for objects (unlike vectors), and provides an efficient representation such that an object is modeled by its components and the existing relations between them. In the last decades, many structural approaches have been proposed [3]. Nevertheless, dealing with graphs suffers, on the one hand from the high complexity of the graph matching problem which is a problem of computing distances between graphs, and on the other hand from the robustness to structural noise which is a problem related to the capability to cope with structural variation. These drawbacks have brought about a lack of suitable methods for classification. However, unlike for structural-based representation, a lot of robust classification approaches have been developed for the feature vector representation such as neural network, support vector machines, k-nearest neighbors, Gaussian mixture model, Gaussian, naive Bayes, decision tree and RBF classifiers [12]. In contrast, as remarked by Riesen et al. [2] the classification of graphs is limited to the use of the nearest-neighbor classifiers using one graph similarity measure. On that account, the community of pattern recognition speaks about a gap between structural and statistical approaches [1,16].

Recently, a few works have been carried out concerning the bridging of this gap between structural and statistical approaches. Their aim is to delineate a mapping between graphs and real vectors, this task is called graph embedding. The embedding techniques were originally introduced for statistical approaches with the objective of constructing low dimensional feature-space embeddings of high-dimensional data sets [6,9,17,22]. In the context of graph embedding, different procedures have been proposed in the literature. De Mauro et al. [4] propose a new method, based on recurrent neural network, to project graphs into vector space. Moreover, Hancock et al. [7,27,14,21] use spectral theory to convert graphs into vectors by means of spectral decomposition into eigenvalues and eigenvectors of the adjacency (or Laplacian) matrix of a graph. Besides, a new category of graph embedding techniques was introduced by Bunke et al. [2,20,19], their method is based on the selection of some prototypes and the computation of the graph edit distance between the graph and the set of prototypes. This method was originally developed for the embedding of feature vectors in a dissimilarity space [16,17]. This technique was also used to project string representations into vector spaces [25].

In this paper, we propose a new graph embedding technique by means of constant shift embedding. Originally, this idea was proposed in order to pairwise data into Euclidean vector spaces [22]. It was used to embed protein sequences into an n-dimensional feature vector space. In the current work, we generalize this method to the domain of graphs. Here, the key issue is to convert general dissimilarity data into metric data. The constant shift embedding increases all dissimilarities by an equal amount to produce a set of Euclidean distances. This set of distances can be realized as the pairwise distances among a set of points in an Euclidean space. In our method, we generalize this method by means of graph similarity measure [11,10]. The experimental results have shown that the proposed graph embedding technique improves the accuracy achieved in the graph domain by the nearest neighbor classifier. Furthermore, the results achieved by our method outperforms some alternative graph embedding techniques.

2 Graph Similarity Measure by Means of Node Signatures

Before introducing the graph embedding approach, let us recall how the dissimilarity in the domain of graphs can be computed. Similarity or (dissimilarity) between two graphs is almost always refered as a graph matching problem. Graph matching is the process of finding a correspondence between nodes and edges of two graphs that satisfies some constraints ensuring that similar substructures in one graph are mapped to similar substructures in the other. Many approaches have been proposed to solve the graph matching problem. In this paper we use a recent technique proposed by Jouili et al. in [10,11]. This approach is based on node signatures notion. In order to construct a signature for a node in an attributed graph, all available information into the graph and related to this node is used. The collection of these informations should be refined into an adequate structure which can provides distances between different node signatures. In this perspective, the node signature is defined as a set composed by four subsets which represent the node attribute, the node degree and the attributes of its adjacent edges and the degrees of the nodes on which these edges are connected. Given a graph G=(V,E,α,β), the node signature of $n_i \in V$ is defined as follows:

$$\gamma(n_i) = \left\{ \alpha_i, \ \theta(n_i), \ \{\theta(n_j)\}_{\forall ij \in E}, \ \{\beta_{ij}\}_{\forall ij \in E} \right\}$$

where

- α_i the attribute of the node n_i.
- $\theta(n_i)$ the degree of n_i.
- $\{\theta(n_j)\}_{\forall ij \in E}$ the degrees set of the nodes adjacent to n_i.
- $\{\beta_{ij}\}_{\forall ij \in E}$ the attributes set of the incident edges to n_i.

Then, to compute a distance between node signatures, the *Heterogeneous Euclidean Overlap Metric* (HEOM) is used. The HEOM uses the *overlap* metric for symbolic attributes and the normalized Euclidean distance for numeric attributes. Next the similarities between the graphs is computed: Firstly, a definition of the distance between two sets of node signatures is given. Subsequently, a matching distance between two graphs is defined based on the node signatures sets. Let S_γ be a collection of local descriptions, the set of node signatures S_γ of a graph g=(V,E,α,β) is defined as :

$$S_\gamma(\ g \ ) = \left\{ \ \gamma(n_i) \mid \forall n_i \in V \ \right\}$$

Let A=(V$_a$,E$_a$) and B=(V$_b$,E$_b$) be two graphs. And assume that $\phi : S_\gamma(A) \rightarrow S_\gamma(B)$ is a function. The distance d between A and B is given by φ which is the distance between $S_\gamma(A)$ and $S_\gamma(B)$

$$d(A, B) = \varphi(S_\gamma(A), S_\gamma(B)) = \min_\phi \sum_{\gamma(n_i) \in S_\gamma(A)} d_{nd}(\gamma(n_i), \phi(\gamma(n_i)))$$

The calculation of the function $\varphi(S_\gamma(A), S_\gamma(B))$ is equivalent to solve an assignment problem, which is one of the fundamental combinatorial optimization problems. It consists of finding a maximum weight matching in a weighted bipartite graph. This assignment problem can be solved by the Hungarian method. The permutation matrix P, obtained by applying the Hungarian method to the cost matrix, defines the optimum matching between two given graphs.

3   Graph Embedding via Constant Shift Embedding

Embedding graph corresponds to finding points in a vector space, such that their mutual distance is as close as possible to the initial dissimilarity matrix with respect to some cost function. Embedding yields points in a vector space, thus making the graph available to numerous machine learning techniques which require vectorial input.

Let $G=\{g_1, ..., g_n\}$ be a set of graph and $d\colon G \times G \to \mathbb{R}$ a graph distance function between pairs of its elements and let $D = D_{ij} = d(g_i, g_j) \in \mathbb{R}^{n \times n}$ be an $n \times n$ dissimilarity matrix. Here, the aim is to yield n vectors x_i in a p-dimensional vector space such that the distance between x_i and x_j is close to the dissimilarity D_{ij} which is the distance between g_i and g_j.

3.1   Dissimilarity Matrix: From Graph Domain to Euclidean Space via Constant Shift Embedding

Before stating the main method, the notion of *centralized matrix* is reminded [22]. Let P be an $n \times n$ matrix and let I_n be the $n \times n$ identity matrix and $e_n=(1,...,1)^\mathsf{T}$. Let $Q_n = I_n - \frac{1}{n}e_n e_n^\mathsf{T}$. Q_n is the projection matrix onto the orthogonal complement of e_n. The *centralized matrix* P^c is given by:

$$P^c = Q_n P Q_n$$

Roth et al. [22] observed that the transition from the original matrix D (in our case: the graph domain) to a matrix $\tilde{D}$ which derives from a squared Euclidean distance, can be achieved by a off-diagonal shift operation without influencing the distribution of the initial data. Hence, $\tilde{D}$ is given by:

$$\tilde{D} = D + d_0(e_n e_n^\mathsf{T} - I_n), \text{ or similarly } (\tilde{D}_{ij} = D_{ij} + d_0, \forall i \neq j)$$

Where d_0 is constant. In addition, since D is a symmetric and zero-diagonal matrix, it can be decomposed [13] by means of a matrix S:

$$D_{ij} = S_{ii} + S_{jj} - 2S_{ij}$$

Obviously, S is not uniquely determined by D. All matrices $S + \alpha e_n e_n^T$ yield the same D, $\forall \alpha \in \mathbb{R}$. However, it is proven that the centralized version of S is uniquely defined by the given matrix D (Lemma 1 in [22]):

$$S^c = -\frac{1}{2}D^c, \text{ with } D^c = Q_n D Q_n$$

From the following important theorem [26], we remark the particularity of the interesting matrix S^c.

Theorem 1. D is squared Euclidean distance matrix if and only if S^c is positive and semidefinite.

In other words, the pairwise similarities given by D can be embedded into an Euclidean vector space if and only if the associated matrix S^c is positive and semidefinite. As far as graph domain, S^c will be indefinite. We use the constant shift embedding [22] to cope this problem. Indeed, by shifting the diagonal elements of S^c it can be transformed into a positive semidefinite matrix $\tilde{S}$ (see Lemma 2. in [22]):

$$\tilde{S} = S^c - \lambda_n(S^c)I_n$$

where $\lambda_n(S^c)$ is the minimal eigenvalue of the matrix S^c. The diagonal shift of the matrix S^c transforms the dissimilarity matrix D in a matrix representing squared Euclidean distances. The resulting embedding of D is defined by:

$$\tilde{D}_{ij} = \tilde{S}_{ii} + \tilde{S}_{jj} - 2\tilde{S}_{ij} \Longleftrightarrow \tilde{D} = D - 2\lambda_n(S^c)(e_n e_n^T - I_n)$$

Since every positive and semidefinite matrix can be thought as representing a dot product matrix, there exists a matrix X for which $\tilde{S}^c = XX^T$. The rows of X are the resulting embedding vectors x_i, so each graph g_i has been embedded in a Euclidean space and is represented by x_i. Then, it can be concluded that the matrix $\tilde{D}$ contains the squared Euclidean distances between these vectors x_i. In the next section, we discuss the extension of this method to the graph embedding.

3.2 From Graphs to Vectors

In this section, an algorithm for constructing embedded vectors is described. This algorithm is inspired from the Principal Component Analysis (PCA) [23]. A pseudo-code description of the algorithm is given in Algorithm 1. From a given dissimilarity matrix D of a set of graphs $G=\{g_1, ..., g_n\}$, the algorithm returns a set of embedded vectors $X=\{x_1, \cdots, x_n\}$ such that x_i embed the graph g_i. Firstly, the squared Euclidean distances matrix $\tilde{D}$ is computed by means of the constant shift embedding (line 1). Next, the matrix $\tilde{S}^c$ is computed, and as stated above it can be calculated as $\tilde{S}^c = -\frac{1}{2}\tilde{D}^c$ (line 2). Since $\tilde{S}^c$ is positive and semidefinite matrix, thus, $\tilde{S}^c = XX^T$. The rows of X are the resulting embedding vectors x_i and they will be recovered by means of an eigendecomposition (line 3). Let note that, due to the centralization, it exists at least one the eigenvalue

$\lambda_i=0$, hence, $p \leq n-1$ (line 3-4). Finally by computing the $n \times p$ map matrix: $X_p=V_p(\Lambda_p)^{1/2}$, the embedded vectors are given by the rows of $X = X_p$, in p-dimensional space.

However, in PCA it is known that small eigenvalues contain the noise. Therefore, the dimensionality p can reduced by choosing $t \leq p$ in line 4 of the algorithm. Consequently, a $n \times t$ map matrix $X_t=V_t(\Lambda_t)^{1/2}$ will be computed instead of X_p, where V_t is the column-matrix of the selected eigenvectors (the first t column vectors of V) and Λ_t the diagonal matrix of the corresponding eigenvectors (the top $t \times t$ sub-matrix of Λ). One can ask how to find the optimal t that yields the better performance of a classifier in the vector space. Indeed, the dimensionality t has a pronounced influence on the performance of the classifier. In this paper, the optimal t is chosen empirically. That means, the optimal t is the one which provides the better classification accuracy from 2 to p.

Algorithm 1. Construction of the embedded vectors

Require: Dissimilarity matrix D of the a set of graphs $G=\{g_1, ..., g_n\}$
Ensure: set of vectors $X=\{x_1, \cdots, x_n\}$ where x_i embed g_i
1: Compute the squared Euclidean distances matrix $\tilde{D}$
2: Compute $\tilde{S}^c = -\dfrac{1}{2} \tilde{D}^c$, where $\tilde{D}^c = Q\tilde{D}Q$.
3: Eigendecomposition of $\tilde{S}^c$, $\tilde{S}^c = V\Lambda V^\mathsf{T}$
   – $\Lambda=\mathrm{diag}(\lambda_1, ...\lambda_n)$ is the diagonal matrix of the eigenvalues
   – $V=\{v_1, ...v_n\}$ is the orthonormal matrix of corresponding eigenvectors v_i.
                          ▷ Note that, $\lambda_1 \geq ... \lambda_p \geq \lambda_{p+1} = 0 = ... = \lambda_n$
4: Compute the $n \times p$ map matrix: $X_p=V_p(\Lambda_p)^{1/2}$, with $V_p=\{v_1, ...v_p\}$ and $\Lambda_p= \mathrm{diag}(\lambda_1, ...\lambda_p)$
5: **Output:** The rows of X_p contain the embedded vectors in p-dimensional space.

---

4   Experiments

4.1   Experimental Setup

To perform the evaluation of the proposed algorithm, we used four data sets :

- **GREC:** The GREC data set [18] consists of graphs representing symbols from architectural and electronic drawings. Here, the ending points (i.e. corners, intersections and circles) are represented by nodes which are connected by undirected edges and labeled as lines or arcs. The graph subset used in our experiments has 814 graphs, 24 classes and 37 graphs per class.
- **Letter:** The Letter data set [18] consists of graphs representing distorted Letter drawings. This data set contains 15 capital letters (A, E, F, H, I, K, L, M, N, T, V, W, X, Y, Z), arbitrarily strong distortions are applied to each letter to obtain large sample sets of graphs. Here, the ending points are represented by nodes which are connected by undirected edges. The graph subset used in our experiments has 1500 graphs, 15 classes and 100 graphs per class.

- **COIL:** The COIL data set [15] consists of graphs representing different views of 3D objects in which two consecutive images in the same class represent the same object rotated by 5°. The images are converted into graphs by feature points extraction using the Harris interest points [8] and Delaunay triangulation. Each node is labeled with a two-dimensional attribute giving its position, while edges are unlabeled. The graph subset used in our experiments has 2400 graphs, 100 classes and 24 graphs per class.
- **Mutagenicity:** The Mutagenicity data set [18] consists of graphs representing molecular compounds, the nodes represent the atoms labeled with the corresponding chemical symbol and edges by valence of linkage. The graph subset used in our experiments has 1500 graphs, 2 classes and 750 graphs per class.

The experiments consist in applying our algorithm for each dataset. Our intention is to show that the proposed graph embedding technique is indeed able to yield embedded vectors that can improve classification results achieved with the original graph representation. We begin by computing the dissimilarities matrix of each data set by means of the graph matching introduced in [10,11] and briefly reviewed in Section 2. Then, since the classification in the graph domain can be performed by only the k-NN classifiers, hence, it is used as reference system in the graph domain. Whereas in vector space the k-NN and the SVM[1] classifiers [24] are used to compare the embedding quality of the vectors resulting from our algorithm and the vectors resulting from the graph embedding approach recently proposed by Bunke et al. [2,19,20]. This method was originally developed for the embedding of feature vectors in a dissimilarity space [16,17] and is based on the selection of some prototypes and the computation of the graph edit distance between the graph and the set of prototypes. That is, let assume that $G=\{g_1, ..., g_n\}$ is a set of graphs and $d: G \times G \to \mathbb{R}$ is some graph dissimilarity measure. Let $PS=\{ps_1, ..., ps_p\}$ be a set of $p \leq n$ selected prototypes from G ($PS \subseteq G$). Now each graph $g \in G$ can be embedded into p-dimensional vector $(d_1, ..., d_p)$, where $d_1 = d(g, ps_1), ..., d_p = d(g, ps_p)$. As remarked in [20], the results of this method depends on the choice of p appropriate prototypes. In this paper, four different prototype selector are used, namely, the k-centers prototype selector (KCPS), the spanning prototype selector (SPS), the target-sphere prototype selector (TPS) and the random prototype selector (RandPS). We refer the reader to [20] for a definition of these selectors. A second key issue of this method is the number p of prototypes that must be selected to obtain the better classification performance. To overcome this problem, we define the optimal p by the same procedure that is applied to determine the optimal t of our algorithm (cf. section 3.2).

4.2   Results

The first experiment consists of applying the k-NN classifier in the graph domain and in the embedded vector space. In Table 1 the classification accuracy rates

---

[1] We used the C-SVM classifier with linear kernel function available in weka software http://www.cs.waikato.ac.nz/ml/weka/

Table 1. K-NN classifier results

| Data-set | Graph domain | Vector space | | | | |
| | | | Bunke with | | | |
		Our Embedding	KCPS	SPS	TPS	RandPS
GREC	98.11%	**99.50%**	97.17%	96.19%	97.78%	97.54%
Letter	79.33%	91.33%	92.4%	**92.66%**	92.46%	92.4%
COIL	50.30%	**52.35%**	48.64%	49.1%	48.6%	48.64%
Mutagenicity	63.70%	63.8%	60.93%	63.73%	64.13%	**64.73%**

for k-NN classifier are given for all data sets. One can remark that the results achieved in the vector space improve the results achieved in the graph domain for all data sets. However, this improvement is not statistically significant since it do not exceed $\tilde{2}\%$ for almost all data sets, except for the Letter data set. In the embedded vector space, the performances of our method and the Bunke's methods are quite similar. This first experiment is essentially to show that the embedding techniques are accurate since using the same classifier we obtain rates better then the graph domain. However, in the vector space we are not limited to a simple nearest neighbor classifier. So, in Table 2 the classification accuracy rates for the k-NN classifier in the graph domain (in Table 1) and the SVM classifier in the embedded vector space are given for all data sets. We can remark that our embedding methods clearly improves the accuracy achieved in the graph domain. Regarding the GREC data set, the best classification accuracy is achieved by our method and the Bunke's method with the KCPS selector. The results improved the k-NN classifier in the graph domain by 1.76%. This improvement has no statistical signification. This is due to the fact that the classification in the graph domain of the GREC data set provides already a very good accuracy 98.11%. For the Letter data set, all the embedding methods clearly improve the accuracy achieved in the graph domain. Regarding the COIL data set, it can be remarked that the accuracy achieved in the graph domain is quite low (50.30%). This accuracy is highly improved by the SVM classifiers in the different vector spaces (by 25.43% using our embedding technique). Finally, the results concerning the Mutagenicity data set show that all the variants of the Bunke's embedding fail to improve the accuracy achieved in the graph domain and provide the worst results. Whereas our graph embedding using constant shift embedding results improve the results of the k-NN classifier achieved in the original graph representation (by 7.5%). Therefore, our method provides more significant embedded vector for the graphs in the Mutagenicity data set.

To summarize, with the graph embedding algorithm proposed in this paper and the SVM classifiers, the classification accuracy rates improve the results achieved in the original graph domain for all data sets used in the experiments. This agrees with our intention to improve classification results achieved in graph domain by means of graph embedding technique. Furthermore, the comparison with the four variants of the Bunke's graph embedding has shown that the proposed technique outperforms these alternatives for almost the data sets.

Table 2. SVM Classifier results

	Graph domain	Vector space			
			Bunke with		
Data-set		Our Embedding	KCPS	SPS	TPS	RandPS
GREC	-	**99.87%**	**99.87%**	99.50%	99.75%	99.75%
Letter	-	**94.87%**	91.60%	91.40%	91.53%	91.53%
COIL	-	**75.73%**	73.29%	73,10%	73.25%	73.23%
Mutagenicity	-	**71.20%**	57.53%	57.31%	57.48%	57.45%

5 Conclusions

In the context of graph-based representation for pattern recognition, there is a lack of suitable methods for classification. However almost the huge part of robust classification approaches has been developed for feature vector representations. Indeed, the classification of graphs is limited to the use of the nearest-neighbor classifiers using one graph similarity measure. In this paper, we proposed a new graph embedding technique based on the constant shift embedding which was originally developed for vector spaces. The constant shift embedding increases all dissimilarities by an equal amount to produce a set of Euclidean distances. This set of distances can be realized as the pairwise distances among a set of points in an Euclidean space. Here, the main idea was to generalize this technique in the domain of graphs. In the experiments, the application of the SVM classifiers on the resulting embedded vectors has shown a significantly statistically improvement.

References

1. Bunke, H., Günter, S., Jiang, X.: Towards bridging the gap between statistical and structural pattern recognition: Two new concepts in graph matching. In: Singh, S., Murshed, N., Kropatsch, W.G. (eds.) ICAPR 2001. LNCS, vol. 2013, pp. 1–11. Springer, Heidelberg (2001)
2. Bunke, H., Riesen, K.: Graph classification based on dissimilarity space embedding. In: da Vitoria Lobo, N., Kasparis, T., Roli, F., Kwok, J.T., Georgiopoulos, M., Anagnostopoulos, G.C., Loog, M. (eds.) S+SSPR 2008. LNCS, vol. 5342, pp. 996–1007. Springer, Heidelberg (2008)
3. Conte, D., Foggia, P., Sansone, C., Vento, M.: Thirty years of graph matching in pattern recognition. IJPRAI 18(3), 265–298 (2004)
4. de Mauro, C., Diligenti, M., Gori, M., Maggini, M.: Similarity learning for graph-based image representations. PRL 24(8), 1115–1122 (2003)
5. Marques de sa, J.: Pattern Recognition: Concepts, Methods and Applications. Springer, Heidelberg (2001)
6. Dhillon, I.S., Modha, D.S., Spangler, W.S.: Class visualization of high-dimensional data with applications. Computational Statistics & Data Analysis 41(1), 59–90 (2002)
7. Emms, D., Wilson, R.C., Hancock, E.R.: Graph embedding using quantum commute times. In: Escolano, F., Vento, M. (eds.) GbRPR 2007. LNCS, vol. 4538, pp. 371–382. Springer, Heidelberg (2007)

8. Harris, C., Stephens, M.: A combined corner and edge detector. In: Proceedings of the 4th Alvey Vision Conference, pp. 147–151 (1988)
9. Iwata, T., Saito, K., Ueda, N., Stromsten, S., Griffiths, T.L., Tenenbaum, J.B.: Parametric embedding for class visualization. In: NIPS (2004)
10. Jouili, S., Mili, I., Tabbone, S.: Attributed graph matching using local descriptions. In: Blanc-Talon, J., Philips, W., Popescu, D., Scheunders, P. (eds.) ACIVS 2009. LNCS, vol. 5807, pp. 89–99. Springer, Heidelberg (2009)
11. Jouili, S., Tabbone, S.: Graph matching based on node signatures. In: Torsello, A., Escolano, F., Brun, L. (eds.) GbRPR 2009. LNCS, vol. 5534, pp. 154–163. Springer, Heidelberg (2009)
12. Kotsiantis, S.B.: Supervised machine learning: A review of classification techniques. Informatica 31(3), 249–268 (2007)
13. Laub, J., Müller, K.R.: Feature discovery in non-metric pairwise data. Journal of Machine Learning Research 5, 801–818 (2004)
14. Luo, B., Wilson, R.C., Hancock, E.R.: Spectral embedding of graphs. Pattern Recognition 36(10), 2213–2230 (2003)
15. Nene, S.A., Nayar, S.K., Murase, H.: Columbia object image library (coil-100). Technical report, Department of Computer Science, Columbia University (1996)
16. Pekalska, E., Duin, R.P.W.: The Dissimilarity Representation for Pattern Recognition. Foundations and Applications. World Scientific Publishing, Singapore (2005)
17. Pekalska, E., Duin, R.P.W., Paclík, P.: Prototype selection for dissimilarity-based classifiers. Pattern Recognition 39(2), 189–208 (2006)
18. Riesen, K., Bunke, H.: Iam graph database repository for graph based pattern recognition and machine learning. In: da Vitoria Lobo, N., Kasparis, T., Roli, F., Kwok, J.T., Georgiopoulos, M., Anagnostopoulos, G.C., Loog, M. (eds.) S+SSPR 2008. LNCS, vol. 5342, pp. 287–297. Springer, Heidelberg (2008)
19. Riesen, K., Bunke, H.: Graph classification based on vector space embedding. IJPRAI 23(6), 1053–1081 (2009)
20. Riesen, K., Neuhaus, M., Bunke, H.: Graph embedding in vector spaces by means of prototype selection. In: Escolano, F., Vento, M. (eds.) GbRPR. LNCS, vol. 4538, pp. 383–393. Springer, Heidelberg (2007)
21. Robles-Kelly, A., Hancock, E.R.: A riemannian approach to graph embedding. Pattern Recognition 40(3), 1042–1056 (2007)
22. Roth, V., Laub, J., Kawanabe, M., Buhmann, J.M.: Optimal cluster preserving embedding of nonmetric proximity data. IEEE Transactions on Pattern Analysis and Machine Intelligence 25(12), 1540–1551 (2003)
23. Schölkopf, B., Smola, A., Müller, K.R.: Nonlinear component analysis as a kernel eigenvalue problem. Neural Computation 10(5), 1299–1319 (1998)
24. Shawe-Taylor, J., Cristianini, N.: Support Vector Machines and other kernel-based learning methods. Cambridge University Press, Cambridge (2000)
25. Spillmann, B., Neuhaus, M., Bunke, H., Pekalska, E., Duin, R.P.W.: Transforming strings to vector spaces using prototype selection. In: Yeung, D.-Y., Kwok, J.T., Fred, A., Roli, F., de Ridder, D. (eds.) SSPR 2006 and SPR 2006. LNCS, vol. 4109, pp. 287–296. Springer, Heidelberg (2006)
26. Torgerson, W.S.: Theory and Methods of Scaling. Johm Wiley and Sons, Chichester (1958)
27. Torsello, A., Hancock, E.R.: Graph embedding using tree edit-union. Pattern Recognition 40(5), 1393–1405 (2007)

A Fuzzy-Interval Based Approach
for Explicit Graph Embedding

Muhammad Muzzamil Luqman[1,2], Josep Lladós[2],
Jean-Yves Ramel[1], and Thierry Brouard[1]

[1] Laboratoire d'Informatique, Université François Rabelais de Tours, France
[2] Computer Vision Center, Universitat Autónoma de Barcelona, Spain
{mluqman,josep}@cvc.uab.es, {ramel,brouard}@univ-tours.fr

Abstract. We present a new method for explicit graph embedding. Our algorithm extracts a feature vector for an undirected attributed graph. The proposed feature vector encodes details about the number of nodes, number of edges, node degrees, the attributes of nodes and the attributes of edges in the graph. The first two features are for the number of nodes and the number of edges. These are followed by w features for node degrees, m features for k node attributes and n features for l edge attributes — which represent the distribution of node degrees, node attribute values and edge attribute values, and are obtained by defining (in an unsupervised fashion), fuzzy-intervals over the list of node degrees, node attributes and edge attributes. Experimental results are provided for sample data of ICPR2010[1] contest GEPR[2].

1 Introduction

The website [2] for the 20th International Conference on Pattern Recognition (ICPR2010) contest Graph Embedding for Pattern Recognition (GEPR), provides a very good description of the emerging research domain of graph embedding. It states and we quote: *"In Pattern Recognition, statistical and structural methods have been traditionally considered as two rather separate worlds. However, many attempts at reducing the gap between these two approaches have been done. The idea inspiring these attempts is that of preserving the advantages of an expressive structural representation (such as a graph), while using most of the powerful, vector-based algorithms from Statistical Pattern Recognition. A possible approach to this issue has been recently suggested by Graph Embedding. The latter is a methodology aimed at representing a whole graph (possibly with attributes attached to its nodes and edges) as a point in a suitable vectorial space. Of course the relevant property is that of preserving the similarity of the graphs: the more two graphs are considered similar, the closer the corresponding points in the space should be. Graph embedding, in this sense, is a real bridge joining the two worlds: once the object at hand has been described in terms of graphs, and the*

---

[1] 20th International Conference on Pattern Recognition.
[2] Graph Embedding for Pattern Recognition.

D. Ünay, Z. Çataltepe, and S. Aksoy (Eds.): ICPR 2010, LNCS 6388, pp. 93–98, 2010.

latter represented in the vectorial space, all the problems of matching, learning and clustering can be performed using classical Statistical Pattern Recognition algorithms."

2   Our Method

Our proposed method for graph embedding encodes the details of an undirected attributed graph into a feature vector. The feature vector is presented in next section, where we discuss it in detail.

2.1   Proposed Vector

The proposed feature vector not only utilizes information about the structure of the graph but also incorporates the attributes of nodes and edges of the graph, for extracting discriminatory information about the graph. Thus yielding a feature vector that corresponds to the structure and/or topology and/or geometry of the underlying content.

Fig.1 presents the feature vector of our method for graph embedding. In the remainder of this section, we provide a detailed explanation of the extraction of each feature of our feature vector.

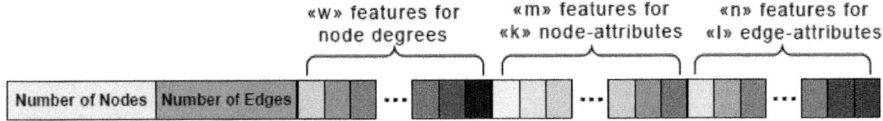

Fig. 1. The vector for graph embedding

Number of Nodes and Number of Edges. The number of nodes and the number of edges in a graph constitutes the basic discriminatory information that can be extracted very easily from a graph. This information helps to discriminate among graphs of different sizes. The first two features of our feature vector are composed of these details on the graph i.e. the number of nodes and the number of connections in the graph.

Features for Node Degrees. The degree of a node refers to the total number of incoming and outgoing connections for the node. The distribution of the node degrees of a graph is a very important information that can be used for extracting discriminatory features from the graph. This information helps to discriminate among graphs which represent different structures and/or topologies and/or geometry.

Our signature contains w features for node degrees. The number of features for node degrees, i.e. w, is computed from the learning dataset and may differ from one dataset to another, thus yielding a variable length feature vector for different datasets. If the learning dataset is not available, our method is capable of learning its parameters directly from the graphs in the test dataset. The w features for node degrees are obtained by defining fuzzy-intervals over list of node degrees of all graphs in the (learning) dataset and then computing the cardinality for each of the w features. In order to avoid repetition we detail the method for defining fuzzy-intervals and computing the corresponding feature values in section 2.2 — as the same method has been used for obtaining the set of features for node and edge attributes.

Features for Node Attributes. The attributes of the nodes in the graph contain very important complementary (and/or supplementary in some cases) information about the underlying content. This information, if used intelligently, can provide discriminating features for the graphs.

Our signature contains m features for k node attributes. The number k refers to the number of node attributes in the graph. Whereas, the number m of the features for node attributes is computed from the learning dataset. Both m (the number of features for node attributes) and k (the number of node attributes) may differ from one dataset to another, thus yielding a variable length of feature vector for different datasets. If the learning dataset is not available, our method is capable of learning its parameters directly from the graphs in the test dataset. The m features for k node attributes are obtained by defining fuzzy-intervals over list of node attributes of all graphs in the (learning) dataset and then computing the cardinality for each of the m features. Each of the k node attributes are processed one by one and can contribute a different number of features to this set of features. The method for obtaining fuzzy-intervals and computing the corresponding feature values is detailed in section 2.2.

Features for Edge Attributes. The attributes of the edges in the graph also contain important complementary (and/or supplementary) information about the underlying content. This information may also provide discriminating features for the graphs.

Our signature contains n features for l edge attributes. The number l refers to the number of edge attributes in the graph. Whereas, the number n of the features for edge attributes is computed from the learning dataset. Both n (the number of features for edge attributes) and l (the number of edge attributes) may differ from one dataset to another, thus yielding a variable length of feature vector for different datasets. If the learning dataset is not available, our method is capable of learning its parameters directly from the graphs in the test dataset. The n features for l edge attributes are obtained by defining fuzzy-intervals over list of edge attributes of all graphs in the (learning) dataset and then computing the cardinality for each of the n features. Each of the l edge attributes are processed one by one and can contribute a different number of features to this set of features. The method for obtaining fuzzy-intervals and computing the corresponding feature values is detailed in section 2.2.

2.2   Defining Fuzzy-Intervals and Computing Feature Values

This section applies to computation of w features for node degrees, m features for k node attributes and n features for l edge attributes – and details our approach for defining a set of fuzzy-intervals for a given data (data here refers to the list of node degrees - or - the list of values for each of the k node attributes - or - the list of values for each of the l edge attributes – as applicable).

We use a histogram based binning technique for obtaining an initial set of intervals for the data. The technique is originally proposed by [1] for discretization of continuous data and is based on use of Akaike Information Criterion (AIC). It starts with an initial m-bin histogram of data and finds optimal number of bins for underlying data. The adjacent bins are iteratively merged using an AIC-based cost function until the difference between AIC-beforemerge and AIC-aftermerge becomes negative.

This set of bins is used for defining fuzzy-intervals i.e. the features and their fuzzy zones (for example Fig.2). The fuzzy-intervals are defined by using a set of 3 bins for defining a feature, set of 3 bins for defining its left fuzzy zone and a set of 3 bins for defining its right fuzzy zone. It is important to note here, and as seen in Fig.2, that the fuzzy zones on left and right of a feature overlaps the fuzzy zones of its neighbors. We have used the sets of 3 bins for defining fuzzy zones in order to be able to assign membership weights for full membership, medium membership and low membership to the fuzzy zones – membership weights of 1.00, 0.66 and 0.33 respectively. Generally an x number of bins can be used for defining the fuzzy zones for the features. For any given value in the data it is ensured that the sum of membership weights assigned to it always equals to 1. The first and the last features in the list have one fuzzy zone each (as seen in Fig.2).

Our method is capable of learning these fuzzy-intervals for a given data, either from a learning dataset (if it is available) or directly from the the graphs in the test dataset. Once the fuzzy-intervals are obtained for a given dataset we get the structure of the feature vector for the dataset i.e. 1 feature for number of nodes, 1 feature for number of edges, w features for node degrees, m features for k node attributes and n features for l edge attributes. We perform a pass on the set of graphs in the dataset and while embedding each graph into a feature vector, we use these fuzzy-intervals for computing the cardinality for each feature in the feature vector. This results into a numeric feature vector for a given undirected attributed graph.

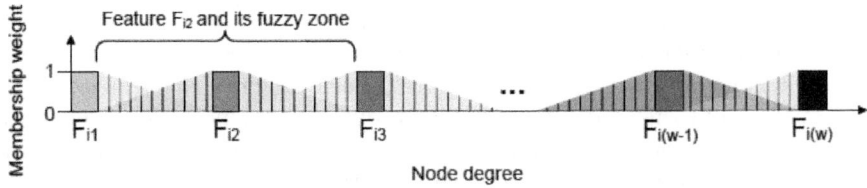

Fig. 2. The fuzzy intervals of features for node degrees

3   Experimentation

Experimental results are presented for sample datasets of 20th International Conference on Pattern Recognition (ICPR2010), contest GEPR (Graph Embedding for Pattern Recognition) [2]. These datasets contain graphs extracted from three image databases: the Amsterdam Library of Object Images (ALOI), the Columbia Object Image Library (COIL), the Object Databank (ODBK). Table 1 provides the details on the graphs in each of these datasets.

Table 1. Dataset details

Dataset	Number of graphs	Node attributes (k)	Edge attributes (l)
ALOI	1800	4	0
COIL	1800	4	0
ODBK	1248	4	0

The 4 node attributes encode the size and the average color of the image area represented by a node.

We have used the sample data for learning parameters for our method (as described in section 2). Table 2 provides the details on the length of feature vector for each image database.

Table 2. Length of feature vectors

Dataset	Length of feature vector
ALOI	1595
COIL	1469
ODBK	1712

And finally, Table 3 presents the performance index, which is a clustering validation index to evaluate the separation between the classes when represented by the vectors, as calculated by the scripts provided by the contest. Further details on the performance index can be found at [2].

Table 3. Performance indexes

Dataset	Performance index
ALOI	0.379169273726
COIL	0.375779781743
ODBK	0.352542741399

The experimentation is performed on Intel Core 2 Duo (T8300 @ 2.4GHz) with 2GB (790MHz) of RAM.

4   Conclusion

We have presented a method for explicit graph embedding. Our algorithm extracts a feature vector for an undirected attributed graph. The proposed feature vector not only utilizes information about the structure of the graph but also incorporates the attributes of nodes and edges of the graph, for extracting discriminatory information about the graph. Thus yielding a feature vector that corresponds to the structure, topology and geometry of the underlying content. The use of fuzzy-intervals, for noise sensitive information in graphs, enables the proposed feature vector to incorporate robustness against the deformations introduced in graphs as a result of noise.

The experimentation on sample datasets shows that the proposed method can be used to extract a huge number of meaningful features from the graphs. An important property of our method is that the number of features could be controlled by using appropriate parameter for the number of bins for defining a feature and its fuzzy zones. One possible future extension to this work is to extend it to directed attributed graphs, which could be done by introducing new features for in-degree and out-degree of nodes.

Acknowledgment

This research was supported in part by PhD grant PD-2007-1/Overseas/FR/ HEC/222 from Higher Education Commission of Pakistan.

References

1. Colot, O., Olivier, C., Courtellemont, P., El-Matouat, A.: Information criteria and abrupt changes in probability laws. In: Signal Processing VII: Theories and Applications, pp. 1855–1858 (1994)
2. http://nerone.diiie.unisa.it/contest/index.shtml (as on May 26, 2010)

The ImageCLEF Medical Retrieval Task at ICPR 2010 — Information Fusion to Combine Visual and Textual Information

Henning Müller[1,2] and Jayashree Kalpathy-Cramer[3]

[1] Geneva University Hospitals and University of Geneva, Switzerland
[2] University of Applied Sciences Western Switzerland, Sierre, Switzerland
[3] Oregon Health and Science University, Portland, OR, USA
henning.mueller@sim.hcuge.ch
http://www.imageclef.org/

Abstract. An increasing number of clinicians, researchers, educators and patients routinely search for medical information on the Internet as well as in image archives. However, image retrieval is far less understood and developed than text–based search. The ImageCLEF medical image retrieval task is an international benchmark that enables researchers to assess and compare techniques for medical image retrieval using standard test collections. Although text retrieval is mature and well researched, it is limited by the quality and availability of the annotations associated with the images. Advances in computer vision have led to methods for using the image itself as search entity. However, the success of purely content–based techniques has been limited and these systems have not had much clinical success. On the other hand a combination of text– and content–based retrieval can achieve improved retrieval performance if combined effectively. Combining visual and textual runs is not trivial based on experience in ImageCLEF. The goal of the fusion challenge at ICPR is to encourage participants to combine visual and textual results to improve search performance. Participants were provided textual and visual runs, as well as the results of the manual judgments from Image-CLEFmed 2008 as training data. The goal was to combine textual and visual runs from 2009. In this paper, we present the results from this ICPR contest.

1 Introduction

Image retrieval is a burgeoning area of research in medical informatics [1,2,3]. With the increasing use of digital imaging in all aspects of health care and medical research, there has been a substantial growth in the number of images being created every day in healthcare settings. An increasing number of clinicians, researchers, educators and patients routinely search for relevant medical information on the Internet as well as in image archives and PACS (Picture Archival and Communication Systems) [1,3,4]. Consequently, there is a critical need to manage the storage and retrieval of these image collections. However, image

D. Ünay, Z. Çataltepe, and S. Aksoy (Eds.): ICPR 2010, LNCS 6388, pp. 99–108, 2010.

retrieval is far less understood and developed than text–based searching. Text retrieval has a long history of evaluation campaigns in which different groups use a common test collection to compare the performance of their methods. The best known such campaign is the Text REtrevial Conference (TREC[1], [5]), which has been running continuously since 1992. There have been several offshoots from TREC, including the Cross–Language Evaluation Forum (CLEF[2]). CLEF operates on an annual cycle, and has produced numerous test collections since its inception in 2000 [6]. While CLEFs focus was originally on cross–language text retrieval it has grown to include multimedia retrieval tracks of several varieties. The largest of these, ImageCLEF[3] , started in 2003 as a response to the need for standardized image collections and a forum for evaluation. It has grown to become todays pre–eminent venue for image retrieval evaluation.

The coming sections will describe the ImageCLEF challenge itself and the details for the fusion task that was organized at ICPR (International Conference on Pattern Recognition). Then, the results and techniques of the participants will be analyzed in more detail and the main lessons learned from this context will be explained.

2   The Annual ImageCLEF Challenge

ImageCLEF is an international benchmark that includes several sub–tracks concerned with various aspects of image retrieval [7]; one of these tracks is the medical retrieval task run since 2004. This task within ImageCLEF enables researchers to assess and compare techniques for medical image retrieval using standard collections. ImageCLEFmed uses the same methodology as information retrieval challenges including TREC. Participants are given a set of topics that represent information needs. They submit an ordered list of runs that contain images that their system believe best meet the information need. Manual judgments using domain experts, typically clinicians, are used to create ground truth. The medical image retrieval tracks test collection began with a teaching database of 8,000 images in 2004. Since then, it has grown to a collection of over 74,000 images from the scientific literature, as well as a set of topics that are known to be well–suited for textual, visual or mixed retrieval methods. A major goal of ImageCLEF has been to foster development and growth of multimodal retrieval techniques: i.e., retrieval techniques that combine visual, textual, and other methods to improve retrieval performance.

Traditionally, image retrieval systems have been text–based, relying on the textual annotations or captions associated with images. Several commercial systems, such as Google Images[4] and Yahoo! images[5], employ this approach. Although text–based information retrieval methods are mature and well researched,

---

[1] http://trec.nist.gov/

[2] http://www.clef-campaign.org/

[3] http://www.imageclef.org/

[4] http://images.google.com/

[5] http://images.yahoo.com/

they are limited by the quality of the annotations applied to the images. Advances in techniques in computer vision have led to a second family of methods for image retrieval: content–based image retrieval (CBIR). In a CBIR system, the visual contents of the image itself are represented by visual features (colors, textures, shape) and compared to similar abstractions of all images in the database. Typically, such systems present the user with an ordered list of images that are visually most similar to the sample (or query) image. The text–based systems typically perform significantly better than purely visual systems at ImageCLEF.

Multimodal systems combine the textual information associated with the image with the actual image features in an effort to improve performance, especially early precision. However, our experience from the ImageCLEF challenge, especially of the last few years has been that these combinations of textual and visual systems can be quite fragile, with the mixed runs often performing worse than the corresponding textual run. We believe that advances in machine learning can be used more effectively to learn how best to incorporate the multimodal information to provide the user with search results that best meet their needs [8]. Thus, the goal of the fusion challenge at ICPR is to encourage participants to effectively combine visual and textual results to improve search performance. Participants were provided textual and visual runs that were submitted to the actual competition, as well as the results of the manual judgments from the ImageCLEFmed 2008 challenge as training data. The goal was to combine similar textual and visual runs from the 2009 challenge for testing. In this paper, we present the preliminary from this ICPR contest.

3   The ImageCLEF Fusion Challenge

In both 2008 and 2009, the Radiological Society of North America (RSNA[6]) made a subset of its journals image collections available for use by participants in ImageCLEF. The 2009 database contains 74,902 images, the largest collection yet [9]. The organizers created a set of 25 search topics based on a user study conducted at Oregon Health & Science University (OHSU) in 2009 [4]. These topics consisted of 10 visual, 10 mixed and 5 semantically oriented topics, as categorized by the organizers based on past experience and nature of the query. During 2008 and 2009, a panel of clinicians, using a web–based interface, created relevance judgments. The manually judged results were used to evaluate the submitted runs using the trec_eval[7] software package. This package provides commonly used information retrieval measures including mean average precision (MAP), recall as well as precision at various levels for all topics.

For the ICPR fusion contest, the goal was to combine the best visual and textual runs that had been submitted previously to improve performance over the purely visual and purely textual runs. After participants registered they were provided access to the training data in early November 2009. The training set consisted of the four best textual and visual runs from different groups in 2008.

---

[6] http://www.rsna.org/
[7] http://trec.nist.gov/trec_eval/

Only one of these groups participated in the fusion challenge, so there was no advantage for any group. These runs were anonymized to remove information about the group. We also provided the qrel, the file that contained the output for the manual judgments as well as the results obtained by the training runs using the trec_eval package. Participants could create fusion runs using combinations of the provided training runs and evaluate the performance using the trec_eval along with the abovementioned qrel file as well as the results of the evaluation measures for the runs.

We released the test runs two weeks later. Again these consisted of the four best textual and four best visual runs, this time from 2009. The ground truth in the form of qrel was not provided at this time. The judgments were released in early January so that the participants could evaluate their runs in time for submission to ICPR 2010. To summarize, the timeline for this contest was as follows:

- 16.11.2009 Release of training data
- 30.11.2009 Release of test data
- 04.01.2010 Submission of results
- 10.01.2010 Release of ground truth data
- 15.01.2010 Conference paper submission

4   Fusion Techniques Used by the Participants

There was quite a variety of techniques relying on either the similarity scores of the supplied runs or the ranks. Early fusion was hardly possible as only the outcome of the system was supplied and no further information, limiting the variety of the approaches.

OHSU used a simple scheme based principally on the number of times that a particular image occurs in the results sets as the main criterion. Two runs use only textual information (fusion2, fusion4) and two runs combine both visual and textual techniques (fusion1, fusion3). Then as a second criterion either the sum of the ranks was used (fusion1, fusion2) or the sum of the scores (fusion3, fusion4).

The *MedGIFT* (Medical projects around the GNU Image Finding Tool) group employed two principal approaches for the fusion described in more detail in [10]. Methods are based on ranks and on the scores. Whereas ranks can be used directly, the scores were normalized to be in the range 0..1 to be better comparable among the submissions. In terms of combination rules a max combination was used where of all systems the maximum was taken (combMax), a sum rule summing up normalized scores or ranks (combSum) and the last rule includes the frequency of the documents into this (CombMnz).

The results of the best system (*SIFT*) in the context are described in [11]. This group uses a probabilistic fusion, where weights are calculated from training data (ProbFuse). The training takes into account that documents retrieved later are generally less relevant and these are subsequently weighted with a learned decrease of the weight (SegFuse). All these techniques can create border effects as the results are grouped in blocks. This can be removed with SlideFuse. SlideFuse most often had the best results.

The *PRISMA* group developed two methods called rankMixer and rankBooster. Both take into account the frequency of an image in the results lists to be combined and its scores. These are used to calculate a function for calculating the similarity score for a particular image.

Finally, the *ISDM* group developed an approach based on a generative statistical model. It uses an attentive reader model, meaning that early documents are weighted high and then attention decreases, in their case with a logarithmic model. The importance of single runs is in a second approach estimated based on population–based incremental learning.

5   Results of the Participants

Table 1 contains the performance of the training runs that were provided. As can be seen, the textual runs perform significantly better than the visual runs for all measures. This has to be taken into account when combining the runs.

Table 1. Results of the training runs

Run	Recall	MAP	P5	P10
Text1	0.63	0.29	0.49	0.46
Text2	0.65	0.28	0.51	0.47
Text3	0.54	0.27	0.51	0.47
Text4	0.61	0.28	0.44	0.41
Visual1	0.06	0.028	0.15	0.13
Visual2	0.24	0.035	0.17	0.17
Visual3	0.17	0.042	0.22	0.17

This performance gap was similarly true for the test runs (Table 2). Overall, the performance was better for the textual runs in 2009 whereas it was worse for the visual runs as can be seen when comparing the two tables.

Table 2. Results of the test runs

Run	Recall	MAP	P5	P10
Text1	0.73	0.35	0.58	0.56
Text2	0.66	0.35	0.65	0.62
Text3	0.77	0.43	0.70	0.66
Text4	0.80	0.38	0.65	0.62
Visual1	0.12	0.01	0.09	0.08
Visual2	0.12	0.01	0.08	0.07
Visual3	0.11	0.01	0.09	0.07
Visual4	0.11	0.01	0.09	0.08

Participants were successful in creating fusion runs that were better than the original text and visual runs, as well being substantially better than the official mixed runs that had been submitted to ImageCLEFmed 2009. None of the officially submitted fusion runs was better than the best text run in the competition.

We received 49 runs from five groups as part of the fusion task. Of the 35 mixed runs that were submitted, 18 had higher MAP compared to the best textual training run and interestingly, 25 had higher MAP compared to the best official mixed run in 2009 as seen in Figure 1. This shows the potential performance gains through fusing varying techniques and it shows how little focus most ImageCLEF participants put into this..

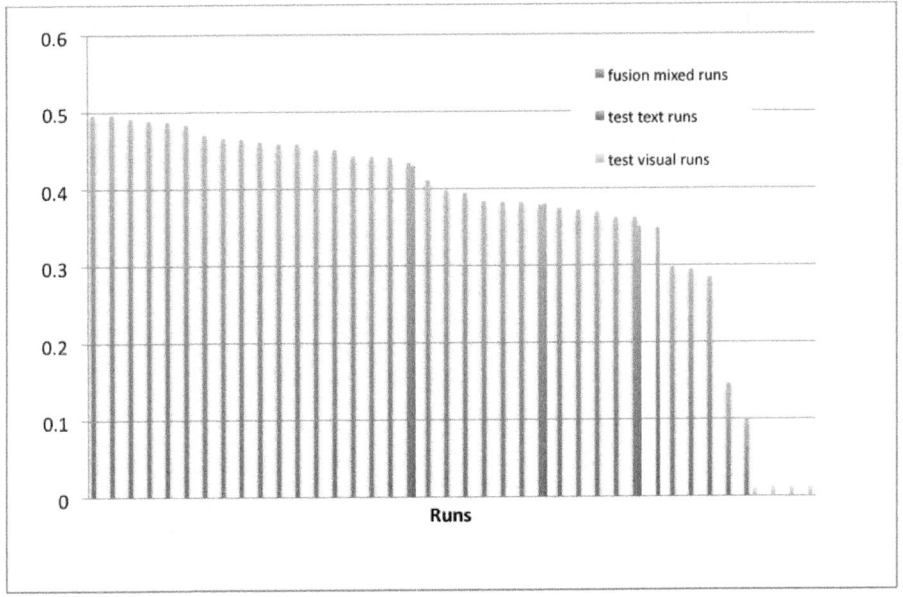

Fig. 1. MAP of all fusion runs and test runs

Figure 2 shows the precisions of the best original runs and the best fusion runs. There is a slight improvement in early precision with the best fusion runs both textual and mixed. However, the fusion runs created using only visual runs performed quite poorly, which is not surprising as the basic results were all very low. Although there was little difference between the best fusion mixed and textual runs for the MAP, the runs with highest early precision used the visual runs in combination with the textual runs. This underlines the importance of visual information, even with a very poor performance, for early precision. This also shows that the information contained in visual and textual retrieval runs is very complementary.

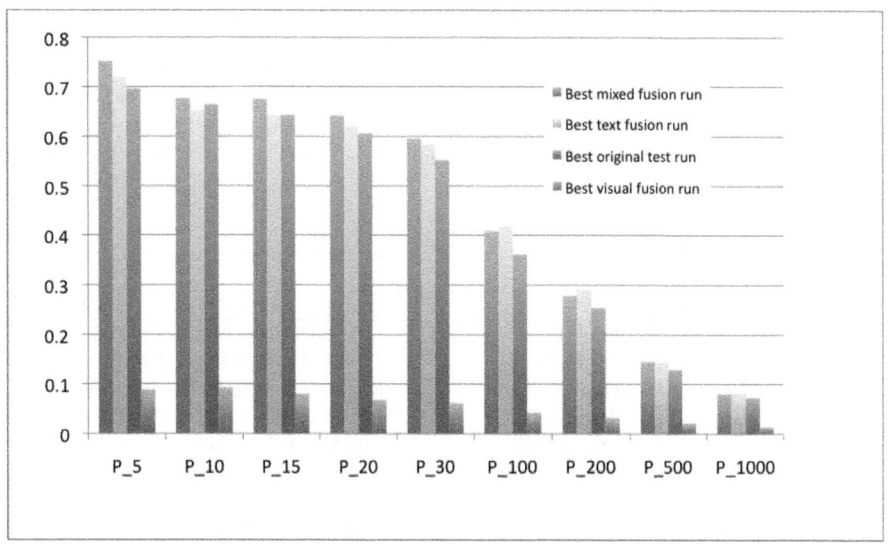

Fig. 2. Early precision (P_X meaning precision after X documents are retrieved) of original text runs and fusion runs

Table 3. Performance metrics for fusion of text runs

Group	Runid	type	map	bpref	P5	P10	P30
SIFT, Ireland	txtOnlySlideFuse	Textual	**0.487**	**0.499**	**0.72**	0.652	0.584
PRISMA, Chile	testt234v	Textual	0.480	0.498	0.704	**0.672**	0.5973
PRISMA, Chile	testt1234v	Textual	0.474	0.487	0.712	0.648	0.596
PRISMA, Chile	testt123v	Textual	0.473	0.489	0.712	0.664	0.584
SIFT, Ireland	txtOnlySegFuse	Textual	0.466	0.472	0.696	0.668	0.577
PRISMA, Chile	testt124v	Textual	0.464	0.474	0.688	0.64	**0.604**
SIFT, Ireland	txtOnlyProbFuse	Textual	0.447	0.454	0.704	0.652	0.556
PRISMA, Chile	testt134v	Textual	0.43	0.444	0.712	0.656	0.563
OHSU, USA	fusion1	Textual	0.300	0.337	0.448	0.376	0.381
OHSU, USA	fusion4	Textual	0.270	0.347	0.28	0.332	0.361
OHSU, USA	fusion3	Textual	0.175	0.235	0.328	0.32	0.24

In Table 3 the results when fusing only the textual runs are shown. The best runs of each performance measure are marked in bold. Best results are obtained with a probabilistic model that learned the importance of specific parts of the results. The best four results are all very close. MAP and early precision are both very well correlated among the runs and the best run regarding MAP also had best early precision. BPref (Binary preference) shows whether a technique has many un–judged images ranked highly and in this case it correlates very closely with MAP, which is not surprising as all runs are based on the exact same runs or basic technqiues.

Table 4. Performance metrics for fusion of visual runs

Group	Runid	type	map	bpref	P5	P10	P30
SIFT, Ireland	visOnlySegFuse	Visual	**0.0179**	0.0353	0.088	**0.092**	0.0613
SIFT, Ireland	visOnlySlideFuse	Visual	0.0175	**0.0354**	**0.104**	0.088	**0.064**
SIFT, Ireland	visOnlyProbFuse	Visual	0.0154	0.0338	0.088	0.08	0.056

Table 5. Performance metrics for fusion mixed runs

Group	Runid	type	map	bpref	P5	P10	P30
SIFT, Ireland	txtimgSlideFuse	Mixed	**0.495**	0.494	0.712	0.66	0.588
SIFT, Ireland	txtimgSlideFuse	Mixed	**0.495**	0.494	0.712	0.66	0.588
PRISMA, Chile	gt841t234v3	Mixed	0.491	**0.497**	**0.76**	**0.696**	**0.611**
medGIFT, CH	combSUMlogRank	Mixed	0.488	0.490	0.712	0.672	0.592
medGIFT, CH	combMNZlogRank	Mixed	0.487	0.489	0.712	0.672	0.592
medGIFT, CH	combSUMByFreqlogRank	Mixed	0.484	0.489	0.712	0.672	0.592
SIFT, Ireland	txtimgSegFuse	Mixed	0.469	0.459	0.696	0.672	0.585
PRISMA, Chile	testt1234v234	Mixed	0.466	0.461	0.752	0.676	0.595
PRISMA, Chile	testt1234v134	Mixed	0.464	0.458	0.744	0.692	0.589
medGIFT, CH	GESUM3MAXLinearRank	Mixed	0.461	0.465	0.720	0.656	0.556
OHSU, USA	fusion2	Mixed	0.458	0.478	0.672	0.628	0.575
medGIFT, CH	SUM3MAXByFreqLinearRank	Mixed	0.458	0.462	0.720	0.656	0.556
PRISMA, Chile	testt1234v123	Mixed	0.451	0.449	0.744	0.688	0.577
PRISMA, Chile	testt1234v124	Mixed	0.450	0.449	0.712	0.656	0.576
medGIFT, CH	combMNZScoreNorm	Mixed	0.442	0.442	0.720	0.656	0.579
medGIFT, CH	combSUMFreqScoreNorm	Mixed	0.442	0.446	0.688	0.692	0.568
medGIFT, CH	combSUMScoreNorm	Mixed	0.441	0.442	0.720	0.656	0.579
SIFT, Ireland	txtimgProbFuse	Mixed	0.434	0.419	0.696	0.652	0.563
PRISMA, Chile	testt1234v1234	Mixed	0.411	0.403	0.720	0.648	0.551
PRISMA, Chile	testt134v234	Mixed	0.398	0.399	0.664	0.664	0.528
PRISMA, Chile	testt134v134	Mixed	0.394	0.395	0.688	0.66	0.528
ISDM, Spain	gen2	Mixed	0.383	0.385	0.688	0.668	0.54
ISDM, Spain	gen5	Mixed	0.382	0.384	0.696	0.652	0.536
ISDM, Spain	gen1	Mixed	0.382	0.385	0.688	0.664	0.54
ISDM, Spain	gen4	Mixed	0.379	0.382	0.704	0.66	0.527
PRISMA, Chile	testt234v124	Mixed	0.374	0.373	0.616	0.604	0.523
ISDM, Spain	gen3	Mixed	0.372	0.375	0.688	0.64	0.521
PRISMA, Chile	testt134v123	Mixed	0.369	0.374	0.648	0.6	0.509
PRISMA, Chile	testt134v124	Mixed	0.362	0.374	0.616	0.596	0.508
PRISMA, Chile	testt124v123	Mixed	0.362	0.357	0.608	0.596	0.517
PRISMA, Chile	testt234v1234	Mixed	0.298	0.304	0.520	0.508	0.431
PRISMA, Chile	testt134v1234	Mixed	0.294	0.304	0.520	0.5 00	0.437
PRISMA, Chile	testt124v1234	Mixed	0.284	0.291	0.496	0.504	0.420
ISDM, Spain	wsum1	Mixed	0.147	0.200	0.528	0.456	0.293
ISDM, Spain	wmnz1	Mixed	0.100	0.142	0.352	0.292	0.216

Table 4 shows the visual fusion results of the participants. Only a single group submitted three runs. Results could be increased over the original results but they remained low as the based results were not performing well at all. Other participants also combined visual runs only without submitting them to the contest but results were very similar to the results presented here.

Table 5 displays all submitted mixed runs. The early precision and the MAP of these runs are clearly superior to all the text runs shown in Table 2. We can also see that the best runs in terms of MAP are not best in terms of early precision, so to understand these a more detailed analysis of the techniques needs to be performed. All among the early results have a very similar score. The first six runs only have an absolute difference in terms of MAP of 1%. When compared to the fusion results using only text it can be seen that MAP is slightly lower but early precision is significantly lower with a much higher margin.

Combinations of only the textual runs delivered similar results to the mixed runs with the best technique (SIFT group) obtaining 0.487, so slightly lower than the combination of the mixed runs. Other groups similarly had slightly better results using the mixed combinations compared to only comparing the text runs. For early precision this was similar but with a stronger difference, obtaining 0.72 compared to 0.76 for the best mixed combination run, with most other groups having a slightly lower early precision for the text only runs.

6 Conclusions

The first fusion challenge to combine visual and textual runs from medical image retrieval was organized for ICPR 2010. The goal of this context was to encourage participants to explore machine learning and other advanced techniques to effectively combine runs from the ImageCLEFmed challenge given a set of training runs and their performance metrics. Five groups submitted a total of 49 runs, many of which demonstrated the effectiveness of a multimodal approach to image retrieval. It was encouraging to note that about half of the submitted runs performed better than all the test runs. On the other hand, a few of the mixed runs that we submitted performed poorly, possibly due to the really poor performance of the visual test runs. The best runs obtained a MAP of 0.495 compared to the best run in the ImageCLEF of 0.43 and the best combined run in ImageCLEF 2009 of even 0.41. Such gains of over 20% show the potential of well combining visual and textual cues for medical image retrieval. The focus of ImageCLEF should be on fostering such developments In the past, particularly the combination of media has been of limited effectiveness in ImageCLEF as most research groups work on either visual or textual retrieval but not the two. The small participation of only five research groups on the other hand also showed that there might be even more potential if successful techniques for fusion are consistently applied and tested.

Acknowledgements

We would like to acknowledge the support of the National Library of Medicine grant 1K99LM009889 and of the BeMeVIS project of the HES–SO.

References

1. Müller, H., Michoux, N., Bandon, D., Geissbuhler, A.: A review of content-based image retrieval systems in medicine–clinical benefits and future directions. IJMI 73(1), 1–23 (2004)
2. Tagare, H.D., Jaffe, C., Duncan, J.: Medical image databases: A content–based retrieval approach. JAMIA 4(3), 184–198 (1997)
3. Hersh, W., Müller, H., Jensen, J., Yang, J., Gorman, P., Ruch, P.: Advancing biomedical image retrieval: Development and analysis of a test collection. JAMIA 13(5), 488–496 (2006)
4. Radhouani, S., Kalpathy-Cramer, J., Bedrick, S., Hersh, W.: Medical image retrieval, a user study. Technical report, Medical Inforamtics and Outcome Research, OHSU, Portland, OR, USA (June 2009)
5. Harman, D.: Overview of the first Text REtrieval Conference (TREC–1). In: Proceedings of the first Text Retrieval Conference (TREC–1), Washington DC, USA, pp. 1–20 (1992)
6. Savoy, J.: Report on CLEF-2001 experiments: Effective combined query-translation approach. In: Peters, C., Braschler, M., Gonzalo, J., Kluck, M. (eds.) CLEF 2001. LNCS, vol. 2406, pp. 27–43. Springer, Heidelberg (2002)
7. Müller, H., Deselaers, T., Kim, E., Kalpathy-Cramer, J., Deserno, T.M., Clough, P., Hersh, W.: Overview of the ImageCLEFmed 2007 medical retrieval and annotation tasks. In: Peters, C., Jijkoun, V., Mandl, T., Müller, H., Oard, D.W., Peñas, A., Petras, V., Santos, D. (eds.) CLEF 2007. LNCS, vol. 5152, pp. 473–491. Springer, Heidelberg (2008)
8. Müller, H., Kalpathy-Cramer, J.: Analyzing the content out of context — features and gaps in medical image retrieval. International Journal on Healthcare Information Systems and Informatics 4(1), 88–98 (2009)
9. Müller, H., Kalpathy-Cramer, J., Eggel, I., Bedrick, S., Said, R., Bakke, B., Kahn Jr., C.E., Hersh, W.: Overview of the CLEF 2009 medical image retrieval track. In: Peters, C., Caputo, B., Gonzalo, J., Jones, G.J.F., Kalpathy-Cramer, J., Müller, H., Tsikrika, T. (eds.) CLEF 2009. LNCS, vol. 6242, pp. 72–84. Springer, Heidelberg (2010)
10. Zhou, X., Depeursinge, A., Müller, H.: Information fusion for combining visual and textual image retrieval. In: International Conference on Pattern Recognition. IEEE Computer Society, Los Alamitos (2010)
11. Zhang, L., Toolan, F., Lillis, D., Collier, R., Dunnion, J.: Probabilistic data fusion for image retrieval. In: International Conference on Pattern Recognition. IEEE Computer Society, Los Alamitos (2010)

ISDM at ImageCLEF 2010 Fusion Task

A. Revuelta-Martínez, I. García-Varea, J.M. Puerta, and L. Rodríguez

Departamento de Sistemas Informáticos
Universidad de Castilla-La Mancha
02071 Albacete, Spain
{Alejandro.Revuelta,Ismael.Garcia,Jose.Puerta,Luis.RRuiz}@uclm.es

Abstract. Nowadays, one of the main problems in information retrieval is filtering the great amount of information currently available. Late fusion techniques merge the outcomes of different information retrieval systems to generate a single result that, hopefully, could increase the overall performance by taking advantage of the strengths of all the individual systems. These techniques have a great flexibility and allow an efficient development of multimedia retrieval systems. The growing interest on these technologies has led to the creation of a subtrack in the ImageCLEF entirely devoted to them: the information fusion task. In this work, Intelligent Systems and Data Mining group approach to that task is presented. We propose the use of an evolutive algorithm to estimate the parameters of three of all the fusion approaches present in the literature.

Keywords: multimodal information retrieval, late fusion, estimation of distribution algorithms.

1 Introduction

Modern technologies allow any user to generate, store and share a large amount of data. As a consequence, one of the most important problems nowadays is filtering data in order to find relevant information. This need motivated the development of Information Retrieval (IR) systems [10,2].

Initially, IR systems were only based on text documents. More recently, however, increasing nontextual information is available thanks to the wide use of cheap multimedia hardware (e.g., digital cameras) and publication options (e.g., the Internet and its social networks). In this scenario, new IR engines were designed for retrieving different kinds of media. Furthermore, several approaches aimed at considering different media in the same query were studied. The rationale behind this idea is that different media usually have complementary information. For example, in text retrieval we try to find documents semantically close to the query while in visual retrieval we try to find documents which contain images presenting similar visual features. This way, considering different media could improve the results and, additionally, the user experience could be enhanced by allowing multimodal interaction.

D. Ünay, Z. Çataltepe, and S. Aksoy (Eds.): ICPR 2010, LNCS 6388, pp. 109–118, 2010.

Several techniques have been considered to deal with this problem which are mainly intended to combine textual and visual retrieval:

- Annotated images and the use of textual retrieval techniques.
- Perform an initial text query and afterwards refine the results using visual features and relevance feedback [15].
- Directly retrieve documents considering both, their textual and visual features [3].
- Perform the same retrieval process using several IR engines and then merge their results, also known as *late fusion* (see section 2 for details).

This work focus on the late fusion approach which presents important advantages: a great flexibility, extensive use of the currently developed technologies and, as well, the possibility to be extended to different media IR systems with little effort. This is an interesting problem that has led to the development of the fusion task presented in ImageCLEF at ICPR 2010[1] (see section 3).

This document presents the Intelligent Systems and Data Mining (ISDM) group approach to ImageCLEF's fusion task (see section 4). Our proposal tries to improve three different methods used in late fusion by applying combinatorial optimization techniques. These fusion methods allow weighting the input IR systems and, in this work, we propose to estimate these weights (in addition to other parameters) using an Estimation of Distribution Algorithm (EDA) [7].

2   Late Fusion in Information Retrieval

As it was previously mentioned, late fusion in information retrieval can be stated as the problem of retrieving of the same query by different and independent IR engines in order to merge their outputs to obtain a final and global result. It is straightforward to see that the constituent systems are not modified since the fusion is performed only on their outcomes. These techniques present, therefore, a high flexibility and scalability. All these advantages may allow the easy development of multimedia retrieval tools as a consequence of permitting the different engines to operate with different information sources.

Several approaches have been considered, so far, for late fusion. In [6], a set of operations to compute the score of a given document d_j by combining the scores provided by each individual IR system is presented. The two operations that yielded the best results were:

$$CombSUM_{score}(d_j) = \sum_{i=1}^{N} \gamma_i score_i(d_j) \qquad (1)$$

$$CombMNZ_{score}(d_j) = |d_j| \sum_{i=1}^{N} \gamma_i score_i(d_j) \qquad (2)$$

---

[1] http://www.imageclef.org/2010/ICPR/Fusion/

where $|d_j|$ is the number of IR systems which retrieve d_j, N is the number of individual IR engines considered, γ_i is the weight assigned to the i-th IR engine and $score_i(d_j)$ is the score of document d_j returned by the i-th IR system. Fusion is performed by merging all the retrieved documents and rescoring them with one of these operations.

However, some engines do not return the actual scores of the retrieved documents. In these cases the previous operations are extended by using a score function based on the position of a document in a ranking [8]:

$$rscore_i(d_j) = 1 - \frac{rank_{i,j} - 1}{retrieved_i} \qquad (3)$$

where $rank_{i,j}$ is the rank of d_j in the i-th retrieval system and $retrieved_i$ is the number of documents retrieved by the i-th engine.

Other rank-based approaches consider late fusion as a voting problem and apply techniques taken from social choice theory to perform the fusion:

- Borda-Fuse [1]: If a given IR engine retrieves D documents, it is considered that D points are assigned to the first document, $D - 1$ points are assigned to the second one, and so on. The fusion is performed by summing up all the points for each document to subsequently sort them.
- Condorcet Fusion [12]: This method considers the results of the individual IR systems as ranked votes so that the best score is assigned to the document that beats or ties with every other document in a pair-wise comparison. Fusion is performed by generalizing this idea to a rank of documents.

Probabilistic techniques have also been applied to fusion problems. In [9], a division of the documents retrieved by each IR system is proposed, which is performed by dividing its outcome in segments of the same size and then using some training data to estimate the probability $Pr(d_{j,k}|m)$ of a document d_j being relevant given that it is retrieved by the m-th model in the k-th segment. These probabilities are later used to score the documents of the fused result.

Finally, an alternative approach tries to estimate a multinomial distribution to rank the documents [4]. This technique will be discussed in detail in section 2.1.

2.1 Generative Model

In order to combine the outputs of the different information retrieval engines a statistical technique (presented in [4]) has been considered. This approach is based on a generative model which will be denoted as GeM from now on.

This model assumes that the documents returned by the IR engines are generated by a multinomial distribution with a parameter vector θ. Under this consideration, a particular document d_j is generated with probability θ_j. This way, the fusion is performed by estimating the parameters of the multinomial distribution from the outcomes of the IR systems and, subsequently, using those parameters to rank the documents in the fused set.

The estimation of parameters follows the idea of the *Impatient Reader Model* which represents a real user's attention while reading a list of documents. The model will assign a probability $Pr(d|r)$ to each document to represent the likelihood of the user reading it. Better scored documents are more likely to be read but user's attention is expected to decrease as ranks get higher. This is modelled by an exponential distribution:

$$Pr(d|r) \sim f(r|\lambda) = \lambda e^{-\lambda r} \tag{4}$$

where $\lambda \geq 0$ is the exponential parameter and r is the rank of the document in the list.

The actual estimation of the θ parameters is performed by taking into account the information provided by the impatient reader model for all the individual IR systems. The overall probability of the impatient reader viewing document d_j is considered to be:

$$Pr(d_j|r_{j1}\cdots r_{jN}) \sim z_j = \frac{1}{N}\sum_{i=1}^{N} f(r_{ji}|\hat{\lambda}_i) \tag{5}$$

where r_{ji} is the rank of document d_j in the i-th considered IR engine output, N is the number of considered IR engines and $\hat{\lambda}_i$ is the estimated value of λ_i.

Using Bayes' rule, the maximum a posteriori estimate for θ_j is obtained by combining the maximum likelihood estimate of θ_j with the conjugate prior for the multinomial distribution (i.e., the Dirichlet distribution):

$$\hat{\theta}_j = \frac{|d_j| + \mu z_j}{|s| + \mu(\sum_i^{|S|} z_i)} \tag{6}$$

where μ is a non-negative hyperparameter that characterizes the posterior distribution, $|s|$ is the total number of documents retrieved by all the IR systems and $|S|$ is the number of documents retrieved by all the IR systems after deleting duplicated ones. Eq. (6) provides the values which will be used to rank the documents in the fused result.

In Eq. (5), all the IR systems have been considered to have the same weight in the final result. However, this assumption is unrealistic, especially in this task, since text-based systems perform much better than visual-based ones (see section 3). The GeM model can be modified to consider different weights rewriting Eq. (5) as:

$$Pr(d_j|r_{j1}\cdots r_{jN}) \sim z_j' = \sum_{i=1}^{N} \gamma_i f(r_{ji}|\hat{\lambda}_i) \tag{7}$$

where γ_i is the weight of the i-th IR system and all of them should satisfy the constraint $\sum_{i=1}^{N} \gamma_i = 1$.

3   Fusion Task

ImageCLEF[2] is the cross-language image retrieval track which is run as part of the Cross Language Evaluation Forum (CLEF). It features several subtracks that

---

[2] http://www.imageclef.org/

have evolved since its first edition. This year, ImageCLEF organizes a challenge as part of the International Conference on Pattern Recognition (ICPR) where a subtrack is devoted to late fusion: the *information fusion task* [13].

The information fusion task is aimed at exploring strategies to combine the outcomes of different visual and text retrieval systems. The data used in this task was obtained from the best four purely visual and the best four purely textual runs of the 2009 ImageCLEF medical retrieval task [14]. The query topics are mainly semantic and, thus, text-based retrieval performs much better than visual-based retrieval in this task.

Participants had to use this data to generate one fused result. Each team could send up to 20 submissions which were evaluated using the well-known *Mean Average Precision* (MAP) metric [10].

4   ISDM Proposal

Three fusion techniques have been selected from the literature which allow the consideration of different weights for each one of the underlying IR engines. The adopted techniques are: GeM (as our main approach), CombSUM and CombMNZ (which are the baseline for most of the published late fusion works).

These methods have several parameters that directly affect the retrieval performance. Here, we propose the use of an EDA [7] to estimate these parameters.

In addition, the considered GeM model has been modified to deal with the document scores from the different IR engines. Therefore, the equation employed by GeM to rank the documents can be rewritten as:

$$\hat{\theta}_j = \frac{\sum_{k=1}^{N} score_k(d_j) + \mu z'_j}{|s| + \mu(\sum_{i}^{|S|} z'_i)} \qquad (8)$$

where $score_k(d_j)$ is the normalized score that the k-th IR system assigns to document d_j and z'_j is computed using the weights returned by the optimization algorithm.

4.1   Parameter Estimation

The GeM model requires the estimation of the μ and λ_i parameters, as well as the different systems weights (γ_i). Our proposal estimates those values using a specific EDA, the *Population-Based Incremental Learning* (PBIL) optimization algorithm [16].

PBIL techniques are based on the idea of a population of suitable solutions. In this case, these solutions are composed of all the parameters to be computed. Hence, an individual I of the population can be expressed as:

$$I = \hat{\gamma}_1, \cdots, \hat{\gamma}_N, \hat{\lambda}_1, \cdots, \hat{\lambda}_N, \hat{\mu} \qquad (9)$$

The overall population is represented as a set of Gaussian distributions, each one of them modelling one of the values of Eq. (9). PBIL algorithm iteratively constructs a population by sampling these Gaussian distributions to subsequently update the Gaussian distribution parameters according to the best individuals.

The LiO [11] toolkit was used to implement the PBIL algorithm whose basic operation mode is:

1. Initialize the means and variances of the Gaussian distributions with arbitrary parameters.
2. Repeat the following steps until reaching a predefined stop criterion[3]:
   (a) Generate a population of M individuals by sampling the Gaussian distributions.
   (b) Compute the MAP, precision and recall metrics for all the individuals.
   (c) Rank the individuals according to their MAP values.
   (d) Delete any individual that is dominated[4] by another one with a higher MAP.
   (e) Update the Gaussian distributions parameters according to the means and variances of the remaining individuals.
3. Finally, the algorithm returns the best individual found so far, as well as the set of non-dominated individuals generated in the (d) step of the last iteration.

At the end, the algorithm returns the best individual found so far, which represents the best considered set of estimated parameters. Furthermore, we can take advantage of the remaining individuals in the final iteration as they represent potentially good solutions and, therefore, they can be used as additional solutions to be returned.

In the cases of CombSUM and CombMNZ, the optimization algorithm is exactly the same although considering only the weights of the different IR systems.

5   Results

5.1   Submitted Runs

All the submitted runs were obtained employing the optimization algorithm presented in section 4.1. Three PBIL runs were performed, considering a different fusion technique (GeM, CombMNZ and CombSUM) in each one of them.

- Runs **gen1.r1**, **gen2.r1**, **gen3.r1**, **gen4.r1** and **gen5.r1** were obtained by fusing the individual IR outcomes applying the GeM technique with the parameters represented by the n-best scored individuals returned by the GeM PBIL execution (**gen1.r1** considering the best individual and **gen5.r1** considering the fifth best scored one).

---

[3] The stop criterion was: performing 1000 iterations, 512000 evaluations, 500 iterations without improving or 256000 evaluations without improving.

[4] One individual is said to dominate a second one if its precision and recall are both higher than those in the second individual and, at least, one of them is strictly greater. This concept, formally named Pareto dominance [5], is widely used in multiobjective optimization. The rationale behind it is to capture the trade-offs among opposing objectives.

- Run **wmnz1.r1** was obtained using the CombMNZ technique to fuse the individual IR outcomes considering, to this end, the weights represented by the best scored individual returned by the CombMNZ PBIL execution.
- Run **wsum1.r1** was obtained using the CombSUM technique considering the weights represented by the best scored individual returned by the CombSUM PBIL execution.

5.2   Evaluation

Table 1 shows the baseline results for all the considered techniques on the provided data sets. Table 2, on the other hand, reports the results obtained with the PBIL-based optimization technique.

Three experiments have been performed with the techniques presented in the previous sections:

1. The first one consisted of executing the PBIL algorithm on the provided training data considering the GeM, CombSUM and CombMNZ fusion techniques. The MAP results of this experiment can be found in the first row of table 2.
2. The second experiment was performed by applying GeM, CombSUM and CombMNZ fusion techniques on the provided test data. The parameters employed were the same ones that were computed in the first experiment. These results are shown in the second row of table 2 and were the ones obtained by the ISDM submissions to the fusion task.
3. After the submission deadline the task organizers released the relevance judgements of the test set. This information was used to perform a new set of PBIL executions (considering GeM, CombSUM and CombMNZ), this time directly on the test set. These results are shown in the third row of table 2.

The results of the first and third experiments are not representative of future performance since there is a significant overlapping between the training and the evaluation data. However, they can provide an indication of the algorithm performance on the different data sets and could be used to compare different fusion techniques.

The results show that the optimization technique indeed improves fusion performance on the datasets. There is a higher improvement on the test set than in the training set, reaching a MAP of 0.477 for the first one and of 0.377 for

Table 1. Baseline MAP results of the fusion task using CombSUM, CombMNZ and GeM without weighting the IR systems and considering arbitrary parameters

	Train	Test
GeM	0.361	0.342
CombMNZ	0.364	0.401
CombSUM	0.367	0.342

Table 2. MAP results of the fusion task obtained by ISDM. Each column corresponds to the techniques employed in the submitted runs described in section 5.1.

	gen1.r1	gen2.r1	gen3.r1	gen4.r1	gen5.r1	wmnz1.r1	wsum1.r1
PBIL—Train	0.373	0.373	0.372	0.371	0.371	0.371	**0.377**
Test	0.382	**0.383**	0.372	0.379	0.382	0.375	0.333
PBIL—Test	0.454	0.454	0.454	0.454	0.454	**0.477**	0.454

the second one. However, evaluation using data not present at the training set achieves significantly lower results.

In the case of GeM, performance is increased, with respect to the baseline, in all the experiments and presents the best score in the second experiment (which is the most representative of future results). On the contrary, CombSUM and CombMNZ do not outperform their baselines in the second experiment.

Results support the idea that CombMNZ and CombSUM have a greater dependence on the queries than GeM and, thus, GeM is more robust and perform better when dealing with new queries. Further research on GeM and the optimization techniques is promising since they would be an important option to perform late fusion if their results could be improved.

Considering all the runs submitted to the contest [13], the other participants top submissions achieved higher scores than ours. Comparing our best submission to the overall best one, there is a MAP difference of 0.11. However, the difference is shorter if early precision is considered (0.064 for precision at 5 and 0.028 for precision at 10). Finally, the third experiment described in this section shows that our approach could be more competitive (achieving a MAP difference of 0.018 with respect to the best submission) by improving the performance of queries not seen in the training stage.

6 Conclusions and Future Work

In this paper, the problem of late fusion in the scope of information retrieval has been firstly introduced. Then, several techniques to cope with this problem, along with the information fusion task presented in the ImageCLEF track at ICPR 2010, have been described. Finally, we have discussed in detail the approach followed by the ISDM group to generate their submissions to the fusion task as well as the experimental results achieved.

Results support that a good configuration of the model parameters can significantly improve retrieval results. In the case of the GeM technique, the baseline results are improved in all the scenarios despite requiring more parameters to be estimated than the other techniques. However, the PBIL-based optimization technique does not improve the results of CombSUM and CombMNZ when retrieving new queries.

Applying optimization techniques to late fusion retrieval systems is a promising area of study and therefore requires future work. It would be interesting to test these techniques on larger training sets, or to reduce the parameters (e.g., using only two weights: the first one for text retrieval systems and the second one for visual retrieval ones) in order to allow a more accurate and better estimation of the parameters. Another future line of work could be the modification of the GeM approach to consider different statistical models while keeping the same underlying ideas of the impatient reader and the generative model. Finally, other optimization techniques could also be considered for this task.

Acknowledgements

Work supported by the EC (FEDER/ESF) and the Spanish government under the MIPRCV "Consolider Ingenio 2010" program (CSD2007-00018), and the Spanish Junta de Comunidades de Castilla-La Mancha regional government under projects PBI08-0210-7127 and PCI08-0048-8577.

References

1. Aslam, J.A., Montague, M.: Models for metasearch. In: 24th Annual International ACM SIGIR Conference on Research and Development in Information Retrieval, pp. 276–284. ACM, New York (2001)
2. Croft, B., Metzler, D., Strohman, T.: Search Engines: Information Retrieval in Practice. Pearson, London (2010)
3. Deselaers, T., Weyand, T., Keysers, D., Macherey, W., Ney, H.: Fire in Image-CLEF 2005: Combining content-based image retrieval with textual information retrieval. In: Peters, C., Gey, F.C., Gonzalo, J., Müller, H., Jones, G.J.F., Kluck, M., Magnini, B., de Rijke, M.(eds.) CLEF 2005. LNCS, vol. 4022, pp. 652–661. Springer, Heidelberg (2006)
4. Efron, M.: Generative model-based metasearch for data fusion in information retrieval. In: 9th ACM/IEEE-CS Joint Conference on Digital Libraries, pp. 153–162. ACM, New York (2009)
5. Fonseca, C.M., Fleming, P.J.: An overview of evolutionary algorithms in multiobjective optimization. Evolutionary Computation 3(1), 1–16 (1995)
6. Fox, E.A., Shaw, J.A.: Combination of multiple searches. In: 2nd Text REtrieval Conference, pp. 243–252 (1994)
7. Larrañaga, P., Lozano, J.: Estimation of Distribution Algorithms. A New Tool for Evolutionary Computation. Kluwer Academic Publishers, Dordrecht (2002)
8. Lee, J.H.: Analyses of multiple evidence combination. In: 20th Annual International ACM SIGIR Conference on Research and Development in Information Retrieval, pp. 267–276. ACM, New York (1997)
9. Lillis, D., Toolan, F., Mur, A., Peng, L., Collier, R., Dunnion, J.: Probability-based fusion of information retrieval result sets. Artificial Intelligence Review 25(1-2), 179–191 (2006)
10. Manning, C.D., Raghavan, P., Schütze, H.: Introduction to Information Retrieval. Cambridge University Press, Cambridge (2008)
11. Mateo, J.L., de la Ossa, L.: LiO: an easy and flexible library of metaheuristics. Tech. Rep. DIAB-06-04-1, Departamento de Sistemas Informáticos de la Universidad de Castilla-La Mancha (2006), http://www.dsi.uclm.es/simd/SOFTWARE/LIO/

12. Montague, M., Aslam, J.A.: Condorcet fusion for improved retrieval. In: 11th International Conference on Information and Knowledge Management, pp. 538–548. ACM, New York (2002)
13. Müller, H., Kalpathy-Cramer, J.: The ImageCLEF medical retrieval task at ICPR 2010 – information fusion to combine visual and textual information. In: Ünay, D., Çataltepe, Z., Aksoy, S. (eds.) ICPR 2010. LNCS, vol. 6388, pp. 101–110. Springer, Heidelberg (2010)
14. Müller, H., Kalpathy–Cramer, J., Eggel, I., Bedrick, S., Radhouani, S., Bakke, B., Kahn Jr., C.E., Hersh, W.: Overview of the CLEF 2009 medical image retrieval track. In: Peters, C., Caputo, B., Gonzalo, J., Jones, G.J.F., Kalpathy-Cramer, J., Müller, H., Tsikrika, T. (eds.) CLEF 2009. LNCS, vol. 6242, pp. 72–84. Springer, Heidelberg (2010)
15. Paredes, R., Deselaers, T., Vidal, E.: A probabilistic model for user relevance feedback on image retrieval. In: Popescu-Belis, A., Stiefelhagen, R. (eds.) MLMI 2008. LNCS, vol. 5237, pp. 260–271. Springer, Heidelberg (2008)
16. Sebag, M., Ducoulombier, A.: Extending population-based incremental learning to continuous search spaces. In: Eiben, A.E., Bäck, T., Schoenauer, M., Schwefel, H.-P. (eds.) PPSN 1998. LNCS, vol. 1498, pp. 418–427. Springer, Heidelberg (1998)

Rank-Mixer and Rank-Booster: Improving the Effectiveness of Retrieval Methods*

Sebastian Kreft and Benjamin Bustos

PRISMA Research Group
Department of Computer Science, University of Chile
{skreft,bebustos}@dcc.uchile.cl

Abstract. In this work, we present two algorithms to improve the effectiveness of multimedia retrieval. One, as earlier approaches, uses several retrieval methods to improve the result, and the other uses one single method to achieve higher effectiveness. One of the advantages of the proposed algorithms is that they can be computed efficiently in top of existing indexes. Our experimental evaluation over 3D object datasets shows that the proposed techniques outperforms the multimetric approach and previously existing rank fusion methods.

Keywords: Multimedia databases, effectiveness, boosting.

1 Introduction

In the last years, we have experienced a phenomenon of multimedia information explosion, where the volume of produced digital data increases exponentially in time. This exponential growth is caused by many factors, like more powerful computing resources, high-speed internet, and the diffusion of the information society all over the world. Additionally, an enormous production of data is attributed to the quick dissemination of cheap devices for capturing multimedia data like audio, video, and photography. Thus, it has become essential to develop effective methods to search and browse large multimedia repositories.

The *content-based retrieval* (CBR) of multimedia data (or of other semantically unstructured-type data) is a widely used approach to search in multimedia collections. CBR performs the retrieval of relevant multimedia data according to the actual content of the multimedia objects, rather than considering an external description (e.g., annotations). Instead of text-based query, the database is queried by an example object to which the desired database objects should be *similar*. This is known as the *query-by-example* retrieval scheme.

Usually, the similarity measure used to compare two multimedia objects is modeled as a metric distance (in the mathematical meaning), which is known as the *metric space approach* [14]. This is because the metric axioms have allowed researchers to design efficient (fast) access methods employed in the similarity search. With this approach, the search can be performed in an efficient way.

---

* Partially funded by Conicyt (Chile), through the Master Scholarship (first author).

D. Ünay, Z. Çataltepe, and S. Aksoy (Eds.): ICPR 2010, LNCS 6388, pp. 119–128, 2010.

However, depending on the particular application domain, the similarity measure may not model 100% correctly the human notion of similarity.

The *effectiveness* of a multimedia research system is related with the quality of the answer returned by the similarity query. In the metric approach, given a distance function, a similarity query corresponds to a search for close objects in some topological space. An effective distance function should treat two similar objects, according to the human concept of similarity, as two close points in the corresponding space. Indeed, the effectiveness of a similarity search system measures its ability to retrieve relevant objects while at the same time holding back non-relevant ones. Improving the effectiveness of a similarity search system is at least as important as improving its efficiency, because effectiveness is directly related to the quality of the answers that the search system returns.

In this paper we present two novel algorithms that improves the effectiveness of similarity measures. The first method, *Rank-Mixer*, merges several results obtained with different similarity measures, and the second method, *Rank-Booster*, improves the quality of the answer obtained using just one similarity measure. Both methods only use the information given by the similarity measure, and they do not rely on training databases like other approaches based on off-line supervised learning. We show how to use these methods over existing index structures, thus the efficiency of the search is not affected. To evaluate the performance of our proposed algorithms, we made an extensive experimental evaluation in a standard reference collection for 3D model retrieval and used the data of the ImageCLEF@ICPR fusion task, showing that Rank-Booster and Rank-Mixer are able to improve the effectiveness of the best available 3D model similarity measures and the best textual methods of ImageCLEF.

2   Related Work

2.1   Similarity Queries

The most common similarity query is the nearest neighbors or k-NN, which returns the k most similar objects of the database with respect to a query object q not necessarily present in the database.

The (dis)similarity is defined as a function that takes two multimedia objects as input and returns a positive value. The value 0 means that the objects are equal. Typically, to compute the (dis)similarity between multimedia objects the metric approach is used. In this approach, the dissimilarity is a distance for which the triangle inequality holds. This is done generally by computing for every object a feature vector (FV) that represents the properties of that object.

2.2   Metric Combination

It has been shown [3,4] that a query dependent combination of metric spaces yields to higher effectiveness of the similarity search. One way to combine several metrics is by mean of multi-metric spaces, where the (dis)similarity function is computed as a linear combination of some selected metrics.

Definition 1. *Multi-metric Space*
Let $\mathcal{X} = \{(\mathbb{X}_i, \delta_i), 1 \leq i \leq n\}$ *a set of metric spaces, the corresponding* Multi-metric space *is defined as the pair* $(\prod_{i=1}^{n} \mathbb{X}_i, \Delta_{\mathbb{W}})$, *where* $\Delta_{\mathbb{W}}$ *is a linear multi-metric, which means*

$$\Delta_{\mathbb{W}}(x, y) = \sum_{i=1}^{n} w_i \delta_i(x_i, y_i), \tag{1}$$

In the above definition, the vector of weights $\mathbb{W} = \langle w_i \rangle$ is not fixed, and is a parameter of Δ. When $\forall i \, w_i \in [0, 1] \wedge \exists i \, w_i > 0$, $\Delta_{\mathbb{W}}$ is also a metric.

2.3 Rank Fusion

In the area of information retrieval and pattern recognition, there are several methods that given different ranks of objects improve the effectiveness of the result by combining them. Here we present some of them, for a more detailed survey on these methods see Suen and Lam [12]. Most of these methods give each element a score and then rank them according to the assigned score.

Borda Count [9], originally developed for voting systems, has been widely used in information retrieval. This method gives each element the score $\sum_{r \in \mathcal{R}} r(d)$.

Reciprocal Rank [6] gives each element the score $\sum_{r \in \mathcal{R}} \frac{1}{k + r(d)}$.

Logistic Regression Method [9] solves the problem of Borda Count that does not take into account the quality of the different ranks. The score assigned by this method is $\sum_{r \in \mathcal{R}} w_r r(d)$ where the weights w_r are computed as a logistic regression. This method is similar to the idea of entropy-impurity [3].

Med-Rank [7] is an aggregation method intended for vector spaces. However, it can also be applied to combine different ranks. In this method, each element gets a score equal to the index i, such that it appears at least in $f_{min}|\mathcal{R}|$ different ranks up to position i. Mathematically, the score is $\min\{i \in \{1, \ldots, n\}/ |\{r(d)/r(d) \leq i\}| > f_{min}|\mathcal{R}|\}$, where f_{min} is a parameter, usually taken as $f_{min} = 0.5$.

3 Improving Effectiveness of Retrieval Methods

3.1 Rank-Mixer

This algorithm combines the answer of different multimedia retrieval methods (not necessarily metrics) and produces a new improved answer. The idea behind this algorithm is that if an object is reported to be similar to the query object by several retrieval methods, then the object should be a relevant one. We give a *score* to the objects according to their position in the rankings. Then, we rank all the objects according to the total score they got.

For each retrieval method, we compute the k-NN. Then we apply a function f^+ to the ranking of each object to assign a *score* to the objects. As we want to give higher scores to the first objects in the ranking, f^+ must be a decreasing function. On the other hand, as we do not have the complete ranking, we need to assign an implicit value of 0 to the unseen objects, thus f^+ must be a positive

$$\begin{aligned}
&\textbf{function } \text{booster}(q, \mathbb{U}, m, k, f^+, k_b)\\
&\quad rank \leftarrow kNN(q, \max\{k, k_b\}, \mathbb{U})
\end{aligned}$$

function mixer(q, $\mathbb{U}$, $\mathcal{M}$, f^+, k)
  for each $m \in \mathcal{M}$ **do**
    $rank \leftarrow kNN(q, \mathbb{U}, k)$
    for each $o \in rank$ **do**
      $mrank[o] \leftarrow mrank[o] + f^+(pos(o, rank))$
  Sort descending $mrank$
  return mrank[1:k]

function booster(q, $\mathbb{U}$, m, k, f^+, k_b)
  $rank \leftarrow kNN(q, \max\{k, k_b\}, \mathbb{U})$
  for each $o \in rank$ **do**
    $trank \leftarrow kNN(o, k_b, \mathbb{U})$
    for each $p \in trank$ **do**
      $brank[p] \leftarrow brank[o] + f^+(pos(p, trank), pos(o, rank))$
  Sort descending $brank$
  $brank \leftarrow selectElements(brank)$
  return [brank,rank-brank][1:k]

Fig. 1. Rank-Mixer and Rank-Booster algorithm

function. Adding a positive constant to f^+ we are able to control how much we "punish" elements not present in all ranks. Then, we add all the scores obtained by the objects in each ranking and rank them according to the final scores. This method is in fact a generalization of both Borda Count and Reciprocal-Rank with a score function $\sum_{r \in \mathcal{R}} f^+(r(d))$. The outline of the algorithm is presented in Fig. 1 (left).

Efficiency. The time needed to perform the query is the time of performing a k-NN query for each retrieval method. These queries can be answered efficiently by using some indexing techniques [2,5].

We could weight each retrieval method, similar to the multi-metric approach, either statically or dynamically at query time. The weighed version of Rank-Mixer has the advantage that the running time does not depend on the weights, opposed to what happens in multi-metric spaces [4].

3.2 Rank-Booster

This algorithm uses a single retrieval method to improve the effectiveness of the answer. This algorithm relies on the fact that good retrieval methods have good results for the first elements. For example, in our experimental evaluation, the nearest-neighbor (NN) is computed correctly with 50-80% and when retrieved the same number of relevant objects (R-Precision) the 50-60% of them are relevant.

In this algorithm, we compute the k-NN for the given query. Then for the first k_b (a parameter of the algorithm) elements of the ranking we perform a k_b-NN query and use a similar strategy as the one used for the mixer method to combine these rankings. The difference between this algorithm and Rank-Mixer, is that we score the objects according to two values. These are the position of each object in the ranking and the position of the object that generated the ranking. Finally, as some objects could get a low score, meaning that they are not good enough, we keep just the first elements of the generated ranking. The outline of this algorithm is presented in Fig. 1 (right).

The function f^+, just as in the fixed Rank-Mixer algorithm, must be positive and decreasing in each coordinate. One variation of the algorithm presented above is to always keep the NN of the original answer.

Efficiency. This algorithm needs no special index to work. It can be built over any existing indexing method. We only need to store the k-NN for each object of the database, thus requiring $(k-1)|\mathbb{U}|$ space (the NN of an object of the database is the object itself, thus we do not need to store it). This space may seem a lot, but in fact is lower than the space used by most FVs. For example, as we will show in Section 4.1, for 3D objects $k_b \leq 15$ gives the the best results. And since the dimensionality of the FVs for 3D objects ranges from 30 to over than 500, the space needed would be around 2%–50% of the space needed by a FV. Besides, this information can be dynamically built at query time, thus requiring less space in practice.

4   Experimental Evaluation

Before using our method in the ImageCLEF@ICPR fusion task, we tested our algorithms with two different 3D models datasets: the first one is the dataset of the SHREC 2009 *"Generic retrieval on new benchmark"* track [1], which comes from the NIST generic shape benchmark [8]. The other dataset is the *test* collection from the Princeton Shape Benchmark (PSB) [11][1].

The SHREC dataset is composed of 720 models and 80 query objects. Both models and queries are classified into 40 different classes, each one having exactly 20 objects (18 in the database and 2 queries). The *test* collection of PSB has 907 objects classified into 92 classes. The classes have between 4 and 50 elements. As the PSB does not provide queries for the dataset, we chose the rounded 10% of each class as queries. Thus the *test* collection now have respectively 810 objects in the database and 97 queries.

As retrieval methods we used different FVs with me metric L_1. One of them, the DSR [13] is itself an optimized metric combination, so we are comparing against it in our tests.

4.1   Experimental Results

Rank-Mixer. The first test we performed was intended to evaluate which function performs better for the mixer method. We considered functions of the following forms: $-\log(x)$, $-x^\alpha$, $1/x^\alpha$. Table 1 shows the complete result and Fig. 2 shows the effectiveness of the Rank-Mixer for some of the functions. The table and the graph show that the $f^+(x) = -\log(x)$ gives the best results and clearly outperforms the Borda Count ($f(x) = -x$) and the Reciprocal Rank ($f(x) = 1/x$ or $f(x) = 1/(x+60)$). In the following tests we will use $f(x) = -\log(x)$ and we will call this method log-rank. The results also show that the log-rank method outperforms MedRank.

---

[1] We actually tested in both PSB train and test; but we omitted some results because of lack of space.

Table 1. Effectiveness of Rank-Mixer for different functions

Function	NN	1T	2T	E	DCG	Function	NN	1T	2T	E	DCG
$-x^{0.1}$	0.850	0.505	0.637	0.442	0.787	$-x^{0.5}$	0.850	0.493	0.622	0.430	0.781
$-\log(x)$	0.863	0.504	0.640	0.443	0.788	$-x$	0.850	0.482	0.603	0.418	0.773
$1/x^{0.1}$	0.863	0.503	0.637	0.444	0.788	$-x^2$	0.838	0.469	0.585	0.409	0.763
$-x^{0.25}$	0.850	0.502	0.629	0.440	0.783	$1/x^2$	0.762	0.113	0.137	0.095	0.506
$-x^{0.2}$	0.850	0.502	0.631	0.440	0.784	$1/x$	0.762	0.113	0.137	0.095	0.506
$-x^{0.333}$	0.850	0.496	0.627	0.438	0.782	$1/(x+60)$	0.025	0.025	0.050	0.032	0.324
$1/x^{0.5}$	0.838	0.495	0.628	0.438	0.784	MedRank	0.850	0.484	0.622	0.433	0.777

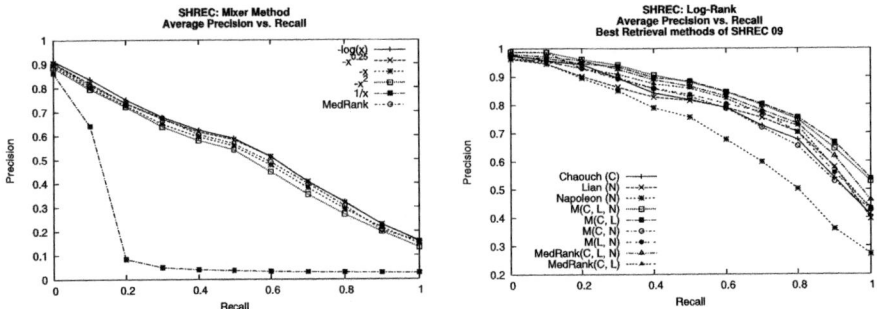

Fig. 2. Left: Effectiveness of Rank-Mixer for different functions, right: Effectiveness of best SHREC descriptors

Table 2. Left: Effectiveness of log-rank on SHREC for best 3D retrieval methods. Right: Effectiveness of Rank-Booster on SHREC.

Method	NN	1T	2T	E	DCG	Method	NN	1T	2T	E	DCG
Chaouch(C)	0.963	0.730	0.848	0.602	0.917	SIL	0.775	0.435	0.582	0.404	0.744
Lian(L)	0.925	0.724	0.844	0.595	0.904	B(SIL)	0.775	0.460	0.590	0.409	0.745
Napoleon(N)	0.950	0.639	0.771	0.541	0.882	DBD	0.825	0.417	0.541	0.377	0.735
M(C, L, N)	0.975	0.781	0.895	0.638	0.945	B(DBD)	0.825	0.453	0.589	0.408	0.739
M(C, L)	0.950	0.788	0.906	0.642	0.938	RSH	0.750	0.384	0.504	0.347	0.705
M(C, N)	0.975	0.728	0.842	0.595	0.924	B(RSH)	0.750	0.412	0.502	0.350	0.698
M(L, N)	0.938	0.733	0.865	0.612	0.922	DSR	0.850	0.546	0.691	0.479	0.819
MedRank(C,L,N)	0.950	0.774	0.891	0.632	0.936	B(DSR)	0.850	0.592	0.717	0.500	0.821
MedRank(C,L)	0.925	0.751	0.865	0.611	0.922						

Figure 3 shows that log-rank is similar to the multimetric approach, it also shows that the proposed method outperforms the DSR and the multimetric approach. Also, in the right figure we present two upper bounds that can be obtained using the log-rank method, the first is obtained using the best static combination of retrieval methods and the second one is obtained using the best possible dynamic combination. Figure 2 compares log-rank method with the best retrieval methods of SHREC 09 [1], these are Aligned Multi-View Depth Line, Composite Shape Descriptor and Multi-scale Contour Representation. The results are detailed in Table 2. The results shows that the log-rank outperforms MedRank and that it increases the effectiveness about 8%.

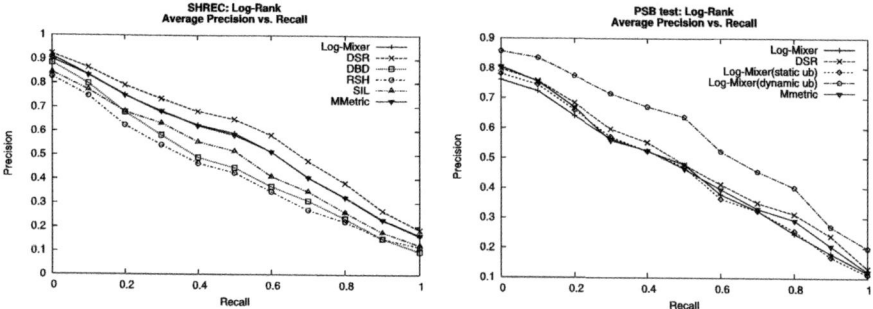

Fig. 3. Effectiveness of log-rank. Left: Shrec Dataset, right: PSB test.

Table 3. Effectiveness of Fixed Rank-Mixer. Left: combination of FVs, right: combination of the results of Chaouch and Lian.

Method	NN	1T	2T	E
log-rank	0.863	0.504	0.640	0.442
Fixed log-rank ($k = 32$)	0.850	0.496	N.A.	0.428
Fixed log-rank ($k = 34$)	0.850	0.495	N.A.	0.425
Fixed log-rank ($k = 36$)	0.850	0.490	0.611	0.427
Fixed log-rank ($k = 38$)	0.850	0.489	0.613	0.427
Fixed log-rank ($k = 40$)	0.863	0.490	0.617	0.428

Method	NN	1T	2T	E
M(Chaouch, Lian)	0.950	0.788	0.906	0.642
Fixed log-rank ($k = 32$)	0.950	0.774	N.A.	0.640
Fixed log-rank ($k = 34$)	0.950	0.774	N.A.	0.639
Fixed log-rank ($k = 36$)	0.950	0.775	0.901	0.641
Fixed log-rank ($k = 38$)	0.950	0.776	0.899	0.639
Fixed log-rank ($k = 40$)	0.950	0.776	0.899	0.638

The above results where computed using the whole rank, that is $k = |\mathbb{U}|$. However, we can compute some statistics for fixed k, such as the Nearest Neighbor or the E-Measure given that $k \geq 32$. If we test it on the SHREC dataset, we could also compute the R-Precision if $k \geq 18$ and the Bull-Eye Percentage if $k \geq 36$. For the PSB, we would have to take $k \geq 50$ just to compute the First Tier. Relying on the same basis of the definition of the E-Measure that a user is interested just in the first screen of results, we will take $k \geq 32$ for our tests, and we will test it on the SHREC dataset. Table 3 shows the results, where "N.A." means not applicable. The function we used in these tests was $f(x) = \log(200) - \log(x)$.

These results show that there is no need of having the complete rank, it is enough to have approximately the 40 first elements to get an improvement close to the one obtained using the whole rank.

Rank-Booster. In our tests, the *selectElements* function of the Rank-Booster algorithm selects the first $k(k-1)/2$ elements. Motivated by the results of the previous experiments, we took $f^{+}(x, y) = \log(2k_b + 1) - \log(x + y + 1)$.

Before performing the tests, we had to compute the best value for k_b. For doing so we computed the R-Precision of the Rank-Booster with different values of k_b. We only used the DSR feature vector because we wanted to choose a value of k_b independent of the used methods. It is important to notice that $k_b = 1$ is the same as not using any improvement method over the descriptor.

Figure 4 shows that $k_b = 13$ is the best choice for SHREC and that every $3 \leq k_b \leq 16$ yields to improvement in the effectiveness of the method. The figure

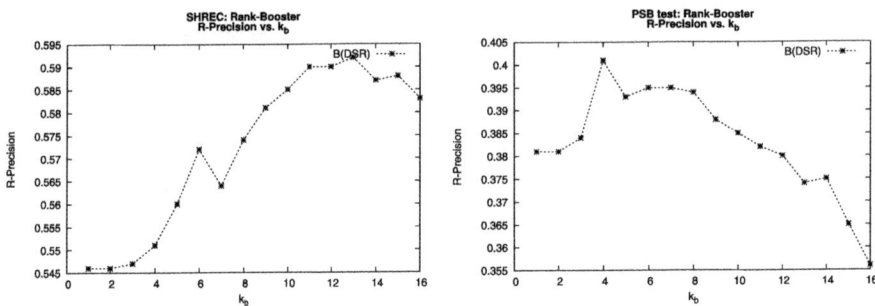

Fig. 4. R-Precision of Rank-Booster. Left: Shrec Dataset, right: PSB test.

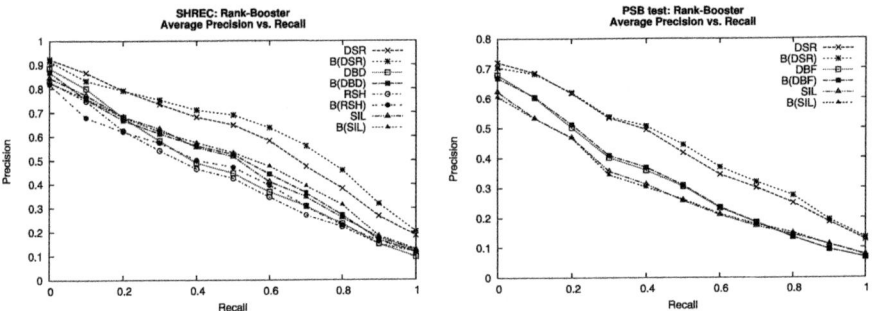

Fig. 5. Effectiveness of Rank-Booster. Left: Shrec Dataset, right: PSB test.

also shows that the best possible value for PSB is $k_b = 6$. It also follows that any $0 < k_b \leq 9$ yields to improvements in the effectiveness or maintains it.

Figure 5, show the precision-recall curves of the retrieval methods used in SHREC and the best methods in PSB. Table 2(right) show the details of the results for the SHREC dataset. These results show that the Rank-Booster method applied over the DSR descriptor gives better results that the ones obtained in SHREC 09 [1] using the global-local approach. Rank-Booster increases effectiveness up to 8%. This is an extremely good result, because this method could be applied to any existing framework, without the need of having several retrieval methods. Although we have devised no way to efficiently compute the optimal value for k_b, it follows from the results that, it suffices to take a small k_b to achieve higher effectiveness.

Combination of Rank-Mixer and Rank-Booster. Table 4 shows that combining Rank-Mixer and Rank-Booster leads to a further improvement of the effectiveness. Combining the methods increases the effectiveness about 2% with respect to the Rank-Mixer.

Table 4. Effectiveness of combining Rank-Mixer and Rank-Booster

Method	NN	1T	2T	E	DCG
Mixer	0.863	0.504	0.640	0.442	0.738
Mixer(B)	0.812	0.511	0.649	0.451	0.786
Booster(Mixer)	0.863	0.517	0.656	0.455	0.782
Booster(Mixer(B))	0.812	0.497	0.633	0.440	0.771

Table 5. Results of ImageCLEF@ICPR Fusion Task. Left: original methods, right: fusion of methods.

Method	MAP	P5	P10
Text1	0.35	0.58	0.56
Text2	0.35	0.65	0.62
Text3	0.43	0.7	0.66
Text4	0.38	0.65	0.62
Visual1	0.01	0.09	0.08
Visual2	0.01	0.08	0.07
Visual3	0.01	0.09	0.07
Visual4	0.01	0.09	0.08

Method	MAP	P5	P10
T(2,3,4) V3 (8.41)	0.491	0.760	0.696
T(2,3,4)	0.480	0.704	0.672
T(1,2,3,4)	0.474	0.712	0.648
T(1,2,3)	0.473	0.712	0.664
T(1,2,3,4) V(234)	0.466	0.752	0.676
T(1,2,3,4) V(134)	0.464	0.744	0.692
T(1,2,4)	0.464	0.688	0.640
T(1,2,3,4) V(1,2,3)	0.451	0.744	0.688

ImageCLEF@ICPR Fusion Task Results. In this task [10] we had to fusion textual and visual results. As not all methods returned the same number of elements we slightly modified our algorithm. We use the function $f^+(x) = T - \log_2(x)$, with $T = 8.0$ fixed for all except the first result. The first result is the best static combination of the methods with the best possible value of $T = 8.41$. The best result yields an improvement of 14% and the best fully automatic combination yields an improvement of 11%.

5   Conclusions

We presented two algorithms for increasing the effectiveness of multimedia retrieval methods. One of these algorithms, the log-rank, outperforms the state of the art rank fusion methods MedRank and Reciprocal Rank. The other algorithm can not be compared against these state of the art methods because it only uses one single method to improve its effectiveness. One important advantage of the proposed methods is that they do not rely on metric retrieval methods, and they can be applied over any method that generates a ranking of the elements given a query object. Another advantage of these methods is that they can be directly applied on top of the indexing scheme of the used methods, without the need of building a custom indexing scheme. An additional advantage of the proposed methods is that one does not need to normalize the databases nor the multimedia descriptors, as required by the multimetric approach.

In the future work, we will study the problem of estimating the parameter k_b of the Rank-Booster method. We will also research how to select the retrieval methods to use in order to get the effectiveness of Rank-Booster and Rank-Mixer closer to the upper bound showed in Section 4.1.

128 S. Kreft and B. Bustos

Acknowledgments

We want to thank to Afzal Godil for kindly giving us the best ranks of the SHREC contest. Finally, we would like to thank to Mohamed Chaouch, Zhouhui Lian, Thibault Napoléon and their respective teams, for letting us to use their results in our investigation.

References

1. Akgül, C., et al.: Shrec 2009 track: Generic shape retrieval. In: Proc. Eurographics 2009 Workshop on 3D Object Retrieval (3DOR), pp. 61–68. Eurographics Association (2009)
2. Böhm, C., Berchtold, S., Keim, D.A.: Searching in high-dimensional spaces: Index structures for improving the performance of multimedia databases. ACM Computing Surveys 33(3), 322–373 (2001)
3. Bustos, B., Keim, D., Saupe, D., Schreck, T., Vranić, D.: Using entropy impurity for improved 3D object similarity search. In: Proc. IEEE International Conference on Multimedia and Expo (ICME), pp. 1303–1306. IEEE, Los Alamitos (2004)
4. Bustos, B., Skopal, T.: Dynamic similarity search in multi-metric spaces. In: Proc. ACM SIGMM International Workshop on Multimedia Information Retrieval (MIR), pp. 137–146. ACM Press, New York (2006)
5. Chávez, E., Navarro, G., Baeza-Yates, R., Marroquin, J.: Searching in metric spaces. ACM Computing Surveys 33(3), 273–321 (2001)
6. Cormack, G.V., Clarke, C.L.A., Büttcher, S.: Reciprocal rank fusion outperforms condorcet and individual rank learning methods. In: Proc. Annual International ACM Conference on Research and Development in Information Retrieval (SIGIR), poster (2009)
7. Fagin, R., Kumar, R., Sivakumar, D.: Efficient similarity search and classification via rank aggregation. In: Proc. ACM SIGMOD International Conference on Management of Data (SIGMOD), pp. 301–312. ACM, New York (2003)
8. Fang, R., Godil, A., Li, X., Wagan, A.: A new shape benchmark for 3D object retrieval. In: Bebis, G., Boyle, R., Parvin, B., Koracin, D., Remagnino, P., Porikli, F., Peters, J., Klosowski, J., Arns, L., Chun, Y.K., Rhyne, T.-M., Monroe, L. (eds.) ISVC 2008, Part I. LNCS, vol. 5358, pp. 381–392. Springer, Heidelberg (2008)
9. Ho, T.K., Hull, J.J., Srihari, S.N.: On multiple classifier systems for pattern recognition. In: Proc. International Conference on Pattern Recognition (ICPR), pp. 66–75 (1992)
10. Müller, H., Kalpathy-Cramer, J.: The ImageCLEF medical retrieval task at ICPR 2010 - information fusion to combine visual and textual information. In: Ünay, D., Çataltepe, Z., Aksoy, S. (eds.) ICPR 2010. LNCS, vol. 6388, pp. 101–110. Springer, Heidelberg (2010)
11. Shilane, P., Min, P., Kazhdan, M., Funkhouser, T.: The princeton shape benchmark. In: Shape Modeling International (2004)
12. Suen, C.Y., Lam, L.: Multiple classifier combination methodologies for different output levels. In: Kittler, J., Roli, F. (eds.) MCS 2000. LNCS, vol. 1857, pp. 52–66. Springer, Heidelberg (2000)
13. Vranic, D.V.: Desire: a composite 3D-shape descriptor. In: IEEE International Conference on Multimedia and Expo. (ICME), pp. 962–965 (2005)
14. Zezula, P., Amato, G., Dohnal, V., Batko, M.: Similarity Search: The Metric Space Approach (Advances in Database Systems). Springer, Heidelberg (2005)

Information Fusion for Combining Visual and Textual Image Retrieval in ImageCLEF@ICPR

Xin Zhou[1], Adrien Depeursinge[1,2], and Henning Müller[1,2]

[1] Geneva University Hospitals and University of Geneva, Switzerland
[2] University of Applied Sciences Western Switzerland, Sierre, Switzerland
`henning.mueller@sim.hcuge.ch`

Abstract. In the ImageCLEF image retrieval competition multimodal image retrieval has been evaluated over the past seven years. For ICPR 2010 a contest was organized for the fusion of visual and textual retrieval as this was one task where most participants had problems. In this paper, classical approaches such as the maximum combinations (combMAX), the sum combinations (combSUM) and the multiplication of the sum and the number of non–zero scores (combMNZ) were employed and the trade–off between two fusion effects (chorus and dark horse effects) was studied based on the sum of n maxima. Various normalization strategies were tried out. The fusion algorithms are evaluated using the best four visual and textual runs of the ImageCLEF medical image retrieval task 2008 and 2009. The results show that fused runs outperform the best original runs and multi–modality fusion statistically outperforms single modality fusion. The logarithmic rank penalization shows to be the most stable normalization. The dark horse effect is in competition with the chorus effect and each of them can produce best fusion performance depending on the nature of the input data.

1 Introduction

In the ImageCLEF image retrieval competition, multimodal image retrieval has been evaluated over the past seven years. For ICPR 2010 a contest was organized in order to investigate the problem of fusing visual and textual retrieval. Information fusion is a widely used technique to combine information from various sources to improve the performance of information retrieval. Fusion improvement relies on the assumption that the heterogeneity of multiple information sources allows self–correction of some errors leading to better results [1]. Medical documents often contain visual information as well as textual information and both are important for information retrieval [2]. The ImageCLEF benchmark addresses this problem and has organized a medical image retrieval task since 2004 [3]. So far it has been observed in ImageCLEF that text–based systems strongly outperformed visual systems, sometimes by up to a factor of ten [4]. It is important to determine optimal fusion strategies allowing overall performance improvement as in the past some groups had combinations leading to poorer results than textual retrieval alone. The ImageCLEF@ICPR fusion task [5] described in this paper is organized to address this goal, making available the four

D. Ünay, Z. Çataltepe, and S. Aksoy (Eds.): ICPR 2010, LNCS 6388, pp. 129–137, 2010.

best visual and the four best textual runs of ImageCLEF 2009 including runs of various participating groups.

Information fusion, which originally comes from multi–sensor processing [6], can be classified by three fusion levels: signal level, feature level, and decision level [7] (also named the raw data level, representation level, and classifier level in [8]). The ImageCLEF@ICPR fusion task focuses on the decision level fusion: the combination of the outputs of various systems [9]. Other related terminologies include evidence combination [7] and rank aggregation [10], which are widely used by meta–search engines [11].

Many fusion strategies have been proposed in the past. Using the maximum combination (*combMAX*), the sum combination (*combSUM*) and the multiplication of the sum and the number of non–zero scores (*combMNZ*) were proposed by [12] and are described in Section 2.3. Ideas such as Borda–fuse [13] and Condorcet–fuse [14] were rather inspired from voting systems. Borda–fuse consists of voting with a linear penalization based on the rank whereas Condorcet–fuse is based on pair–wise comparisons. Others strategies exist such as Markov models [10] and probability aggregation [15]. A terminology superposition also exists. For example, the round–robin strategy as analyzed in [6] is equivalent to the *combMAX* strategy, the Borda–fuse strategy, despite the idea being inspired from voting, is in fact the *combSUM* strategy with descending weights for ranks. First proposed in 1994, *combMAX*, *combSUM*, and *combMNZ* are still the most frequently used fusion strategies and were taken as the base of our study. However, these three methods have limitations. On the one hand, *combMAX* favors the documents highly ranked in one system (*Dark Horse Effect* [16]) and is thus not robust to errors. On the other hand, *combSUM* and *combMNZ* favor the documents widely returned to minimize the errors (*Chorus Effect*) but relevant documents can obtain high ranks when they are returned by few systems. In this paper, we investigate a trade–off between these methods while using the sum of n maximums: *combSUM(n)MAX*.

Two other important issues of information fusion are the normalization of the input scores [17,18] and the tuning of the respective weights (i.e contribution) given to each system [16,19]. The normalization method proposed by Lee [17] consists of mapping the score to [0;1]. It was declared to perform best in [18]. Our study reused this normalization method, which is based on a topic basis or a run basis to produce normalized scores.

2   Methods

2.1   Dataset

The test data of the ImageCLEF@ICPR fusion task consist of 8 runs submitted to the ImageCLEF 2009 medical image retrieval task. 4 runs of the best textual retrieval systems and 4 representing the best visual systems were made available[1]. There are 25 query topics in ImageCLEFmed 2009. For each topic, a

---

[1] For more details about the retrieval systems, please visit ImageCLEF working notes available at http://www.clef-campaign.org/

maximum of 1000 images can be present in a run. The format of each run follows the requirements of trec_eval.

Ranks and scores of the 8 runs are available as well as the ground truth for evaluation. Training data consists of the 4 best textual runs and the 3 best visual runs for the same task in 2008. The 7 runs in the training data and the 8 runs in the test data were not produced by the same systems. Therefore, weight selection on a run basis can not be applied to the test data.

2.2 Rank Penalty vs. Score Normalization

To enable the combination of heterogeneous data, each image must be mapped to a value V (e.g. score, rank) that is normalized among all systems. Symbols employed are $\overline{V}$ for normalized values, and S and R for scores and ranks given by the input system.

The scores given by the input systems are not homogeneous and require normalization. The normalization method proposed by Lee [17] is used:

$$\overline{V(S)} = \frac{S - S_{min}}{S_{max} - S_{min}} \tag{1}$$

with S_{max} and S_{min} the highest and lowest score found. Two groups of normalized values were produced by either applying this method on a run or topic basis: $\overline{V_{run}(S)}$ and $\overline{V_{topic}(S)}$.

The rank is always between 1 and 1000. However, low ranks need to be penalized as less relevant. Linear normalized rank values are obtained:

$$\overline{V_{linear}(R)} = N_{images} - R, \tag{2}$$

where N_{images} equals the lowest rank (1000 in our case). Experiments have shown that for most information retrieval systems, performance tends to decreases in a logarithmic manner [16]. As a consequence a logarithmic penalization function was tried:

$$\overline{V_{log}(R)} = \ln N_{images} - \ln R, \tag{3}$$

In the rest of the paper, $\overline{V}$ generally refers to one of these four groups of normalized values: $\overline{V_{topic}(S)}$, $\overline{V_{run}(S)}$, $\overline{V_{linear}(R)}$ and $\overline{V_{log}(R)}$.

2.3 Combination Rules

In this section, the various combination rules evaluated are made explicit.

combMAX. *combMAX* computes the value for a result image i as the maximum value obtained over all N_k runs:

$$V_{\text{combMAX}}(i) = \arg \max_{k=1:N_k} (\overline{V_k(i)}). \tag{4}$$

combSUM. *combSUM* computes the associated value of the image i as the sum of the $\overline{V}(i)$ over all N_k runs:

$$V_{\text{combSUM}}(i) = \sum_{k=1}^{N_k} \overline{V_k(i)}. \tag{5}$$

combMNZ. *combMNZ* aims at giving more importance to the documents retrieved by several systems:

$$V_{\texttt{combMNZ}}(i) = F(i) \sum_{k=1}^{N_k} \overline{V_k(i)},\tag{6}$$

where $F(i)$ is the frequency, obtained by counting the number of runs that retrieved the image i. Images that obtain identical values were arbitrarily ordered.

combSUM(n)MAX. The *combMAX* and *combSUM* rules both have drawbacks. *CombMAX* is not robust to errors as it is based on a single run for each image. *CombSUM* has the disadvantage of being based on all runs and thus includes runs with low performance. As a trade–off the sum of N_{max} maxima rule combSUM(n)MAX is proposed:

$$V_{\texttt{combSUM}(n)\texttt{MAX}}(i) = \sum_{j=1}^{n} \arg\max_{k \in \mathcal{E}_{N_k} \setminus \mathcal{E}_j} (\overline{V_k(i)}),\tag{7}$$

with n the number of maxima to be summed and $\mathcal{E}_{N_k} \setminus \mathcal{E}_j$ the ensemble of N_k runs minus the j runs with maximum value for the image i. When $n = 1$, only 1 maximum is taken, which is equivalent to combMAX. Summing $n > 1$ maxima increases the stability of combMAX. When $n = N_k$, this strategy sums up all maximums and is equivalent to combSUM. $n < N_k$ can potentially avoid runs with low performance if assuming that runs with maximum scores or ranks have higher confidence and thus allow best retrieval performance.

$\{F(i) : V(i)\}$. As *combMNZ* proved to perform well, integrating the frequency is expected to improve performance. Instead of using a multiplication between the sum of values and the frequency, images are separated into pairs $\{F(i) : V(i)\}$ and are sorted hierarchically. Images with high frequency are ranked higher. Images with the same frequency are then ranked by value $V(i)$, where $V(i)$ can be obtained by any combination rules.

3   Results

An analysis was performed to analyze the distribution of the relevant documents in the training data. Each run in the training data contains 30 topics. Within each topic there are 1000 ranked images. The 1000 ranks were divided into 100 intervals, and the number of relevant images were counted in each interval. As some topics contain few relevant images, all 30 topics were summed to obtain a more stable curve. Two curves containing the average numbers for all visual systems as well as all textual systems are shown in Figure 1. In Table 1, the performance of the best fused runs is compared with the best runs of ImageCLEF 2009. The retrieval performance is measured using the mean average precision (MAP). MAP obtained with various combination methods are shown in Figure 2 and 3.

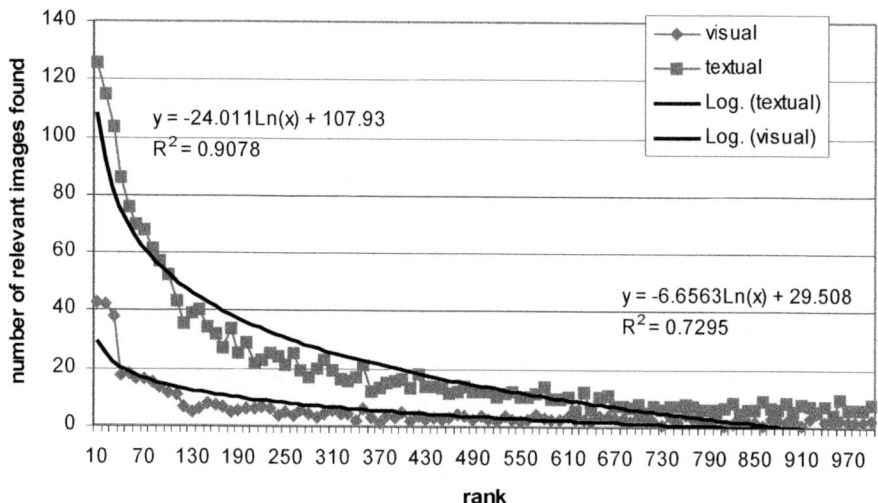

Fig. 1. Distribution of the relevant documents in the training data

Table 1. Best original vs. fused runs

Run	2008	2009
best original textual run	0.2881	0.4293
best original visual run	0.0421	0.0136
best textual fusion run	0.3611	0.4766
best visual fusion run	0.0611	0.0198
best mixed fusion	0.3654	0.488

4 Interpretation

Two trends using logarithmic regression were calculated to analyze the distribution of relevant documents in Figure 1. Two observations can be made: 1) for both modalities the number of relevant images decreases logarithmically ($R^2 > 0.7$), which confirms Vogt [16]; 2) the quality of text retrieval is constantly 4 times better than that of visual. Fixed weights w were applied to combine the two modalities with $w = 0.8$ for text systems and $w = 0.2$ for visual systems.

Fusion results using w are shown in Table 1. Two observations highlight the benefits of heterogeneity: 1) for both modalities the best fused runs outperform all original runs; 2) multi–modal fusion outperformed the best run obtained with single modality fusion. Two–tailed paired t tests were performed in order to study the statistical significance of the two observations. Observation 2) is significant with both training ($p_{train} < 0.012$) and test ($p_{test} < 0.0116$) data. Observation 1) is significant with test ($p_{test} < 0.0243$) but not training ($p_{train} < 0.4149$) data.

The comparative analysis of *combMAX*, *combSUM* and *combMNZ* as well as *combSUM(n)MAX* is shown in Figure 2 and   3. Operators based on few

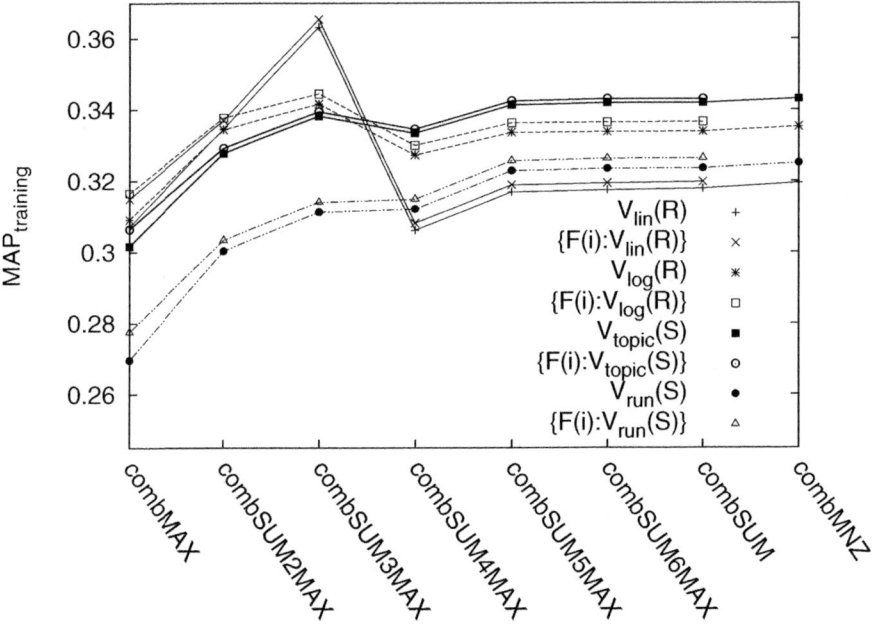

Fig. 2. MAP on training data

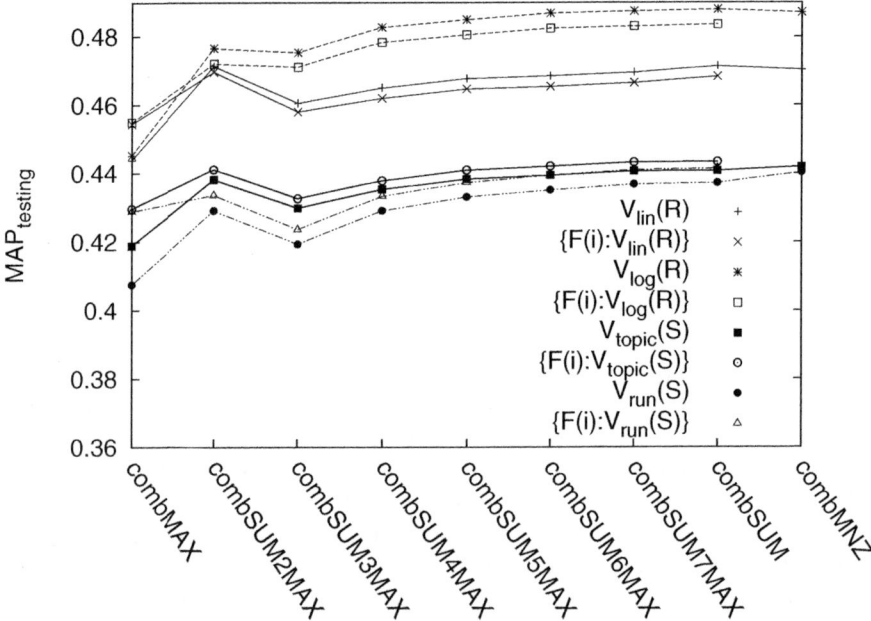

Fig. 3. MAP on test data

maxima (left side of the graph: *combMAX*, *combSUM(*n*)MAX* with small values of *n* favor the *Dark Horse Effect* whereas those based on several runs (*combSUM*, *combMNZ*) favor the *Chorus Effect*. None of the performance curves is monotone, thus neither *Chorus* or *Dark Horse Effect* can be declared best. The presence of coincident local minimum MAP for all techniques is due to the absence of both mentioned effects.

For training data, maximum MAP was obtained with linear rank penalization using *combSUM3MAX* whereas for test data, log rank penalization using *combSUM* gave the best results. With logarithmic rank penalization, the behavior of MAP is the most stable among all techniques on both training and test data. This is in accordance with the descriptive analysis of the data where the relevance of images decreases with $\log(R)$ (Figure 1).

The normalized scores of 2008 produced very close performance to rank-based methods whereas same technique in 2009 only provided poor results. The performance using normalized scores for fusion depends highly on score definition of each run and is thus less stable.

Frequency $F(i)$ favors images returned by numerous systems, which slightly improved all techniques with runs of 2008. However with runs of 2009 using frequency slightly decreased the performance of those techniques using rank penalization. Performance margins between visual and textual runs are larger in 2009 than in 2008. Results can be interpreted as frequency of images in runs with poor quality can provide noise.

Comparing with results produced by the other groups [5], our best run combining both textual runs and visual runs (named mixed run) is ranked 3rd best among 49 runs. The best mixed run outperformed our best mixed run in MAP but has poorer early precision. The second best mixed run slightly outperformed our best mixed run both in terms of MAP and early precision. Two other comparisons on single modality fusion showed that our best run fusing only textual runs is ranked at 3rd best. Only one group submitted fusion of purely visual runs. Our best fusion run of viusal runs outperforms all submitted fusion runs of visual runs. As technical details are not available yet, further comparisons remain furture work. However, our best fused runs of the test data (both single modality or multi–modality) are obtained by using *combSUM* with logarithmic rank normalization. As *combSUM* (equivalent to Borda–fuse) is one of the most straightforward fusion strategies, the major difference of performance is mostly due to the logarithmic rank normalization.

5 Conclusions

In this paper, we studied the fusion of textual and visual retrieval systems. Fused runs outperform the original runs and combining visual information with text can significantly improve fusion performance. It was observed that the logarithmic rank penalisation was more stable than linear penalization. As *Dark Horse Effect* oriented operators, *combSUM(*n*)MAX* outperforms *combMAX*, whereas *combSUM* and *combMNZ* give often close results on favoring *Chorus Effect*. In our experiments, no significant differences in performance were observed between the two effects and neither *Chorus* or *Dark Horse Effect* can be declared best. The improvement depends on the nature of the data to be fused.

References

1. Dietterich, T.G.: Ensemble methods in machine learning. In: Kittler, J., Roli, F. (eds.) MCS 2000. LNCS, vol. 1857, pp. 1–15. Springer, Heidelberg (2000)
2. Müller, H., Michoux, N., Bandon, D., Geissbuhler, A.: A review of content-based image retrieval systems in medicine–clinical benefits and future directions. International Journal of Medical Informatics 73(1), 1–23 (2004)
3. Hersh, W., Müller, H., Kalpathy-Cramer, J., Kim, E., Zhou, X.: The consolidated ImageCLEFmed medical image retrieval task test collection. Journal of Digital Imaging 22(6), 648–655 (2009)
4. Müller, H., Kalpathy–Cramer, J., Eggel, I., Bedrick, S., Radhouani, S., Bakke, B., Kahn Jr., C.E., Hersh, W.: Overview of the CLEF 2009 medical image retrieval track. In: Peters, C., Caputo, B., Gonzalo, J., Jones, G.J.F., Kalpathy-Cramer, J., Müller, H., Tsikrika, T. (eds.) CLEF 2009. LNCS, vol. 6242, pp. 72–84. Springer, Heidelberg (2010)
5. Müller, H., Kalpathy-Cramer, J.: The ImageCLEF medical retrieval task at icpr 2010 — information fusion to combine viusal and textual information. In: Ünay, D., Çataltepe, Z., Aksoy, S. (eds.) ICPR 2010. LNCS, vol. 6388, pp. 101–110. Springer, Heidelberg (2010)
6. Wu, S., Mcclean, S.: Performance prediction of data fusion for information retrieval. Information Processing & Management 42(4), 899–915 (2006)
7. Valet, L., Mauris, G., Bolon, P.: A statistical overview of recent literature in information fusion. IEEE Aerospace and Electronic Systems Magazine 16(3), 7–14 (2001)
8. Croft, W.B.: Combining approaches to information retrieval. In: Advances in Information Retrieval, pp. 1–36. Springer US, Heidelberg (2000)
9. Kludas, J., Bruno, E., Marchand-Maillet, S.: Information fusion in multimedia information retrieval. In: Proceedings of 5th International Workshop on Adaptive Multimedia Retrieval (AMR), vol. 4918, pp. 147–159. ACM, New York (June 2008)
10. Dwork, C., Kumar, R., Naor, M., Sivakumar, D.: Rank aggregation methods for the web. In: WWW 2001: Proceedings of the 10th International Conference on World Wide Web, New York, NY, USA, pp. 613–622 (2001)
11. Renda, E.M., Straccia, U.: Web metasearch: rank vs. score based rank aggregation methods. In: SAC 2003: Proceedings of the 2003 ACM Symposium on Applied Computing, pp. 841–846. ACM Press, New York (2003)
12. Fox, E.A., Shaw, J.A.: Combination of multiple searches. In: Text REtrieval Conference, pp. 243–252 (1993)
13. Aslam, J.A., Montague, M.: Models for metasearch. In: SIGIR 2001: Proceedings of the 24th Annual International ACM SIGIR Conference on Research and Development in Information Retrieval, pp. 276–284. ACM, New York (2001)
14. Montague, M., Aslam, J.A.: Condorcet fusion for improved retrieval. In: CIKM 2002: Proceedings of the Eleventh International Conference on Information and Knowledge Management, pp. 538–548. ACM, New York (2002)
15. Lillis, D., Toolan, F., Collier, R., Dunnion, J.: Probfuse: a probabilistic approach to data fusion. In: SIGIR 2006: Proceedings of the 29th ACM SIGIR Conference on Research and Development in Information Retrieval, New York, USA, pp. 139–146 (2006)

16. Vogt, C.C., Cottrell, G.W.: Fusion via a linear combination of scores. Information Retrieval 1(3), 151–173 (1999)
17. Lee, J.H.: Analyses of multiple evidence combination. In: SIGIR 1997: Proceedings of the 20th Annual International ACM SIGIR Conference on Research and Development in Information Retrieval, pp. 267–276. ACM, New York (1997)
18. Wu, S., Crestani, F., Bi, Y.: Evaluating score normalization methods in data fusion. In: Information Retrieval Technology, AIRS 2006, pp. 642–648 (2006)
19. Wu, S., Bi, Y., Zeng, X., Han, L.: Assigning appropriate weights for the linear combination data fusion method in information retrieval. Information Processing & Management 45(4), 413–426 (2009)

Overview of the Photo Annotation Task in ImageCLEF@ICPR

Stefanie Nowak

Audio-Visual Systems, Fraunhofer IDMT, Ilmenau, Germany
stefanie.nowak@idmt.fraunhofer.de
http://www.imageclef.org/2010/ICPR/PhotoAnnotation

Abstract. The Photo Annotation Task poses the challenge for automated annotation of 53 visual concepts in Flickr photos and was organized as part of the ImageCLEF@ICPR contest. In total, 12 research
teams participated in the multilabel classification challenge while initially 17 research groups were interested and got access to the data. The
participants were provided with a training set of 5,000 Flickr images
with annotations, a validation set of 3,000 Flickr images with annotations and the test was performed on 10,000 Flickr images. The evaluation
was carried out twofold: first the evaluation per concept was conducted
by utilizing the Equal Error Rate (EER) and the Area Under Curve
(AUC) and second the evaluation per example was performed with the
Ontology Score (OS). Summarizing the results, an average AUC of 86.5%
could be achieved, including concepts with an AUC of 96%. The classification performance for each image ranged between 59% and 100% with
an average score of 85%. In comparison to the results achieved in Image-
CLEF 2009, the detection performance increased for the concept-based
evaluation by 2.2% EER and 2.5% AUC and showed a slight decrease
for the example-based evaluation.

1   Introduction

The automated annotation and detection of concepts in photos and videos is
an important field in multimedia analysis and received a lot of attention from
the multimedia and machine learning community during the last years. Adopted
methods are often difficult to compare as they are evaluated regarding different
datasets, concepts or evaluation measures. Benchmarking campaigns cope with
this problem and decide on the evaluation cycle including the definition of tasks,
the access to data and the evaluation and presentation of the results. Evaluation initiatives in multimedia got popular with the text-based evaluations of
TREC, the video analysis evaluation of TRECVid and the multimodal, cross-
lingual evaluation efforts of CLEF. ImageCLEF is since 2003 a part of the CLEF
evaluation initiative. It focuses on the evaluation of multimodal image retrieval
approaches in the consumer and medical domain. In 2010, the ImageCLEF consortium posed several image-related task in the form of an ICPR contest, including the general annotation of photos with visual concepts, a robot vision

D. Ünay, Z. Çataltepe, and S. Aksoy (Eds.): ICPR 2010, LNCS 6388, pp. 138–151, 2010.

task, a fusion task and an interactive retrieval session. In this paper, we present the results of the ImageCLEF Photo Annotation Task. Researchers of 12 teams accepted the challenge to annotate a set of 10,000 images with multiple labels. The task is described in Section 2 detailing the dataset, annotations and submission format. In Section 3 the evaluation methodology is outlined. Section 4 presents the submissions of the participating groups. The results of the benchmark are illustrated in Section 5. Section 6 analyses the results in comparison to the results of the similar Photo Annotation Task of ImageCLEF 2009. Finally, Section 7 concludes the paper.

2   Task Description, Dataset and Annotations

The Photo Annotation Task poses the challenge of multilabel classification in consumer photos. A subset of the MIR Flickr 25,000 image dataset [1] was chosen as database. This collection contains 25,000 photos from the Flickr platform, that were collected based on the interestingness rating of the community and the creative commons copyright of the images.

For the classification challenge a set of 53 visual concepts was defined. The visual concepts are oriented on the holistic impression of the images. They contain concepts concerning the scene description, the representation of photo content and its quality. For each photo, it was manually assessed which of the concepts are present. The assessment of the photos was carried out by 43 expert annotators on an image-based level as presented in Figure 1. Some concepts were

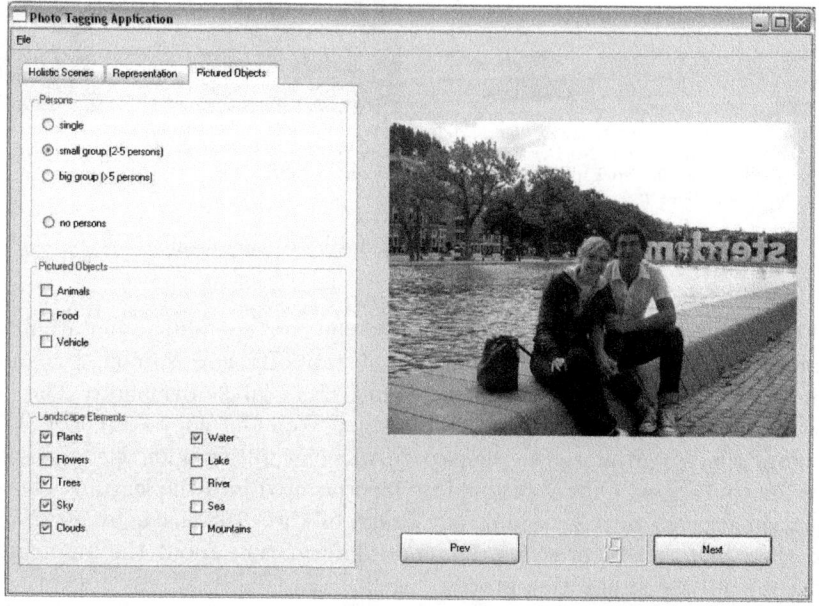

Fig. 1. Relevance assessment tool for the Photo Annotation Task

judged independently like `Animals` while others had to be chosen from a group of concepts, e.g. `small group of persons`. The number of photos that were annotated by one person varied between 30 and 2,500 images. Following the annotation step, three persons performed a validation by screening all photos on a concept basis.

Partly Blurred, Indoor, Macro, Neutral Illumination, No Persons, Food, Aesthetic Impression, No Visual Season, No Visual Time

Day, Sunset-Sunrise, Citylife, Outdoor, Overall Quality, Aesthetic Impression, Fancy, No Persons, Neutral Illumination, No Blur, Water, Lake, Clouds, Sky, Building-Sights, Landscape, No Visual Season

Day, Sunny, Landscape, Outdoor, Overall Quality, Aesthetic Impression, No Persons, Vehicle, Neutral Illumination, No Blur, Plants, Clouds, Sky, No Visual Season

Family-Friends, No Visual Season, Outdoor, Day, Neutral Illumination, Partly Blurred, Small Group, Overall Quality

Fig. 2. Example images from the Photo Annotation Task

Figure 2 shows four example photos and the corresponding annotations. The number of annotations per photo varied substantially. For example, the annotations of the photos in Figure 2 range from 8 to 17 labels per photo. The visual concepts are additionally organized in a small Web Ontology Language (OWL) ontology which was offered to the participants as additional knowledge resource. The Photo Tagging Ontology could be incorporated into the learning and classification process by e.g. taking advantage of the relations between concepts and their hierarchical ordering. For more information about the ontology, the concepts and the annotation process see [2].

Finally, the task of the participants was to annotate a set of 10,000 photos with 53 visual concepts. Each photo should be annotated with confidence values describing the presence or absence of the concepts. For the training of the algorithms, 5,000 photos with annotations and Exchangeable Image File Format (EXIF) data were provided. A validation set consisting of 3,000 photos and its annotations was available for tuning the system parameters. The participants needed to sign a license agreement to get access to the data and annotations. In total 17 research groups signed this agreement and downloaded the data. Summarizing, the challenge of the annotation task consists in coping with the unbalanced amount of data per concept, with the subjectivity of the presence of some concepts as well as with the diversity of images belonging to the same concept class. Approaches that try to adopt the ontology in the learning process are appreciated, as the question is still open whether incorporating real-world knowledge leads to superior results in contrast to applying purely machine learning approaches.

3   Evaluation Methodology

Structured evaluation of the quality of information retrieval systems has a long tradition. Several evaluation measures were investigated, e.g. [3], [4], that can be classified into two main categories. The first category is called *example-based* evaluation. For each media item (example) the actual set of labels (ground truth) is compared with the predicted set using set-theoretic operations. A score is generated for each item and then averaged over all items. The second category stands for *concept-based* evaluation measures which groups any known measure for binary evaluation. The annotations are subdivided according to each concept and a single evaluation per concept is performed. Finally, the results are averaged over all concepts.

The two evaluation paradigms were followed in the analysis of the results of the Photo Annotation Task. For the concept-based evaluation the Equal Error Rate (EER) and the Area Under Curve (AUC) were calculated as explained in Section 3.1. The evaluation on example basis was performed with the Ontology Score (OS) as illustrated in Section 3.2.

3.1   Evaluation per Concept

The concept-based measures AUC and EER can be calculated from the Receiver Operating Characteristics (ROC) curve and are common measures used in the evaluation of classification tasks, e.g. [5], [6]. A ROC curve graphically plots the true-positive rate against the false-positive rate. The EER is measured at the break-even point of the ROC curve, where the true-positive rate of a system is equal to the false-positive rate. AUC describes the overall quality of a classification system independent from an individual threshold configuration. It is calculated by integration of the ROC curve. In the benchmark, the EER and the AUC of the ROC curves summarize the performance of the individual runs, by taking the average values over all concepts.

3.2   Evaluation per Example

The evaluation per example is assessed with the OS measure [7] and is based on the Photo Tagging Ontology. The OS considers partial matches between system output and ground truth by calculating misclassification costs for each missing or wrongly annotated concept per image. The cost function is based on the level of the concept in the hierarchy of the ontology in which lower costs are assigned to deeper levels. In an optimization procedure incorrect annotated concepts are matched to the ground truth and the costs are summed up. Violations against real-world knowledge, e.g. the simultaneous annotation of mutually exclusive concepts, are penalized. All in all, the score is based on structure information (distance between concepts in the hierarchy), relationships from the ontology and the agreement among annotators on a concept.

4   Participation

Researchers from 12 teams situated in seven countries participated in the benchmark. The participating groups are listed in the following together with a short description of their methods. This information was extracted from the submission questionnaire that is filled out during the submission process. A summary of the approaches detailing the descriptors and classifiers used is presented in Table 1.

AVEIR: The AVEIR consortium is a group consisting of the four research institutes MRIM, UPMC, CNRS and LSIS. They submitted four runs derived by a rank average of a varying number of AVEIR runs.

CNRS|Telecom ParisTech: The institute TELECOM from Paris, France submitted five runs. They use global features that reflect the colour and texture information and local features that describe local shape information and combine this information in a late fusion approach. For classification a SVM and boosting is applied.

CVSSPRet: The team of the University of Surrey, UK proposed five runs. It adopts various combinations of sampling strategies, extract different SIFT features, partly using also their spatial information, and soft assignment for histograms. They experiment with multiple kernel fisher discriminant analysis or kernel discriminant analysis using spectral regression and combinations at classifier-level and kernel-level for classification [8].

IJS: The team of the Department of Knowledge Technologies, Jozef Stefan Institute, Slovenia proposes a hierarchical multilabel classification approach based on random forests of predictive clustering trees. Several SIFT features, local binary patterns, colour histogram and GIST features are utilized. They submitted four runs.

ISIS: The Intelligent Systems Lab of the University of Amsterdam, The Netherlands submitted two runs based on a retrained model of the ImageCLEF 2009

Table 1. Results of the ICPR Photo Annotation Task. The table lists the best results for each measure per group, the number of runs submitted and the descriptors and classifiers applied. It is sorted ascending due to the EER measure. Please note that the best run of a group for one measure is not necessarily the best run evaluated with the other measures.

Group ID	#	Descriptor	Classifier	EER	Rank	∅ Rank	AUC	Rank	∅ Rank	OS	Rank	∅ Rank
CVSSPRet	5	various SIFT	spectral regression	0.21	1	3.6	0.86	1	3.6	0.69	5	16.2
ISIS	2	various SIFT	SVM	0.22	4	5.0	0.86	4	5.0	0.78	1	1.5
IJS	4	global and local	random forests	0.24	8	9.5	0.83	8	9.5	0.71	3	5.5
CNRS	5	global and local	SVM, boosting	0.28	12	14.0	0.79	12	14.0	0.42	23	25.0
AVEIR	4	global and local	SVM	0.29	17	21.5	0.79	17	21.5	0.56	12	15.5
MMIS	5	various features	non-parametric density estimation / MRF	0.31	19	26.6	0.76	19	26.8	0.50	17	32.6
LSIS	4	various features	SVM + reranking	0.31	21	27.0	0.75	21	26.8	0.51	16	24.0
UPMC	5	SIFT	SVM	0.34	28	33.0	0.72	29	32.4	0.40	28	26.1
ITI	5	local and global	NN	0.37	30	38.8	0.59	37	39.2	0.40	30	32.0
MRIM	3	colour, texture, feature points	SVM + Fusion	0.38	31	32.0	0.64	30	31.0	0.58	9	23.0
TRS2008	1	SIFT	SVM	0.42	34	34.0	0.62	33	33.0	0.33	38	38.0
UAIC	1	face detection, EXIF	NN + default values	0.48	38	38.0	0.14	43	43.0	0.68	6	6.0

approach [9]. They use a sampling strategy that combines a spatial pyramid approach and saliency points detection, extract different SIFT features, perform a codebook transformation and classify with a SVM approach.

ITI: The team of the Institute of Computer Technology, Polytechnic University of Valencia, Spain submitted five runs in which they experiment with different feature combinations and classifiers. As basis they use a local feature dense grid extraction with patch colour histogram representation and random projection and combine this with different colour and GIST features. A nearest neighbour approach in a discriminant subspace and a fast linear discriminant classifier are applied as classifiers.

LSIS: The Laboratory of Information Science and Systems, France proposes four runs. These runs contain a reranking of SVM outputs. SVM was trained with features of 200 dimensions as described in ImageCLEF 2009 proceedings [10]. The reranking is sequentially done by an asymmetric normalisation. It improved the results from 0.71 to 0.74 AUC.

MMIS: The team of the Knowledge Media Institute, Open University, UK proposes two approaches and submitted five runs in total. The first one is based on a non-parametric density estimation using a Laplacian kernel in which they compare a baseline run with another that utilizes web-based keyword correlation. The second approach relies on Markov Random Fields and presents two different models, one that explores the relations between words and image features and a final one that incorporates the relation between words in the model.

MRIM: The Multimedia Information Modelling and Retrieval group at the Laboratoire Informatique de Grenoble, Grenoble University, France submitted three runs. They linearly fused the results of four approaches that are based on colour similarity, SVMs on feature points and a SVM on colour and textures. The hierarchy of concepts was additionally incorporated in their runs.

TRS2008: The group of Beijing Information Science and Technology University, China submitted one run. They use 128 dimensional SIFT features. The images are classified with a SVM model.

UAIC: The team of the Faculty of Computer Science of Alexandru Ioan Cuza University, Romania participated with one run. Similar to the ImageCLEF 2009 approach [11], they use a four step mechanism consisting of face detection, a clustering process, utilizing EXIF data and finally incorporating default values as fall-back solution.

UPMC: The group of the University Pierre et Marie Curie in Paris, France submitted five runs to the Photo Annotation Task. They investigate SVM classification and SVM ranking approaches on PCA reduced features or SIFT features and experiment with incorporating the hierarchy of concepts.

5   Results

Solutions from 12 research groups were submitted to the Photo Annotation Task in altogether 44 run configurations. Table 1 summarizes the performance for each group for the measures EER, AUC and OS. The best results per measure are listed with the rank information of this run and the average rank for all runs per group. A complete list with the results of all runs is provided in the appendix in Table 3 and can also be found at the benchmark website[1].

For the concept-based evaluation, the best run achieved an EER of 21.4% and an AUC of 86% per concept in average. Two other groups got close results with an AUC score of 85.7% and 83.2%. The best annotation quality for a concept achieved by any run is in average 20.8% EER and 86.5% AUC. The concepts *Sunset-Sunrise* (AUC: 96.2%), *Clouds* (AUC: 96.2%) and *Sky* (AUC: 95.9%) are the easiest detectable concepts. The worst concept detection quality can be found for the concepts *Fancy* (AUC: 61.4%), *Overall Quality* (AUC: 66.1%) and *Aesthetic* (AUC: 67.1%). This is not surprising due to the subjective nature of these concepts. Analysing the methods of the first five groups in the EER and

---

[1] www.imageclef.org/2010/ICPR/PhotoAnnotation

AUC ranking, it is obvious that all groups applied discriminative approaches with local features. Some used a fusion of local and global features.

The example-based evaluation reveals that the best run is able to annotate a photo in average 78.4% correctly. Taking into account the best annotation quality per photo out of all runs, the photos can be annotated in average with 85.1% quality, ranging between 59.3% and 100%. The ranking of the runs is different than with the concept-based measures. The best five groups in the ranking of the OS again applied discriminative approaches with local or combined local and global features. One model-free approach (UAIC) considering a combination of methods could achieve good results.

The OS measure needs a binary decision about the presence or absence of concepts. The participants were asked to provide a threshold or scale the confidence values in the way that 0.5 is an adequate threshold to map them into binary decisions. Five groups proposed a threshold for their runs. Two of them asked for different thresholds per concept.

Figure 3 depicts the label density (LD) plotted against the OS results for all 44 runs. The LD is a means to describe the characteristics of an annotated dataset. It defines how many concepts are assigned to each photo in average divided by the total number of available concepts. The LD for the ground truth of the test set is 0.17, which roughly equals the LD for the training and validation set with 0.164 and 0.172, respectively. The provided thresholds were used for calculating the LD of the submitted runs. It can be seen from the figure that an

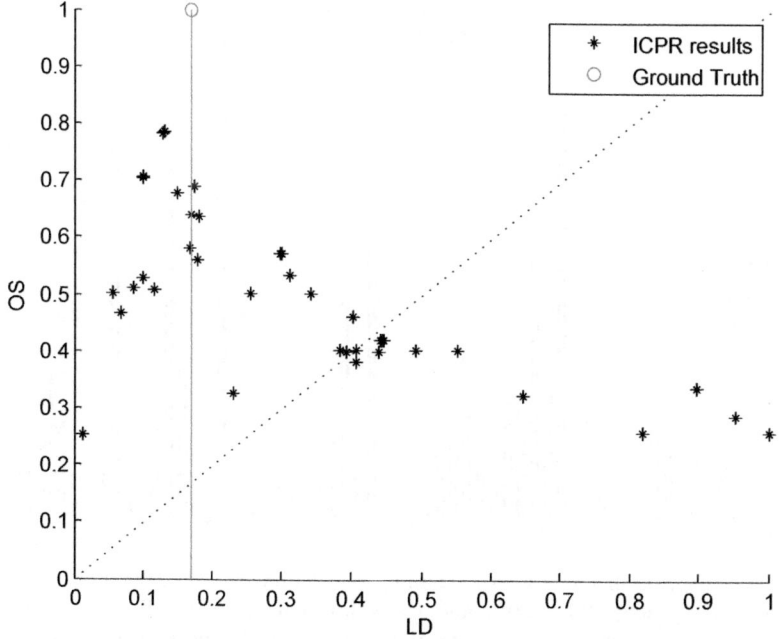

Fig. 3. OS scores plotted against label density

over-annotation of the photos does not lead to good results in the OS. The best results are assigned to runs that annotated close to the LD of the ground truth.

6   Comparison to Results of 2009

The Photo Annotation Task in ImageCLEF 2009 posed a similar problem as the annotation challenge in the ICPR contest. In both evaluation cycles, the participants were asked to annotate a set of Flickr images with 53 visual concepts. While the participants in 2009 were provided with a training set of 5,000 images with annotations and EXIF data and tested on 13,000 images, in 2010 an additional validation set of 3,000 images and annotations was provided. These 3,000 images belonged to the test set of 2009 and in 2010 the test was conducted on the remaining 10,000 test images. Therefore a comparison of the annotation performance on the 10,000 images of the test set that were used in both evaluation cycles can be made. Nevertheless, one has to keep in mind, that in 2010 the algorithms could be trained with ∼ 40% more training data. An increase in detection is therefore not necessarily caused by better annotation systems. Figure 4 illustrates the percentage of occurrence of each concept in the datasets. It can be seen that most concepts are equally distributed in the different sets while the percentage of occurrence between concepts varies significantly.

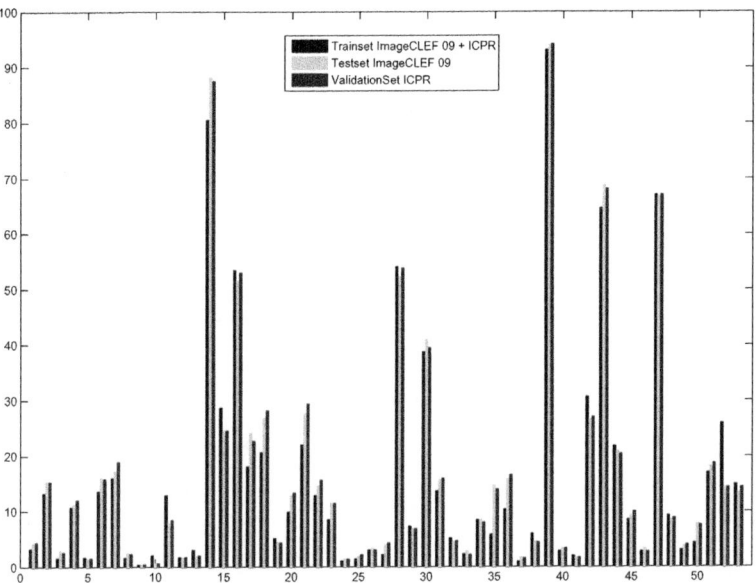

Fig. 4. Concept occurrences for the training set of ImageCLEF 2009 and ICPR 2010, the test set of ImageCLEF 2009 and the validation set of ICPR 2010. The x-axis denotes the number of the concept and the y-axis shows the occurrence of the concept in percent in the dataset.

Table 2. Overview of concepts and results per concept in terms of the best EER and best AUC per concept and the name of the group which achieved these results. The results for the Photo Annotation task in 2009 are illustrated in the middle and the ones for 2010 on the right.

No.	Concept	Group	Best AUC 09	Best EER 09	Group	Best AUC 10	Best EER 10
0	Partylife	ISIS	0.84	0.24	CVSSPRet	0.87	0.22
1	Family-Friends	ISIS	0.83	0.24	CVSSPRet	0.86	0.22
2	Beach Holidays	ISIS	0.91	0.16	CVSSPRet	0.93	0.13
3	Building-Sights	ISIS	0.88	0.20	CVSSPRet	0.90	0.18
4	Snow	LEAR	0.85	0.22	CVSSPRet	0.89	0.19
5	Citylife	ISIS	0.83	0.24	CVSSPRet	0.85	0.22
6	Landscape	ISIS	0.95	0.13	CVSSPRet / ISIS	0.95	0.12
7	Sports	FIRST	0.72	0.33	CVSSPRet	0.78	0.29
8	Desert	ISIS	0.89	0.18	ISIS / CVSSPRet	0.92	0.18
9	Spring	FIRST	0.83	0.24	IJS	0.86	0.20
10	Summer	ISIS	0.81	0.25	CVSSPRet	0.83	0.23
11	Autumn	ISIS	0.86	0.21	ISIS	0.88	0.18
12	Winter	ISIS	0.84	0.23	CVSSPRet	0.88	0.21
13	No-Visual-Season	ISIS	0.80	0.26	CVSSPRet	0.82	0.25
14	Indoor	ISIS	0.83	0.25	CVSSPRet / ISIS	0.84	0.24
15	Outdoor	ISIS	0.90	0.19	ISIS	0.91	0.18
16	No-Visual-Place	ISIS	0.79	0.29	CVSSPRet	0.81	0.27
17	Plants	ISIS	0.88	0.21	CVSSPRet	0.90	0.18
18	Flowers	ISIS / FIRST	0.87	0.21	CVSSPRet	0.89	0.19
19	Trees	ISIS	0.90	0.18	CVSSPRet	0.92	0.16
20	Sky	ISIS	0.95	0.12	ISIS	0.96	0.10
21	Clouds	ISIS	0.96	0.11	ISIS	0.96	0.10
22	Water	ISIS	0.90	0.18	CVSSPRet	0.91	0.16
23	Lake	ISIS	0.90	0.17	CVSSPRet / ISIS	0.91	0.16
24	River	ISIS	0.90	0.17	CVSSPRet	0.93	0.14
25	Sea	ISIS	0.94	0.12	CVSSPRet	0.95	0.12
26	Mountains	ISIS	0.94	0.14	ISIS	0.95	0.12
27	Day	ISIS	0.85	0.24	CVSSPRet	0.87	0.22
28	Night	LEAR	0.91	0.17	IJS	0.92	0.16
29	No-Visual-Time	ISIS	0.84	0.25	CVSSPRet / ISIS	0.86	0.23
30	Sunny	LEAR / FIRST	0.77	0.30	CVSSPRet	0.81	0.27
31	Sunset-Sunrise	ISIS	0.96	0.11	ISIS / CVSSPRet	0.96	0.08
32	Canvas	I2R / XRCE	0.83	0.24	CVSSPRet	0.85	0.22
33	Still-Life	ISIS	0.83	0.25	CVSSPRet	0.86	0.22
34	Macro	ISIS	0.81	0.27	ISIS	0.84	0.24
35	Portrait	XRCE / ISIS	0.87	0.21	CVSSPRet	0.91	0.18
36	Overexposed	LSIS / UPMC	0.81	0.25	ISIS / CNRS	0.83	0.24
37	Underexposed	I2R	0.89	0.18	AVEIR / ITI	0.89	0.19
38	Neutral-Illumination	LEAR	0.80	0.26	IJS	0.81	0.26
39	Motion-Blur	ISIS	0.75	0.31	CVSSPRet / IJS	0.79	0.28
40	Out-of-focus	LEAR	0.82	0.25	CNRS / CVSSPRet	0.84	0.24
41	Partly-Blurred	LEAR	0.86	0.22	ISIS / CVSSPRet	0.87	0.21
42	No-Blur	LEAR	0.85	0.23	ISIS	0.86	0.22
43	Single-Person	ISIS / LEAR	0.80	0.27	CVSSPRet	0.83	0.25
44	Small-Group	ISIS	0.80	0.28	CVSSPRet	0.83	0.25
45	Big-Group	ISIS	0.88	0.20	CVSSPRet	0.91	0.17
46	No-Persons	ISIS	0.86	0.22	CVSSPRet / ISIS	0.87	0.21
47	Animals	ISIS	0.84	0.24	CVSSPRet	0.87	0.21
48	Food	ISIS	0.90	0.19	CVSSPRet	0.92	0.17
49	Vehicle	ISIS	0.83	0.24	CVSSPRet	0.86	0.23
50	Aesthetic-Impression	ISIS	0.66	0.39	CVSSPRet / ISIS	0.67	0.38
51	Overall-Quality	ISIS	0.66	0.39	CVSSPRet / ISIS	0.66	0.39
52	Fancy	LEAR / ISIS	0.59	0.44	ISIS / IJS	0.61	0.42

Table 2 lists the annotation performance per concept in terms of EER and AUC for the evaluation cycles in 2009 and 2010. Additionally the group which could achieve these results is depicted. The AUC for the concepts in 2010 is at least greater or equal to the results in 2009. Also the EER in 2009 was greater or equal to the one in 2010. That means there was no decline in the annotation performance for a concept. In numbers, the concepts could be detected best by any run with an EER of 23% and an AUC of 84% in 2009. This improved to an annotation performance of 20.8% EER and 86.5% AUC per concept in 2010.

Worst detection performance in 2009 with 68.7%. Detection score in 2010: 68.6%   Worst detection performance in 2009 with 68.9%. Detection score in 2010: 67.5%

Worst detection performance in 2010 with 59%. Detection score in 2009: 79.1%   Worst detection performance in 2010 with 60.5%. Detection score in 2009: 82.5%

Fig. 5. Images with the lowest detection rate in terms of OS in 2009 and 2010

Evaluated on an example basis, the photos could be annotated correctly by 85.1% considering the best result per photo of any run in 2010. This is a small decrease by 4.5% in comparison to 2009. Also the best run in 2010 for the example-based evaluation has a lower score (78.4%) than the one in 2009 (81%). In 2010 the classification performance for each image ranged between 59% and 100%, while it ranged between 68.7% and 100% in 2009. Figure 5 shows two images for both evaluation cycles with the lowest detection rate in terms of OS and the annotation rate in the other year for the same photo. It can be seen that the images that could not be annotated well in 2009, were annotated with a similar quality in 2010 while the ones that were not annotated well in 2010 were annotated much better in 2009.

Summarizing, the comparison shows that there was an improvement in the annotation performance evaluated on a concept basis, while the results got slightly worse when evaluated on an example basis. Although if this seems contradictory, these results are reasonable. In contrast to the concept-based measures, the OS is directly dependent on the threshold for mapping the confidence values into a binary decision. If this threshold is not chosen carefully, the results are directly influenced while the threshold does not affect the EER and AUC results. Further, the OS considers all labels per image. If there are a few concepts that could not be annotated with reasonable results, these low scores major influence the average annotation behaviour. In addition, the OS penalizes annotations that violate real-world knowledge provided in the ontology. Some annotation systems do not take into account concepts that condition each other through the hierarchy or concepts that exclude each other. While the concept-based evaluation does not consider the relations between concepts, the example-based evaluation with the OS assigns violation costs in these cases. It seems like the participants of 2009 set a higher value on these cases than in the ICPR benchmark.

7   Conclusion

This paper summarises the results of the Photo Annotation Task. Its aim was to automatically annotate photos with 53 concepts in a multilabel scenario. The results of 12 teams show that the task could be solved reasonably well with the best system achieving an average AUC of 86%. The concepts could be annotated in average with an AUC of 86.5% considering all runs. The classification performance for each image ranged between 59% and 100% with an average score of 85%. Most groups applied discriminative approaches with local or combined local and global features. In comparison to the results achieved in 2009, the detection performance increased for the concept-based evaluation by 2.2% and 2.5% for EER and AUC, respectively. The example-based evaluation showed a slight decrease in performance.

Acknowledgment

This work was supported by grant 01MQ07017 of the German research program THESEUS and a German Academic Exchange Service (DAAD) Scholarship. It was partly performed at the KMI at Open University, UK.

References

1. Huiskes, M.J., Lew, M.S.: The MIR Flickr Retrieval Evaluation. In: Proc. of the ACM Intern. Conference on Multimedia Information Retrieval (2008)
2. Nowak, S., Dunker, P.: A Consumer Photo Tagging Ontology: Concepts and Annotations. In: Dunker, P., Liebetrau, J., Nowak, S., Müller, H., Panagiotis, V. (eds.) Proc. of THESEUS-ImageCLEF Pre-Workshop, Corfu, Greece (2009)
3. Manning, C., Raghavan, P., Schütze, H.: An Introduction to Information Retrieval [Draft], ch. 8. Cambridge University Press, Cambridge (2009)

4. Tsoumakas, G., Vlahavas, I.: Random k-labelsets: An ensemble method for multilabel classification. In: Proc. of the European Conference on Machine Learning (2007)
5. Everingham, M., Zisserman, A., Williams, C., Van Gool, L., Allan, M., et al.: The 2005 PASCAL Visual Object Classes Challenge. In: Quiñonero-Candela, J., Dagan, I., Magnini, B., d'Alché-Buc, F. (eds.) MLCW 2005. LNCS (LNAI), vol. 3944, pp. 117–176. Springer, Heidelberg (2006)
6. Deselaers, T., Hanbury, A.: The visual concept detection task in ImageCLEF 2008. In: Peters, C., Deselaers, T., Ferro, N., Gonzalo, J., Jones, G.J.F., Kurimo, M., Mandl, T., Peñas, A., Petras, V. (eds.) Evaluating Systems for Multilingual and Multimodal Information Access. LNCS, vol. 5706, pp. 531–538. Springer, Heidelberg (2009)
7. Nowak, S., Lukashevich, H., Dunker, P., Rüger, S.: Performance measures for multilabel evaluation: a case study in the area of image classification. In: MIR 2010: Proceedings of the International Conference on Multimedia Information Retrieval, New York, USA, pp. 35–44 (2010)
8. Tahir, M.A., Fei, Y., Barnard, M., Awais, M., Mikolajczyk, K., Kittler, J.: The University of Surrey Visual Concept Detection System at ImageCLEF 2010: Working Notes. In: Ünay, D., Çataltepe, Z., Aksoy, S. (eds.) ICPR 2010. LNCS, vol. 6388, pp. 164–171. Springer, Heidelberg (2010)
9. van de Sande, K., Gevers, T., Smeulders, A.: The University of Amsterdam's Concept Detection System at ImageCLEF 2009. In: Peters, C., Caputo, B., Gonzalo, J., Jones, G.J.F., Kalpathy-Cramer, J., Müller, H., Tsikrika, T. (eds.) CLEF 2009. LNCS, vol. 6242, pp. 261–268. Springer, Heidelberg (2010)
10. Dumont, E., Zhao, Z.Q., Glotin, H., Paris, S.: A new TFIDF Bag of Visual Words for Concept Detection. In: Peters, C., Caputo, B., Gonzalo, J., Jones, G.J.F., Kalpathy-Cramer, J., Müller, H., Tsikrika, T. (eds.) CLEF 2009. LNCS, vol. 6242. Springer, Heidelberg (2010)
11. Iftene, A., Vamanu, L., Croitoru, C.: UAIC at ImageCLEF 2009 Photo Annotation Task. In: Peters, C., Caputo, B., Gonzalo, J., Jones, G.J.F., Kalpathy-Cramer, J., Müller, H., Tsikrika, T. (eds.) CLEF 2009. LNCS, vol. 6242, pp. 283–286. Springer, Heidelberg (2010)

A   Results for All Submissions

Table 3. The table shows the results for all submitted runs in alphabetical order

Run ID	EER	AUC	OS
AVEIR_1262856077602_sorted_run1_AVEIR_sim_rank_seuil05.txt	0.3377	0.7159	0.5602
AVEIR_1262856132951_sorted_run2_AVEIR_sim_rank_seuil05.txt	0.3135	0.7476	0.4673
AVEIR_1262856175981_sorted_run3_AVEIR_sim_rank_seuil05.txt	0.2848	0.7848	0.5273
AVEIR_1262856207702_sorted_run4_AVEIR_sim_rank_seuil05.txt	0.2858	0.7799	0.5106
CNRS_1262433353127__TotalCombineSiftPCAScore060.txt	0.2748	0.7927	0.4195
CNRS_1262433442878__TotalCombineSiftPCAScore063.txt	0.2751	0.7928	0.4199
CNRS_1262433562325__TotalCombineSiftPCAScore065.txt	0.2749	0.7926	0.4203
CNRS_1262433641320__TotalCombineSiftPCAScore067.txt	0.2752	0.7923	0.4204
CNRS_1262434028322__TotalCombineSiftPCAScore070.txt	0.2758	0.7915	0.4199
CVSSPRet_1262719488645_run2_format.txt	0.2216	0.8547	0.5328
CVSSPRet_1262719758659_run1_format.txt	0.2136	0.8600	0.5709
CVSSPRet_1262727440131_fei_run_2.txt	0.2206	0.8534	0.2575
CVSSPRet_1262777799929_fei_run_1.txt	0.2138	0.8588	0.5724
CVSSPRet_1262781593747__fei_run_3.txt	0.2162	0.8572	0.6899
IJS_1262100955676__ijs_feit_run2_1.txt	0.2425	0.8321	0.6374
IJS_1262101154419__ijs_feit_run2.txt	0.2425	0.8321	0.7066
IJS_1262101315087__ijs_feit_run1_1.txt	0.2504	0.8214	0.6395
IJS_1262101509972__ijs_feit_run1.txt	0.2504	0.8214	0.7033
ISIS_1262376995364__uva-isis-both2-4sift.txt	0.2214	0.8538	0.7812
ISIS_1262377180771__uva-isis-bothdenseallharris-4sift.txt	0.2182	0.8568	0.7837
ITI_1262190868772__icprlocalhisto.txt	0.3656	0.5917	0.4017
ITI_1262191191579__scoresall.txt	0.5066	0.4953	0.4002
ITI_1262191384462__scoresconcatlhf.txt	0.4847	0.5184	0.4023
ITI_1262191518097__scoreslh.txt	0.4789	0.5269	0.4007
ITI_1262191649206__scoreslhretrain.txt	0.4806	0.5246	0.4001
LSIS_1262858899140__sorted_run3_lsis_sim_rank_seuil05.txt	0.3106	0.7490	0.4599
LSIS_1262858975795__sorted_run4_lsis_sim_rank_seuil05.txt	0.3106	0.7490	0.5014
LSIS_1262859069185__sorted_run1_lsis_sim_rank_seuil05.txt	0.4964	0.5042	0.3226
LSIS_1262859144313__sorted_run2_lsis_sim_rank_seuil05.txt	0.3106	0.7490	0.5067
MMIS_1261257618173__test_NPDE_baseline2010.txt	0.3049	0.7566	0.5027
MMIS_1261257806321__test_NPDE_GOOGLE_2010.txt	0.3049	0.7566	0.5027
MMIS_1261258062082__test_MRF_baseline2010.txt	0.3281	0.7199	0.2529
MMIS_1261258281151__test_MRF_GOOGLE_2010.txt	0.3281	0.7199	0.2529
MMIS_1261427169687__test_MRF2_nor2_2010.txt	0.4998	0.0000	0.2602
MRIM_1262676461569__res_ICPR_LIG3.txt	0.3831	0.6393	0.5801
MRIM_1262698052599__LIG2_ICPR_sum1.txt	0.4108	0.6209	0.5004
MRIM_1262698323023__LIG3_ICPR_sumsr.txt	0.4082	0.6262	0.2881
TRS2008_1261716477070__testingSetAnnotationsICPR.txt	0.4152	0.6200	0.3270
UAIC_1262539903396__run1.txt	0.4762	0.1408	0.6781
UPMC_1262281196445__run1_pca5000_score01_unbalanced_lossfunction10.txt	0.3377	0.7159	0.4034
UPMC_1262425276587__run2_pca5000_score01_unbalanced_svmonly.txt	0.4159	0.6137	0.3366
UPMC_1262426941383__run4_sift5000_unbalanced_lossfunc10_file_01.txt	0.4331	0.5933	0.3806
UPMC_1262462158758__run3_hierImplications.txt	0.3377	0.7159	0.4034
UPMC_1262467131292__run5_Fusion_sift_pca_Implication.txt	0.4331	0.5933	0.3806

Detection of Visual Concepts and Annotation of Images Using Ensembles of Trees for Hierarchical Multi-Label Classification

Ivica Dimitrovski[1,2], Dragi Kocev[1], Suzana Loskovska[2], and Sašo Džeroski[1]

[1] Department of Knowledge Technologies, Jožef Stefan Institute
Jamova cesta 39, 1000 Ljubljana, Slovenia
[2] Department of Computer Science, Faculty of Electrical Engineering and
Information Technology
Karpoš bb, 1000 Skopje, Macedonia
ivicad@feit.ukim.edu.mk, Dragi.Kocev@ijs.si, suze@feit.ukim.edu.mk,
Saso.Dzeroski@ijs.si

Abstract. In this paper, we present a hierarchical multi-label classification system for visual concepts detection and image annotation. Hierarchical multi-label classification (HMLC) is a variant of classification where an instance may belong to multiple classes at the same time and these classes/labels are organized in a hierarchy. The system is composed of two parts: feature extraction and classification/annotation. The feature extraction part provides global and local descriptions of the images. These descriptions are then used to learn a classifier and to annotate an image with the corresponding concepts. To this end, we use predictive clustering trees (PCTs), which are able to classify target concepts that are organized in a hierarchy. Our approach to HMLC exploits the annotation hierarchy by building a single predictive clustering tree that can simultaneously predict all of the labels used to annotate an image. Moreover, we constructed ensembles (random forests) of PCTs, to improve the predictive performance. We tested our system on the image database from the ImageCLEF@ICPR 2010 photo annotation task. The extensive experiments conducted on the benchmark database show that our system has very high predictive performance and can be easily scaled to large number of visual concepts and large amounts of data.

1 Introduction

An ever increasing amount of visual information is becoming available in digital form in various digital archives. The value of the information obtained from an image depends on how easily it can be found, retrieved, accessed, filtered and managed. Therefore, tools for efficient archiving, browsing, searching and annotation of images are a necessity.

A straightforward approach, used in some existing information retrieval tools for visual materials, is to manually annotate the images by keywords and then

D. Ünay, Z. Çataltepe, and S. Aksoy (Eds.): ICPR 2010, LNCS 6388, pp. 152–161, 2010.

to apply text-based query for retrieval. However, manual image annotation is an expensive and time-consuming task, especially given the large and constantly growing size of image databases.

The image search provided by major search engines, such as Google, Bing, Yahoo! and AltaVista, relies on textual or metadata descriptions of images found on the web pages containing the images and the file names of the images. The results from these search engines are very disappointing when the visual content of the images is not mentioned, or properly reflected, in the associated text.

A more sophisticated approach to image retrieval is automatic image annotation: a computer system assigns metadata in the form of captions or keywords to a digital image [5]. These annotations reflect the visual concepts that are present in the image. This approach begins with the extraction of feature vectors (descriptions) from the images. A machine learning algorithm is then used to learn a classifier, which will then classify/annotate new and unseen images.

Most of the systems for detection of visual concepts learn a separate model for each visual concept [7]. However, the number of visual concepts can be large and there can be mutual connections between the concepts that can be exploited. An image may have different meanings or contain different concepts: if these are organized into a hierarchy (see Fig. 2), hierarchical multi-label classification (HMLC) can be used for obtaining annotations (i.e., labels for the multiple visual concepts present in the image) [7]. The goal of HMLC is to assign to each image multiple labels, which are a subset of a previously defined set (hierarchy) of labels.

In this paper, we present a system for detection of visual concepts and annotation of images, which exploits the semantic knowledge about the inter-class relationships among the image labels organized in hierarchical structure. For the annotation of the images, we propose to exploit the annotation hierarchy in image annotation by using predictive clustering trees (PCTs) for HMLC. PCTs are able to handle target concepts that are organized in a hierarchy, i.e., to perform HMLC. To improve the predictive performance, we use ensembles (random forests) of PCTs for HMLC. For the extraction of features, we use several techniques that are recommended as most suitable for the type of images at hand [7].

We tested the proposed approaches on the image database from the ICPR 2010 photo annotation task [9]. The concepts used in this annotation task are from the personal photo album domain and they are structured in an ontology. Fig. 2 shows a part of the hierarchical organization of the target concepts.

The remainder of this paper is organized as follows. Section 2 presents the proposed large scale visual concept detection system. Section 3 explains the experimental design. Section 4 reports the obtained results. Conclusions and a summary are given in Section 5.

2 System for Detection of Visual Concepts

2.1 Overall Architecture

Fig. 1 presents the architecture of the proposed system for visual concepts detection and image annotation. The system is composed of a feature extraction part and a classification/annotation part. We use two different sets of features to describe the images: visual features extracted from the image pixel values and features extracted from the exchangeable image file format (EXIF) metadata files. We employ different sampling strategies and different spatial pyramids to extract the visual features (both global and local) [4].

As an output of the feature extraction part, we obtain several sets of descriptors of the image content that can be used to learn a classifier to annotate the images with the visual concepts. First, we learn a classifier for each set of descriptors separately. The classifier outputs the probabilities with which an image is annotated with the given visual concepts. To obtain a final prediction, we combine the probabilities output from the classifiers for the different descriptors by averaging them. Depending on the domain, different weights can be used for the predictions of the different descriptors.

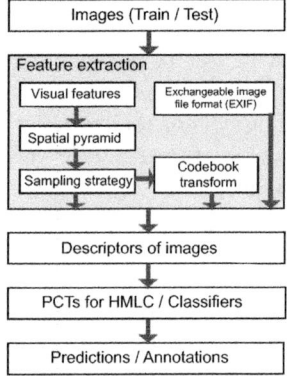

Fig. 1. Architecture of the proposed system for detection of visual concepts and annotation of images

2.2 The Task of HMLC

Hierarchical multi-label classification is a variant of classification were a single example may belong to multiple classes at the same time and these classes are organized in a hierarchy. An example that belongs to some class automatically belongs to all its super-classes, as implied by the hierarchical constraint. Example problems of this kind can be found in several domains including text classification, functional genomics, and object and scene classification. For more detail overview of the possible application areas we refer the reader to [11].

The predefined set of labels can be organized in a semantic hierarchy (see Fig. 2 for an example). Each image is represented with: (1) a set of descriptors (in this example, the descriptors are histograms of five types of edges encountered in the image) and (2) labels/annotations. A single image can be annotated with multiple labels at different levels of the predefined hierarchy. For example, the image in the third row in the Table from Fig. 2 is labeled with clouds and sea. Note that this image is also labeled with the labels: sky, water and landscape because these labels are in the upper levels of the hierarchy.

The data, as presented in the Table from Fig. 2, are used by a machine learning algorithm to train a classifier. The testing set of images contains only the set of descriptors and has no *a priori* annotations.

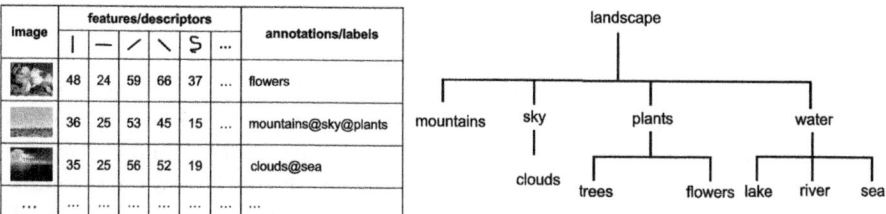

Fig. 2. A fragment of the hierarchy for image annotation. The annotations are part of the hierarchical classification scheme for the ICPR 2010 photo annotation task (right).The table contains set of images with their visual descriptors and annotations (left).

2.3   Ensembles of PCTs for HMLC

In the PCT framework [1], a tree is viewed as a hierarchy of clusters: the top-node corresponds to one cluster containing all data, which is recursively partitioned into smaller clusters while moving down the tree. Note that the hierarchical structure of the PCT does not necessary reflect the hierarchical structure of the annotations.

PCTs are constructed with a standard "top-down induction of decision trees" (TDIDT) algorithm. The heuristic for selecting the tests is the reduction in variance caused by partitioning the instances, where the variance $Var(S)$ is defined by equation (1) below. Maximizing the variance reduction maximizes cluster homogeneity and improves predictive performance.

A leaf of a PCT is labeled with/predicts the prototype of the set of examples belonging to it. With appropriate variance and prototype functions, PCTs can handle different types of data, e.g., multiple targets [3] or time series [12]. A detailed description of the PCT framework can be found in [1]. The PCT framework is implemented in the CLUS system, which is available for download at http://www.cs.kuleuven.be/~dtai/clus.

To apply PCTs to the task of HMLC, the example labels are represented as vectors with Boolean components. Components in the vector correspond to labels in the hierarchy traversed in a depth-first manner. The i-th component of the vector is 1 if the example belongs to class c_i and 0 otherwise. If $v_i = 1$, then $v_j = 1$ for all v_j's on the path from the root to v_i.

The variance of a set of examples (S) is defined as the average squared distance between each example's label v_i and the mean label $\bar{v}$ of the set, i.e.,

$$Var(S) = \frac{\sum_i d(v_i, \bar{v})^2}{|S|} \qquad (1)$$

We consider the higher levels of the hierarchy more important: an error at the upper levels costs more than an error at the lower levels. Considering this, a weighted Euclidean distance is used:

$$d(v_1, v_2) = \sqrt{\sum_i w(c_i)(v_{1,i} - v_{2,i})^2} \qquad (2)$$

where $v_{k,i}$ is the i'th component of the class vector v_k of an instance x_k, and the class weights $w(c_i)$. The class weights decrease with the depth of the class in the hierarchy, $w(c_i) = w_0 \cdot w(c_j)$, where c_j is the parent of c_i and $0 < w_0 < 1$.

Each leaf in the tree stores the mean $\bar{v}$ of the vectors of the examples that are sorted in that leaf. Each component of $\bar{v}$ is the proportion of examples $\bar{v}_i$ in the leaf that belong to class c_i. An example arriving in the leaf can be predicted to belong to class c_i if $\bar{v}_i$ is above some threshold t_i. The threshold can be chosen by a domain expert.

For a detailed description of PCTs for HMLC the reader is referred to [15]. Next, we explain how PCTs are used in the context of an ensemble classifier, namely ensembles further improve the performance of PCTs.

Random Forests of PCTs. To improve the predictive performance of PCTs, we use ensemble methods. An ensemble classifier is a set of classifiers. Each new example is classified by combining the predictions of each classifier from the ensemble. These predictions can be combined by taking the average (for regression tasks) or the majority vote (for classification tasks) [2]. In our case, the predictions in a leaf are the proportions of examples of different classes that belong to it. We use averaging to combine the predictions of the different trees. As for the base classifiers, a threshold should be specified to make a prediction.

We use random forests as an ensemble learning technique. A random forest [2] is an ensemble of trees, obtained both by bootstrap sampling, and by randomly changing the feature set during learning. More precisely, at each node in the decision tree, a random subset of the input attributes is taken, and the best feature is selected from this subset (instead of the set of all attributes). The number of attributes that are retained is given by a function f of the total number of input attributes x (e.g., $f(x) = x, f(x) = \sqrt{x}, f(x) = \lfloor \log_2 x \rfloor + 1$).

2.4 Feature Extraction

We use different commonly used types of techniques for feature extraction from images. We employ three types of global image descriptors: gist features [10], local binary patterns (LBP) [13] and a color histogram, with 8 bins in each color channel for the RGB color space. The LBP operator is computed in a spatial arrangement where the image is split into 4x4 sub-regions.

Local features include scale-invariant feature transforms (SIFT) extracted densely on a multi-scale grid or around salient points obtained from a Harris-Laplace detector [6]. The dense sampling and Harris-Laplace detectors give an equal weight to all key-points, independent of their spatial location in the image. To overcome this limitation, one can use spatial pyramids of 1x1, 2x2 and 1x3 regions [14].

We computed six different sets of SIFT descriptors over the following color spaces: RGB, opponent, normalized opponent, gray, HUE and HSV. For each set of SIFT descriptors, we use the codebook approach to avoid using all visual features of an image [14].

The generation of the codebook begins by randomly sampling 50 key-points from each image and extracting SIFT descriptors in each key-point (i.e., each key-point is described by a vector of numerical values). Then, to create the codewords, we employ k-means clustering on the set of all key-points. We set the number of clusters to 4000, thus we define a codebook with 4000 codewords (a codeword corresponds to a single cluster and a codebook to the set of all clusters). Afterwards, we assign the key-points to the discrete codewords predefined in the codebook and obtain a histogram of the occurring visual features. This histogram will contain 4000 bins, one for each codeword. To be independent of the total number of key-points in an image, the histogram bins are normalized to sum to 1.

The number of key-points and codewords (clusters) are user defined parameters for the system. The values used above (50 key-points and 4000 codewords) are recommended for general images [14].

An image can have an associated text file with metadata information in EXIF (EXchangeable Image File) format [16]. The metadata can be used to construct features that describe certain aspects of the imaging technique and the technical specification of the used camera. These describe, for example the image quality (resolution, focal length, exposure time) and when the picture was taken.

3   Experimental Design

3.1   Definition and Parameter Settings

We evaluated our system on the image database from the ImageCLEF@ICPR 2010 photo annotation task. The image database consists of training (5000), validation (3000) and test (10000) images. The images are labeled with 53 visual concepts organized in a tree-like hierarchy [9]. The goal of the task is to predict which of the visual concepts are present in each of the testing images.

We generated 15 sets of visual descriptors for the images: 12 sets of SIFT local descriptors (2 detectors, Harris-Laplace and dense sampling, over 6 different color spaces) with 32000 bins for each set (8 sub-images, from the spatial pyramids: 1x1, 2x2 and 1x3, 4000 bins each). We also generated 3 sets of global descriptors

(LBP histogram with 944 bins, gist features with 960 bins and RGB color histogram with 512 bins). From the EXIF metadata, we selected the most common tags as features, such as: software, exposure time, date and time (original), exposure bias, metering mode, focal length, pixelXDimension, pixelY-Dimension etc. Since the PCTs can handle missing values, these values for the images without EXIF tags were set to 'unknown' or '?'.

The parameter values for the random forests were as follows: we used 100 base classifiers and the size of the feature subset was set to 10% of the number of descriptive attributes. The weights for the PCTs for the HMLC (w_0) were set to 1: each of the classes from the hierarchy has equal influence on the heuristic score.

3.2   Performance Measures

The evaluation of the results is done using three measures of performance suggested by the organizers of the challenge [3]: area under the ROC curve (AUC), equal error rate (EER) and average ontology score (AOS). The first two scores evaluate the performance for each visual concept, while the third evaluates the performance for each testing image.

The ROC curve is widely used evaluation measure (see Fig. 3). It plots the true positive rate (TPR) vs. false positive rate (FPR). The area between the curve and the axis with FPR (AUC) is the probability that a randomly chosen positive example will be ranked higher than a randomly chosen negative example. The EER is the threshold value at which the TPR and FPR are equal. Hence, the EER balances the probability of error with the probability of false rejection. Lower EER means better predictive performance. The hierarchical AOS measure calculates the misclassification cost for each missing or wrongly annotated concept per image. The AOS score is based on structure information (distance between concepts in the hierarchy), relationships from the ontology and the agreement between annotators for a concept [8].

4   Results and Discussion

We present results from two different experiments (see Table 1). In the first experiment, we use just the training images for learning the classifier. For the second experiment, we merge the training and validation set into a single dataset which we then use to learn the classifier. The results show that by using both datasets (training and validation together) we get better scores.

If we focus on the prediction scores for the individual visual concepts, we can note that we predict best the presence of landscape elements (see Table 2); the best predicted concept is 'Sunrise or Sunset' (from the parent-concept 'Time of day'). The worst predicted concepts are from the 'Aesthetics' group of concepts ('Aesthetic Impression', 'Overall quality' and 'Fancy'). But, this is to be expected because the agreement of human annotators on these concepts is only about 75% [8].

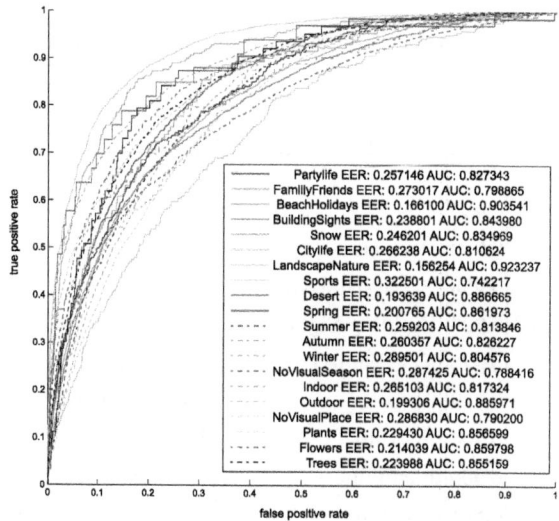

Fig. 3. ROC curves for a subset of the visual concepts

The system has low predictive power when we are predicting the absence of a concept (e.g., 'No persons', 'No visual season' ...). The hierarchy should not include these concepts. These concepts should be assigned after post-processing the results. We also have to predict mutually exclusive concepts (for example: Indoor, Outdoor and No visual place). The notation of HMLC, however, does not account for mutually exclusive concepts. To solve this issues one must re-engineer the hierarchy of the concepts.

Further improvements can be expected if different weighting schemes are used (to combine the predictions of the various descriptors). For instance, the SIFT descriptors are invariant to color changes, and they do not predict well concepts where illumination is important. Thus, the weight of the SIFT descriptors in the combined predictions for those concepts should be decreased.

Let us compare the results of our system with the results from the other participating groups at the ImageCLEF@ICPR 2010 photo annotation task. Our system ranks second by the hierarchical AOS score. By the EER and AUC score it ranks third. Thus, relatively speaking, it performs better under the hierarchical performance measure.

Table 1. Results of the experiments evaluated using Equal Error Rate, Area under Curve and Average Ontology Score

	EER	AUC	AOS
Train and Validation	0.242	0.832	0.706
Train	0.250	0.821	0.703

Table 2. Results per concept for our best run in the Large-Scale Visual Concept Detection Task using the Area Under the Curve. The concepts are ordered by their highest score.

Concept	AUC	Concept	AUC	Concept	AUC
Sunset-Sunrise	0.951	Trees	0.855	Citylife	0.811
Clouds	0.946	Day	0.853	Winter	0.805
Sea	0.939	Portrait	0.849	Out-of-focus	0.803
Sky	0.933	Partly-Blurred	0.848	Animals	0.803
Landscape-Nature	0.923	Building-Sights	0.844	Familiy-Friends	0.799
Night	0.923	No-Visual-Time	0.840	Sunny	0.799
Mountains	0.919	Snow	0.835	No-Persons	0.794
Beach-Holidays	0.904	No-Blur	0.829	Vehicle	0.791
Lake	0.900	Partylife	0.827	No-Visual-Place	0.790
River	0.891	Autumn	0.826	No-Visual-Season	0.788
Food	0.890	Canvas	0.825	Motion-Blur	0.779
Desert	0.887	Indoor	0.817	Single-Person	0.761
Outdoor	0.886	Still-Life	0.817	Small-Group	0.752
Water	0.877	Macro	0.816	Sports	0.742
Underexposed	0.876	Summer	0.814	Aesthetic-Impression	0.661
Spring	0.862	Overexposed	0.812	Overall-Quality	0.657
Flowers	0.860	Big-Group	0.811	Fancy	0.613
Plants	0.857	Neutral-Illumination	0.811	Average	0.832

5   Conclusion

Hierarchical multi-label classification (HMLC) problems are encountered increasingly often in image annotation. However, flat classification machine learning approaches are predominantly applied in this area. In this paper, we propose to exploit the annotation hierarchy in image annotation by using ensembles of trees for HMLC. Our approach to HMLC exploits the annotation hierarchy by building a single classifier that simultaneously predicts all of the labels in the hierarchy.

Applied on the ImageCLEF@ICPR 2010 photo annotation benchmark task our approach was ranked second for the hierarchical performance measure and third for the equal error rate and area the under the curve, out of 12 competing groups. The results were worst for predicting the absence of concepts. This suggests the need for re-engineering the hierarchy or for post processing the predictions to appropriately handle such concepts.

The system we presented is general. It can be easily extended with new feature extraction methods, and it can thus be easily applied to other domains, types of images and other classification schemes. In addition, it can handle arbitrarily sized hierarchies organized as trees or directed acyclic graphs.

References

1. Blockeel, H., De Raedt, L., Ramon, J.: Top-down induction of clustering trees. In: Proc. of the 15th ICML, pp. 55–63 (1998)
2. Breiman, L.: Random Forests. Machine Learning 45, 5–32 (2001)
3. Kocev, D., Vens, C., Struyf, J., Džeroski, S.: Ensembles of Multi-Objective Decision Trees. In: Kok, J.N., Koronacki, J., Lopez de Mantaras, R., Matwin, S., Mladenič, D., Skowron, A. (eds.) ECML 2007. LNCS (LNAI), vol. 4701, pp. 624–631. Springer, Heidelberg (2007)
4. Lazebnik, S., Schmid, C., Ponce, J.: Beyond bags of features: Spatial pyramid matching for recognizing natural scene categories. In: CVPR, pp. 2169–2178 (2006)
5. Li, J., Wang, J.Z.: Real-Time Computerized Annotation of Pictures. IEEE Trans. on Pattern Analysis and Machine Intelligence 30(6), 985–1002 (2008)
6. Lowe, D.G.: Distinctive Image Features from Scale-Invariant Keypoints. International Journal of Computer Vision 60(2), 91–110 (2004)
7. Nowak, S., Dunker, P.: Overview of the CLEF 2009 Large-Scale Visual Concept Detection and Annotation Task. In: Peters, C., Caputo, B., Gonzalo, J., Jones, G.J.F., Kalpathy-Cramer, J., Müller, H., Tsikrika, T. (eds.) CLEF 2009. LNCS, vol. 6242, pp. 94–109. Springer, Heidelberg (2010)
8. Nowak, S., Lukashevich, H.: Multilabel classification evaluation using ontology information. In: Workshop on IRMLeS, Heraklion, Greece (2009)
9. Nowak, S.: ImageCLEF@ICPR Contest: Challenges, Methodologies and Results of the PhotoAnnotation Task. In: Ünay, D., Çataltepe, Z., Aksoy, S. (eds.) ICPR 2010. LNCS, vol. 6388, pp. 140–153. Springer, Heidelberg (2010)
10. Oliva, A., Torralba, A.: Modeling the Shape of the Scene: A Holistic Representation of the Spatial Envelope. International Journal of Computer Vision 42(3), 145–175 (2001)
11. Silla, C., Freitas, A.: A survey of hierarchical classification across different application domains. Data Mining and Knowledge Discovery (in press, 2010)
12. Slavkov, I., Gjorgjioski, V., Struyf, J., Džeroski, S.: Finding explained groups of time-course gene expression profiles with predictive clustering trees. Molecular BioSystems 6(4), 729–740 (2010)
13. Takala, V., Ahonen, T., Pietikainen, M.: Block-Based Methods for Image Retrieval Using Local Binary Patterns. In: Kalviainen, H., Parkkinen, J., Kaarna, A. (eds.) SCIA 2005. LNCS, vol. 3540, pp. 882–891. Springer, Heidelberg (2005)
14. Van de Sande, K., Gevers, T., Snoek., C.: A comparison of color features for visual concept classification. In: CIVR, pp. 141–150 (2008)
15. Vens, C., Struyf, J., Schietgat, L., Dzeroski, S., Blockeel, H.: Decision trees for hierarchical multi-label classification. Machine Learning 73(2), 185–214 (2008)
16. Exchangeable image file format, http://en.wikipedia.org/wiki/EXIF

The University of Surrey Visual Concept Detection System at ImageCLEF@ICPR: Working Notes

M.A. Tahir, F. Yan, M. Barnard, M. Awais, K. Mikolajczyk, and J. Kittler

Centre for Vision, Speech and Signal Processing
University of Surrey
Guildford, GU2 7XH, UK
{m.tahir,f.yan,mark.barnard,m.rana,k.mikolajczyk,j.kittler}@surrey.ac.uk

Abstract. Visual concept detection is one of the most important tasks in image and video indexing. This paper describes our system in the ImageCLEF@ICPR Visual Concept Detection Task which ranked *first* for large-scale visual concept detection tasks in terms of Equal Error Rate (EER) and Area under Curve (AUC) and ranked *third* in terms of hierarchical measure. The presented approach involves state-of-the-art local descriptor computation, vector quantisation via clustering, structured scene or object representation via localised histograms of vector codes, similarity measure for kernel construction and classifier learning. The main novelty is the classifier-level and kernel-level fusion using Kernel Discriminant Analysis with RBF/Power Chi-Squared kernels obtained from various image descriptors. For 32 out of 53 individual concepts, we obtain the best performance of all 12 submissions to this task.

1 Introduction

ImageCLEF@ICPR PhotoAnnotation [1,2] is an evaluation initiative that aims at comparing image-based approaches in the consumer photo domain. It consists of two main tasks: the visual concept detection and annotation tasks. The aim of this paper is to present our system in the Large-Scale Visual Concept Detection Task which ranked *first* in terms of EER and AUC and ranked *third* in terms of hierarchical measure. For the concepts, an average AUC of 86% could be achieved, including concepts with an AUC as high as 96%. For 32 out of 53 individual concepts, we obtained the best performance of all 12 submissions addressing this task.

The rest of paper is organised as follows. Section 2 describes the system followed by a description of the methods submitted in Section 3. Experiments and the results are discussed in Section 4. Section 5 concludes the paper.

2 Visual Concept Detection System

The visual concept detection problem can be formulated as a two class pattern recognition problem. The original data set is divided into N data sets where

D. Ünay, Z. Çataltepe, and S. Aksoy (Eds.): ICPR 2010, LNCS 6388, pp. 162–170, 2010.

$Y = \{1, 2, ..., N\}$ is the finite set of concepts. The task is to learn one binary classifier $h_a : X \rightarrow \{\neg a, a\}$ for each concept $a \in Y$. We may choose various visual feature extraction methods to obtain X. Figure 1 shows the visual concept detection system adopted in this paper. It follows the standard bag-of-words model [3] that has become the method of choice for visual categorisation [4,5,6]. The system consists of six main components. Each component is implemented via state-of-the-art techniques. These components are described below.

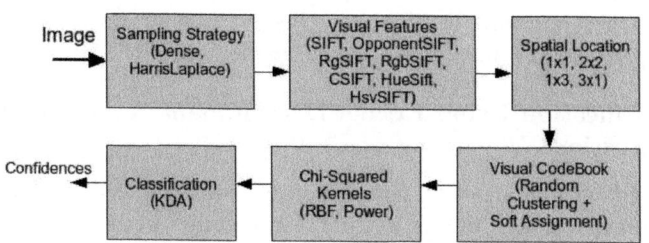

Fig. 1. Visual Concept Detection System

2.1 Sampling Strategy

The model first extracts specific points in an image using a point sampling strategy. Two methods have been chosen: Dense sampling, and Harris-Laplace. Dense sampling selects points regularly over the image at fixed pixel intervals. Typically, around 10,000 points are sampled per image at an interval of 6 pixels. The Harris-Laplace salient point detector [7] uses the Harris corner detector to find potential feature locations and then selects a subset of these points for which the Laplacian-of-Gaussians reaches a maximum over scale.

2.2 Visual Feature Extraction

To describe the area around the sampled points, we use the SIFT descriptor [8], HSV Sift, HUE Sift, two extensions of SIFT [7] and four extensions of SIFT to colour [4]: OpponentSIFT, RGSIFT, C-SIFT, RGB-SIFT. These descriptors have specific invariance properties with respect to common changes in illumination conditions and have been shown to improve visual categorisation accuracy [4].

2.3 Spatial Location and Visual Codebook

In order to create a representation for each image we employ the commonly used bag of visual words technique. All the descriptors in the training set are clustered using the kmeans algorithm into 4000 clusters. This is a hierarchical process, first the data is clustered into 10 high level clusters and then 400 lower level clusters. A histogram is then produced for each image in the training set. This 4000 bin histogram is populated using the *Codeword Uncertainty* method presented by Van Gemert et al [9] where the histogram entry of each visual codeword w is given by

$$UNC(w) = \frac{1}{n}\sum_{i=1}^{n}\frac{K_\sigma(D(w,r_i))}{\sum_{j=1}^{|V|}K_\sigma(D(w_j,r_i))},\tag{1}$$

where n is the number of descriptors in the image, $D(w,r_i)$ is the Euclidean distance between the descriptor r_i and its cluster centre on codeword w, K is a Gaussian kernel with smoothing factor σ and V is the visual vocabulary containing the codeword W. This method of histogram generation has been shown to perform well in the visual concept detection [4].

2.4   Classification Using Kernel Discriminant Analysis and Spectral Regression

Kernel based learning methods are commonly regarded as a solid choice in order to learn robust concept detectors from large-scale visual codebooks. In recent work [5], we have successfully used kernel discriminant analysis using spectral regression (SRKDA), initially introduced by Cai et al [10], for large-scale image and video classification problems. This method combines the spectral graph analysis and regression for an efficient large matrix decomposition in KDA. It has been demonstrated in [10] that it can achieve an order of magnitude speedup over the eigen-decomposition while producing smaller error rate compared to state-of-the-art classifiers. Later in [5], we have shown the effectiveness of SRKDA for large scale concept detection problems. In addition to superior classification results when compared to existing approaches, it can provide an order of magnitude speed-up over support vector machine. The main computationally intensive operation is Cholesky decomposition, which is actually independent of the number of labels. For more details please refer to [5].

The total computational cost of SRKDA for all concepts in visual concept detection is $\frac{1}{6}m^3+m^2Nc$ flams where flam is a compound operation consisting of one addition and one multiplication and m is the number of samples. Compared to the cost of ordinary KDA for VCD, $(N\times(\frac{9}{2}m^3+m^2c))$ flams, SRKDA achieves an order of magnitude ($27N$ times) speed-up over KDA which is massive for large scale image/video datasets.

3   Submitted Runs

We have submitted five different runs described below. All runs use 72 kernels generated from different visual feature representations (2 sampling strategies, 9 different descriptor types and 4 spatial location grids). In this paper, we use only visual information. Future research includes usage of EXIF metadata provided for the photos. The main novelty is the classifier-level and kernel-level fusion using SRKDA with RBF/Power Chi-Squared kernels obtained from various image descriptors. It is worth mentioning that we have also evaluated the performance using SVM with the same kernels and based on the results from validation set, KDA is superior to SVM. These runs are described below:

3.1 RUN1: Classifier-Level Fusion Using RBF Kernels (CLF-KDA)

In general, the discriminatory power of kernel classifiers comes directly from the complexity of the underlying kernels. In this run, we have used standard RBF kernel with Chi-squared distance metric: $k(\boldsymbol{F}, \boldsymbol{F}') = e^{-\frac{1}{A}dist_{\chi^2}(\boldsymbol{F},\boldsymbol{F}')}$ where A is a scalar which normalises the distances. Following [6], A is set to the average χ^2 distance between all elements of the kernel matrix. Each kernel is then trained using SR-KDA with the regularization parameter, δ, tuned using the validation set. The output from each classifier is then combined using the AVG rule [11]. It is worth noting that for this run we have tried various combination rules such as MAX, MIN, MEDIAN. The best result on the validation set is obtained by the AVG rule and is reported here.

3.2 RUN2: Kernel-Level Fusion Using RBF Kernels (KLF-KDA)

In this run, the same RBF kernels with χ^2 distance as in RUN1 are used. However, instead of classifier level fusion, this run uses kernel level fusion with uniform weighting. This corresponds to taking the Cartesian product of the features spaces of the base kernels. Once the kernels are combined, kernel Fisher discriminant analysis is applied as the classifier.

3.3 RUN3: Stacked KDA

This run uses the classifier in RUN2 as a base classifier for each of the 53 concepts to produce 53 scores. These scores are used as feature vectors and another RBF kernel is built with these features. Note however, for some concepts, not all 53 scores are used for building this kernel. In cases where we have information about the correlation of the concepts, for example, for the disjoint concepts "single person", "small group", "big group", and "no persons", only the scores of the base classifiers for these 4 concepts are used. The new kernel is then added to the set of kernels and kernel FDA classifiers are trained in a second round.

3.4 RUN4: Classifier-Level Fusion Using Power Kernels (CLF-KDA-Power)

Conditional positive definite kernels have also drawn attention during the last decade and proved successful in image recognition using SVM [12]. In recent work [13], we have modified SRKDA to support conditional positive definite kernels such as power kernels. The main idea is to use LDL^T decomposition instead of Cholesky decomposition. For more details, please refer to [13]. In this run, we have used Power kernel with Chi-squared distance metric: $k(\boldsymbol{F}, \boldsymbol{F}') = -dist_{\chi^2}(\boldsymbol{F}, \boldsymbol{F}')^{\beta}$ (Conditional Positive Definite if $0 < \beta < 2$). Each power kernel is then trained using modified SRKDA with the regularization parameters δ and β tuned using the validation set. The output from each classifier is then combined using the AVG rule.

3.5 RUN5

Based on the performance on the validation set, this run selects the best of RUN2 and RUN3 for each concept.

4 Experimental Results

The ImageCLEF@ICPR dataset consists of 18000 images of 53 different object classes such as animals, vehicles, etc. The dataset is divided into a predefined "trainval" set (8000 images) and "test" set (10000 images). The "trainval" dataset is further divided for validation purpose into a training set containing 5000 images and a validation set containing 3000 images. The ground truth for the test sets is not released to avoid over-fitting of classifiers.

The Equal Error Rate (EER) and the Area under Curve (AUC) are used as measures for large-scale visual concept detection while an hierarchical measure is used to provide a score for the annotation performance for each image.

4.1 Results on Validation Set

We first evaluate the classifiers performance on the validation set using different techniques and then compare it to the state-of-the art systems that produced the top results in ImageCLEF@ICPR Challenge. Table 1 shows the performance of our runs including the best and worst descriptors. It is clear from the table that fusion of information either at classifier-level or kernel-level has significantly improved the performance. It is interesting to observe that while RBF-CLF has the best performance both in terms of mean AUC and EER, this run ranked top in only few concepts when compared to other submitted runs. Further, it should be noted that we have also tried to select the best combination of descriptors using search techniques such as Sequential Forward Search but were unable to get any improvement at all on the validation set. Since all of the classifiers contain complementary information, we have used all 9 descriptors with four spatial locations and 2 sampling strategies in our experiments.

Table 1. Comparison of different runs on ImageCLEF@ICPR Validation Set. Ind. Best Descriptor = DS-SIFT-1x1 for AUC, HS-SIFT-2x2 for EER. Ind. Worst Descriptor = DS-HSVSIFT-1x1 for AUC, DS-HSVSIFT-3x1 for EER.

Method	AUC	#WINs	EER	#WINs
Ind. Best	0.7843	-	0.2811	-
Ind. Worst	0.7347	-	0.3236	-
CLF-KDA	0.8424	5	0.2319	10
CLF-KDA-Power	0.8379	12	0.2348	15
KLF-KDA	0.8423	23	0.2319	13
Stacked KDA	0.8400	13	0.2324	15

4.2 Results on Test Set

Table 2 shows the performance of best run of each team evaluated independently by the organizers. The best performance using EER and AUC is achieved by our method based on classifier-level fusion using RBF Kernels. In fact the top 2 methods are clearly significantly better than all the other methods. Table 2 also shows the performance using the hierarchical measure in which our method (RUN5) ranked *third*. Technical details of the approaches by other groups are summarised in [2], ISIS approach is an extension of the system proposed in [4,14] where SIFT features are extracted in different colour spaces. The learning step is based on SVM with χ^2 kernel which differs from our system mainly where RBF/Power kernels with KDA is used in the classification stage. For 32 out of 53 individual concepts, we obtain the best performance of all submissions to this task when AUC is used as the evaluation criterion; more than twice when compared with second best method. For EER, the best performance is obtained in 29 out of the 53 individual concepts. These results clearly show the effectiveness of our system for large-scale visual concept detection. Future research aims to use ontology information and multi-label classification techniques that considers correlation among concepts to further improve the performance.

Table 2. The team runs of ImageCLEF@ICPR Photo Annotation Task (from the official evaluations). HM = Hierarchical measure.

Group	EER	#WINs	AUC	#WINs	HM
CVSSP	**0.2136**	29	**0.8600**	32	0.6900
ISIS	0.2182	17	0.8568	15	**0.7836**
IJS	0.2425	5	0.8321	3	0.7065
CNRS	0.2748	1	0.7927	2	0.4204
AVEIR	0.2848	0	0.7848	1	0.5602
MMIS	0.3049	0	0.7566	0	0.5027
LSIS	0.3106	0	0.7490	0	0.5067
UPMC/LIP6	0.3377	0	0.7159	0	0.4034
ITI	0.3656	1	0.5917	0	0.4023
MRIM	0.3831	0	0.6393	0	0.5801
TRS2008	0.4152	0	0.6200	0	0.3270
UAIC	0.4762	0	0.1408	0	0.6781

Table 3 shows the performance of our runs in terms of AUC on a few individual concepts. It is observed that the performance may vary in different concepts. The results indicate that RBF kernels perform quite well when class imbalance is not severe (for example in Day, No-Blur etc). On the other hand, in many highly unbalanced categories like Desert, Lake etc., Power Kernel performs quite well. In some concepts, stacking also has significant effect on the performance e.g. Fancy approx. a 4% improvement over the best run. It is observed that fusion at decision-level or feature-level yields very similar performance on this dataset with the results showing slightly in favour of the classifier-level fusion both in terms of EER and AUC. But the kernel-level fusion has speed advantage over the classifier-level fusion as only one classifier is required to train while the classifier-level fusion requires separate classifiers for the individual descriptors. The results also indicate that RBF-CLF (RUN1) ranked top in the majority of the concepts over other runs indicating that other runs may have overfitted during parameter optimization on the validation set. For RBF-CLF, the same regularisation

Table 3. Comparison of AUC for some individual concepts in ImageCLEF@ICPR Test Set. GT = Ground Truth.

Concept	GT	RUN1	RUN2	RUN3	RUN4
Desert	31	0.8752	0.8762	0.8762	**0.8977**
Lake	90	0.8991	0.8959	0.8959	**0.9122**
Overexposed	93	0.8165	0.8127	0.8201	**0.8276**
Spring	118	0.8257	0.8284	0.8284	**0.8410**
Snow	128	0.8925	**0.8846**	**0.8846**	0.8819
River	132	0.9210	0.9173	0.9173	**0.9264**
Autumn	136	0.8751	0.8724	0.8602	**0.8800**
Sports	146	0.7777	0.7741	0.7577	**0.7813**
Out-of-focus	148	0.8323	0.8279	0.8139	**0.8353**
Beach-Holidays	155	**0.9288**	0.9285	0.9285	0.9236
Canvas	178	**0.8522**	0.8503	0.8503	0.8467
Winter	210	**0.8749**	0.8712	0.8712	0.8633
Big-Group	222	0.9113	**0.9120**	**0.9120**	0.8966
Mountains	233	**0.9414**	0.9408	0.9212	0.9378
Sea	238	**0.9471**	0.9470	0.9247	0.9427
Motion-Blur	241	0.7842	0.7793	0.7806	**0.7849**
Food	269	**0.9156**	**0.9156**	**0.9156**	0.9089
Partylife	293	0.8587	0.8565	0.8565	**0.8647**
Flowers	382	0.8889	**0.8894**	0.8802	0.8792
Sunset-Sunrise	394	0.9617	**0.9619**	0.9587	0.9595
Underexposed	427	0.8682	0.8643	0.8609	**0.8730**
Vehicle	437	0.8548	**0.8550**	**0.8550**	0.8438
Night	566	**0.9071**	0.9063	0.9063	0.9046
Still-Life	659	**0.8584**	0.8575	0.8447	0.8476
Macro	705	**0.8395**	0.8379	0.8303	0.8360
Small-Group	721	**0.8273**	0.8270	0.8257	0.8195
Animals	727	**0.8722**	**0.8722**	0.8617	0.8536
Water	763	**0.9136**	0.9129	0.9001	0.9018
Summer	891	**0.8332**	0.8307	0.8307	0.8297
Trees	891	**0.9157**	0.9155	0.9155	0.9054
Building-Sights	896	**0.9028**	0.9025	0.9025	0.8922
Portrait	1007	0.9050	**0.9051**	**0.9051**	0.8945
Clouds	1107	0.9586	0.9597	**0.9598**	0.9505
Familiy-Friends	1115	0.8572	**0.8578**	**0.8578**	0.8451
Citylife	1151	0.8495	**0.8500**	**0.8500**	0.8378
Sunny	1159	**0.8118**	0.8090	0.8091	0.8027
Fancy	1174	0.5881	0.5839	**0.6100**	0.6051
Landscape-Nature	1362	**0.9517**	0.9511	0.9511	0.9460
Aesthetic-Impression	1408	0.6620	0.6582	0.6673	**0.6706**
No-Visual-Place	1578	**0.8108**	0.8101	0.8079	0.8020
Single-Person	1701	0.8184	0.8192	**0.8342**	0.8019
Overall-Quality	1719	0.6518	0.6483	**0.6605**	**0.6605**
Plants	1872	0.9028	**0.9031**	**0.9031**	0.8841
Sky	1977	0.9582	0.9582	**0.9587**	0.9475
Indoor	2162	0.8412	0.8412	**0.8436**	0.8243
Partly-Blurred	2337	**0.8674**	0.8681	0.8658	0.8540
No-Visual-Time	3121	**0.8559**	0.8557	0.8557	0.8406
Outdoor	4260	0.9018	0.9022	**0.9041**	0.8824
Day	4313	**0.8660**	0.8656	0.8600	0.8509
No-Blur	5274	**0.8578**	0.8573	0.8573	0.8432
No-Persons	5357	0.8718	**0.8739**	**0.8739**	0.8495
No-Visual-Season	6645	**0.8221**	0.8217	0.8217	0.8100
Neutral-Illumination	7480	0.7974	0.7950	0.7893	**0.7991**
Mean		**0.8600**	0.8588	0.8534	0.8547
#WINs		23	12	16	14

parameter, $\delta = 0.1$, is used for all concepts while for RBF-KLF/Stacking, δ is tuned for every concept. Similarly, for power kernel, β is also tuned along with δ on the validation set.

5 Conclusions

Our focus on machine learning methods for concept detection in ImageCLEF@ICPR has proved successful. Our method ranked top for the large-scale visual concept detection task in terms of both EER and AUC. For 32 out of 53 individual concepts, we obtained the best performance of all submissions addressing this task. The main novelty is the use of classifier-level and kernel-level fusion with Kernel Discriminant Analysis employing RBF/Power Chi-Squared kernels obtained from various image descriptors. Future work aims to combine ontology (hierarchy and relations) with visual information to improve the performance.

Acknowledgements

This work was supported by EU Vidi-Video project.

References

1. Nowak, S., Dunker, P.: Overview of the CLEF 2009 large scale visual concept detection and annotation task. In: Peters, C., Caputo, B., Gonzalo, J., Jones, G.J.F., Kalpathy-Cramer, J., Müller, H., Tsikrika, T. (eds.) CLEF 2009. LNCS, vol. 6242, pp. 94–109. Springer, Heidelberg (2010)
2. Nowak, S.: Imageclef@ICPR contest: Challenges, methodologies and results of the photoannotation task. In: Ünay, D., Çataltepe, Z., Aksoy, S. (eds.) ICPR 2010. LNCS, vol. 6388, pp. 140–153. Springer, Heidelberg (2010)
3. Sivic, J., Zisserman, A.: Video google: a text retrieval approach to object matching in videos. In: Proc. of the ICCV (2003)
4. van de Sande, K.E.A., Gevers, T., Snoek, C.G.M.: Evaluating color descriptors for object and scene recognition. PAMI (in press, 2010)
5. Tahir, M.A., Kittler, J., Mikolajczyk, K., Yan, F., van de Sande, K.E.A., Gevers, T.: Visual category recognition using spectral regression and kernel discriminant analysis. In: Proc. of the 2nd International Workshop on Subspace 2009, In Conjunction with ICCV 2009, Kyota, Japan (2009)
6. Zhang, J., Marszałek, M., Lazebnik, S., Schmid, C.: Local features and kernels for classification of texture and object categories: A comprehensive study. IJCV 73(2), 213–238 (2007)
7. Mikolajczyk, K., Schmid, C.: A performance evaluation of local descriptors. PAMI 27(10), 1615–1630 (2005)
8. Lowe, D.G.: Distinctive image features from scale-invariant keypoints. IJCV 60(2), 91–110 (2004)
9. van Gemert, J.C., Veenman, C.J., Smeulders, A.W.M., Geusebroek, J.M.: Visual word ambiguity. PAMI (2009) (in press)

10. Cai, D., He, X., Han, J.: Efficient kernel discriminat analysis via spectral regression. In: Proc. of the ICDM (2007)
11. Kittler, J., Hatef, M., Duin, R.P.W., Matas, J.: On combining classifiers. PAMI 20(3), 226–239 (1998)
12. Boughorbel, S., Tarel, J.P., Boujemaa, N.: Conditionally positive definite kernels for SVM based image recognition. In: Proc. of ICME, Amsterdam, The Netherlands (2005)
13. Tahir, M.A., Kittler, J., Yan, F., Mikolajczyk, K.: Kernel discriminant analysis using triangular kernel for semantic scene classification. In: Proc. of the 7th International Workshop on CBMI, Crete, Greece (2009)
14. van de Sande, K., Gevers, T., Smeulders, A.: The University of Amsterdams Concept Detection system at ImageCLEF 2009. In: Peters, C., Caputo, B., Gonzalo, J., Jones, G.J.F., Kalpathy-Cramer, J., Müller, H., Tsikrika, T. (eds.) CLEF 2009. LNCS, vol. 6242, pp. 261–268. Springer, Heidelberg (2010)

Overview of the ImageCLEF@ICPR 2010 Robot Vision Track*

Andrzej Pronobis[1], Henrik I. Christensen[2], and Barbara Caputo[3]

[1] Centre for Autonomous Systems, The Royal Institute of Technology,
Stockholm, Sweden
pronobis@kth.se
[2] Georgia Institute of Technology, Atlanta, GA, USA
hic@cc.gatech.edu
[3] Idiap Research Institute, Martigny, Switzerland
bcaputo@idiap.ch
http://www.imageclef.org/2010/ICPR/RobotVision

Abstract. This paper describes the robot vision track that has been proposed to the ImageCLEF@ICPR2010 participants. The track addressed the problem of visual place classification. Participants were asked to classify rooms and areas of an office environment on the basis of image sequences captured by a stereo camera mounted on a mobile robot, under varying illumination conditions. The algorithms proposed by the participants had to answer the question "where are you?" (I am in the kitchen, in the corridor, etc) when presented with a test sequence imaging rooms seen during training (from different viewpoints and under different conditions), or additional rooms that were not imaged in the training sequence. The participants were asked to solve the problem separately for each test image (obligatory task). Additionally, results could also be reported for algorithms exploiting the temporal continuity of the image sequences (optional task). A total of eight groups participated to the challenge, with 25 runs submitted to the obligatory task, and 5 submitted to the optional task. The best result in the obligatory task was obtained by the Computer Vision and Geometry Laboratory, ETHZ, Switzerland, with an overall score of 3824.0. The best result in the optional task was obtained by the Intelligent Systems and Data Mining Group, University of Castilla-La Mancha, Albacete, Spain, with an overall score of 3881.0.

Keywords: Place recognition, robot vision, robot localization.

---

* We would like to thank the CLEF campaign for supporting the ImageCLEF initiative. B. Caputo was supported by the EMMA project, funded by the Hasler foundation. A. Pronobis was supported by the EU FP7 project ICT-215181-CogX. The support is gratefully acknowledged.

D. Ünay, Z. Çataltepe, and S. Aksoy (Eds.): ICPR 2010, LNCS 6388, pp. 171–179, 2010.

1 Introduction

ImageCLEF[1] [1, 2, 3] started in 2003 as part of the Cross Language Evaluation Forum (CLEF[2], [4]). Its main goal has been to promote research on multi-modal data annotation and information retrieval, in various application fields. As such it has always contained visual, textual and other modalities, mixed tasks and several sub tracks.

The robot vision track has been proposed to the ImageCLEF participants for the first time in 2009. The track attracted a considerable attention, with 19 inscribed research groups, 7 groups eventually participating and a total of 27 submitted runs. The track addressed the problem of visual place recognition applied to robot topological localization. Encouraged by this first positive response, the track has been proposed for the second time in 2010, within the context of the ImageCLEF@ICPR2010 initiative. In this second edition of the track, participants were asked to classify rooms and areas on the basis of image sequences, captured by a stereo camera mounted on a mobile robot within an office environment, under varying illumination conditions. The system built by the participants had to be able to answer the question "where are you?" when presented with a test sequence imaging rooms seen during training (from different viewpoints and under different conditions) or additional rooms, not imaged in the training sequence.

The image sequences used for the contest were taken from the previously unreleased COLD-Stockholm database. The acquisition was performed in a subsection of a larger office environment, consisting of 13 areas (usually corresponding to separate rooms) representing several different types of functionality. The appearance of the areas was captured under two different illumination conditions: in cloudy weather and at night. Each data sample was then labeled as belonging to one of the areas according to the position of the robot during acquisition (rather than contents of the images).

The challenge was to build a system able to localize semantically (I'm in the kitchen, in the corridor, etc.) when presented with test sequences containing images acquired in the previously observed part of the environment, or in additional rooms that were not imaged in the training sequences. The test images were acquired under different illumination settings than the training data. The system had to assign each test image to one of the rooms that were present in the training sequences, or to indicate that the image comes from a room that was not included during training. Moreover, the system could refrain from making a decision (e.g. in the case of lack of confidence).

We received a total of 30 submission, of which 25 were submitted to the obligatory task and 5 to the optional task. The best result in the obligatory task was obtained by the Computer Vision and Geometry Laboratory, ETHZ, Switzerland. The best result in the optional task was obtained by the Intelligent

---

[1] http://www.imageclef.org/
[2] http://www.clef-campaign.org/

Systems and Data Mining Group (SIMD) of the University of Castilla-La Mancha, Albacete, Spain.

This paper provides an overview of the robot vision track and reports on the runs submitted by the participants. First, details concerning the setup of the robot vision track are given in Section 2. Then, Section 3 presents the participants and Section 4 provides the ranking of the obtained results. Conclusions are drawn in Section 5. Additional information about the task and on how to participate in the future robot vision challenges can be found on the ImageCLEF web pages.

2 The RobotVision Track

This section describes the details concerning the setup of the robot vision track. Section 2.1 describes the dataset used. Section 2.2 gives details on the tasks proposed to the participants. Finally, section 2.3 describes briefly the algorithm used for obtaining a ground truth and the evaluation procedure.

2.1 Dataset

Three datasets were made available to the participants. Annotated training and validation data were released when the competition started. Unlabeled testing set was released two weeks before the results submission deadline. The training, validation and test sets consisted of a subset of the previously unreleased COLD-Stockholm database. The sequences were acquired using the MobileRobots PowerBot robot platform equipped with a stereo camera system consisting of two Prosilica GC1380C cameras (Figure 1). In order to facilitate the participation of those groups not familiar with stereo images, we allowed the participants to sue monocular as well as stereo image data. The acquisition was performed in a subsection of a larger office environment, consisting of 13 areas (usually corresponding to separate rooms) representing several different types of functionality (Figure 2).

The appearance of the areas was captured under two different illumination conditions: in cloudy weather and at night. The robot was manually driven through the environment while continuously acquiring images at a rate of 5fps. Each data sample was then labeled as belonging to one of the areas according to the position of the robot during acquisition, rather than according to the content of the images.

Four sequences were selected for the contest: two training sequences having different properties, one sequence to be used for validation and one sequence for testing. Each of these four sequences had the following properties:

- *training-easy.* This sequence was acquired in 9 areas, during the day, under cloudy weather. The robot was driven through the environment following a similar path as for the test and validation sequences and the environment was observed from many different viewpoints (the robot was positioned at multiple points and performed 360 degree turns).

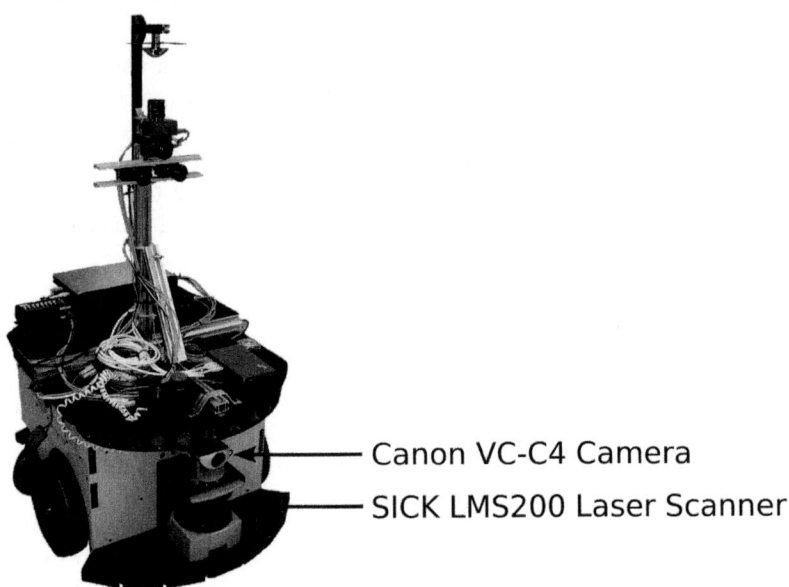

Canon VC-C4 Camera

SICK LMS200 Laser Scanner

Fig. 1. The MobileRobots PowerBot mobile robot platform used for data acquisition

- *training-hard.* This sequence was acquired in 9 areas, during the day, under cloudy weather. The robot was driven through the environment in a direction opposite to the one used for the training-easy sequence, without making additional turns.
- *validation.* This sequence was acquired in 9 areas, at night. A similar path was followed as for the training-easy sequence, without making additional turns.
- *testing.* This sequence was acquired in similar conditions and following similar path as in case of the validation sequence. It contains four additional areas,for a total of 13, that were not imaged in the training or validation sequences: elevator, workshop, living room and laboratory. Exemplar images for these rooms are shown in Figure 3.

As an additional resource, we made available to participants the camera calibration data for the stereo image sequences.

2.2 The Task

The overall goal of the robot vision track is to stimulate research on semantic place recognition for mobile robot localization. The problem can be mapped to an image annotation task, where participants have to recognize the room type (kitchen, a corridor') on the basis of images acquired with a stereo camera, mounted on a mobile robot platform.

Fig. 2. Example pictures of nine rooms used for the robot vision task at ICPR 2010. From left to right, top: corridor, kitchen, large office 1. From left to right, middle: large office 2, student office, printer area. From left to right, bottom: elevator 1, small office 2, large office 2.

Fig. 3. Example pictures of the four additional rooms in the test sequence, used for the robot vision task at ICPR 2010. From left to right: elevator2, workshop, living room and laboratory.

All data, consisting of training, validation and test sequences, were recorded using a mobile robot, manually driven through several rooms, under fixed illumination conditions. The environment of choice was a standard academic office environment. Images in the sequences were annotated according to the position of the robot, as opposed to their informative content. For instance, an image acquired in the room 'corridor', taken when the robot was facing the entrance of the room 'kitchen', is labeled as 'corridor' even if it shows mostly part of the 'kitchen'.

The test sequences were acquired under different illumination conditions. They imaged the same rooms contained into the training and validation sequences, plus some additional rooms not seen before. Therefore participants had to address at the same time two challenges: (a) recognizing correctly rooms seen before, and (b) recognizing as 'unknown' the new rooms in the test sequence.

We consider two separate tasks, *task 1* (obligatory) and *task 2* (optional). In task 1, the algorithm had to be able to provide information about the location of the robot separately for each test image, without relying on information contained in any other image (e.g. when only some of the images from the test sequences are available or the sequences are scrambled). This corresponds to the problem of global topological localization. In task 2, the algorithm was allowed to exploit continuity of the sequences and rely on the test images acquired before the classified image, with the constraint that images acquired after the classified image could be used. The same training, validation and testing sequences were used for both tasks. The reported results were compared separately.

The tasks employed two sets of training, validation and testing sequences. The first, easier set contained sequences with constrained viewpoint variability. In this set, training, validation and testing sequences were acquired following similar path through the environment. The second, more challenging set contained sequences acquired following different paths (e.g. the robot was driven in the opposite direction). The final score for each task was calculated based on the results obtained for both sets.

The competition started with the release of annotated training and validation data. Moreover, the participants were given a tool for evaluating performance of their algorithms. The test image sequences were released later and were acquired in the same environment, under different conditions. They also contained additional rooms that were not imaged previously.

2.3   Ground Truth and Evaluation

The image sequences used in the competition were annotated with ground truth. The annotations of the training and validation sequences were available to the participants, while the ground truth for the test sequence was released after the results were announced. Each image in the sequences was labelled according to the position of the robot during acquisition as belonging to one of the rooms used for training or as an unknown room. The ground truth was then used to calculate a score indicating the performance of an algorithm on the test sequence. The following rules were used when calculating the overall score for the whole test sequence:

- 1 point was granted for each correctly classified image.
- Correct detection of an unknown room was regarded as correct classification.
- 0.5 points was subtracted for each misclassified image.
- No points were granted or subtracted if an image was not classified (the algorithm refrained from the decision).

A script was available to the participants that automatically calculated the score for a specified test sequence given the classification results produced by an algorithm. Each of the two test sequences consisted of a total of 2551 features. Therefore, according to the rules listed above, the maximum possible score is of 2551, both for the easy and the hard test sequences, with a maximum overall score of 5102.

3 Participation

In 2010, 28 groups registered to the Robot Vision task. 8 of them submitted at least one run, namely:

- CVG: Computer Vision and Geometry laboratory, ETH Zurich, Switzerland;
- TRS2008: Beijing Information Science and Technology University, Bejing, China;
- SIMD: Intelligent Systems and Data Mining Group, University of Castilla-La Mancha, Albacete, Spain;
- CAS IDIAP: Center for Autonomous Systems, KTH, Stockholm, Sweden and Idiap Research Institute, Martigny, Switzerland;
- PicSOM TKK: Helsinki University of Technology, TKK Department of Information and Computer Science, Helsinki, Finland;
- Magrit: INRIa Nancy, France;
- RIM at GT: Georgia Institute of Technology, Atlanta, Georgia, USA;

A total of 30 runs were submitted, with 25 runs submitted to the obligatory task and 5 runs submitted to the optional task. In order to encourage participation, there was no limit to the number of runs that each group could submit.

4 Results

This section presents the results of the robot vision track of ImageCLEF@ ICPR2010. Table 1 shows the results for the obligatory task, while Table 2 shows the result for the optional task. Scores are presented for each of the submitted runs that complied with the rules of the contest.

Table 1. Results obtained by each group in the obligatory task. The maximum overall score is of 5102, with a maximum score of 2551 for both the easy and the had sequence.

#	Group	Overall Score	Score Easy	Score Hard
1	CVG	3824.0	2047.0	1777.0
2	TRS2008	3674.0	2102.5	1571.5
3	SIMD	3372.5	2000.0	1372.5
4	CAS IDIAP	3344.0	1757.5	1372.5
5	PicSOM TKK	3293.0	2176.0	1117.0
6	Magrit	3272.0	2026.0	1246.0
7	RIM at GT	2922.5	1726.0	1196.5
8	UAIC	2283.5	1609.0	674.5

Table 2. Results obtained by each group in the optional task. The maximum overall score is of 5102, with a maximum score of 2551 for both the easy and the had sequence.

#	Group	Overall Score	Score Easy	Score Hard
1	SIMD	3881.0	2230.5	1650.5
2	TRS2008	3783.5	2135.5	1648.0
3	CAS IDIAP	3453.5	1768.0	1685.5
4	RIM at GT	2822.0	1589.5	1232.5

We see that the majority of runs were submitted to the obligatory task. A possible explanation is that the optional task requires a higher expertise in robotics that the obligatory task, which therefore represents a very good entry point. The same behavior was noted at the first edition of the robot vision task in 2009.

Concerning the obtained results, we notice that all groups perform considerably better on the easy sequence, compared to the hard sequence. This is true for both the obligatory and the optional task. For the obligatory task, the best performance on the easy sequence was obtained by the PicSOM TKK group with a score of 2176.0. This is only 375 points lower than the maximum possible score of 2551, also considering that the images in the sequence are annotated according to the robots position, but they are classified according to their informative content. As opposed to this, we see that the best performance on the easy sequence was obtained by the CVG group, with a score of 1777.0. This is 774 points lower than the maximum possible score of 2551, more than twice the difference in score between the nest performance and the maximum one for the easy sequence.

This pattern is replicated in the optional task, indicating that the temporal continuity between image frames does not seem to alleviate the problem. We see that the best performance for the easy sequence is obtained by the SIMD group, with a score of 2230.5. For the hard sequence, the best performance (obtained by the CAS IDIAP group) drops to 1685.5.

These results indicate quite clearly that the capability to recognize visually a place under different viewpoints is still an open challenge for mobile robots. This is a strong motivations towards proposing similar tasks to the community in the future editions of the robot vision task.

5   Conclusions

The robot vision task at ImageCLEF@ICPR2010 attracted a considerable attention and proved an interesting complement to the existing tasks. The approach presented by the participating groups were diverse and original, offering a fresh take on the topological localization problem. We plan to continue the task in the next years, proposing new challenges to the participants. In particular, we plan to focus on the problem of place categorization and use objects as an important source of information about the environment.

References

1. Clough, P., Müller, H., Deselaers, T., Grubinger, M., Lehmann, T.M., Jensen, J., Hersh, W.: The CLEF 2005 cross–language image retrieval track. In: Peters, C., Gey, F.C., Gonzalo, J., Müller, H., Jones, G.J.F., Kluck, M., Magnini, B., de Rijke, M., Giampiccolo, D. (eds.) CLEF 2005. LNCS, vol. 4022, pp. 535–557. Springer, Heidelberg (2006)
2. Clough, P., Müller, H., Sanderson, M.: The CLEF cross–language image retrieval track (ImageCLEF) 2004. In: Peters, C., Gey, F.C., Gonzalo, J., Müller, H., Jones, G.J.F., Kluck, M., Magnini, B., de Rijke, M., Giampiccolo, D. (eds.) CLEF 2005. LNCS, vol. 4022, pp. 597–613. Springer, Heidelberg (2006)
3. Müller, H., Deselaers, T., Kim, E., Kalpathy-Cramer, J., Deserno, T.M., Clough, P., Hersh, W.: Overview of the imageCLEFmed 2007 medical retrieval and medical annotation tasks. In: Peters, C., Jijkoun, V., Mandl, T., Müller, H., Oard, D.W., Peñas, A., Petras, V., Santos, D. (eds.) CLEF 2007. LNCS, vol. 5152, pp. 473–491. Springer, Heidelberg (2008)
4. Savoy, J.: Report on CLEF-2001 experiments. In: Peters, C., Braschler, M., Gonzalo, J., Kluck, M. (eds.) CLEF 2001. LNCS, vol. 2406, pp. 27–43. Springer, Heidelberg (2002)

Methods for Combined Monocular and Stereo Mobile Robot Localization

Friedrich Fraundorfer[1], Changchang Wu[2], and Marc Pollefeys[1]

[1] Department of Computer Science, ETH Zürich, Switzerland
{fraundorfer,marc.pollefeys}@inf.ethz.ch
[2] Department of Computer Science,University of North Carolina at Chapel Hill, USA
ccwu@cs.unc.edu

Abstract. This paper describes an approach for mobile robot localization using a visual word based place recognition approach. In our approach we exploit the benefits of a stereo camera system for place recognition. Visual words computed from SIFT features are combined with VIP (viewpoint invariant patches) features that use depth information from the stereo setup. The approach was evaluated under the ImageCLEF@ICPR 2010 competition[1]. The results achieved on the competition datasets are published in this paper.

1 Introduction

The ImageCLEF@ICPR 2010 competition was established to provided a common testbed for vision based mobile robot localization, to be able to evaluate different approaches against each other. For the competition image datasets of a realistic indoor scenario were created and manually labeled to get ground truth data. The mobile robot was equipped with a stereo vision system, that generates an image pair for each location instead of a single image only. This availability of image pairs from a stereo vision system allowed us to design an approach that combines monocular and stereo vision cues. The approach we designed is based on a place recognition system using visual words [7,3]. For one part visual words are computed from SIFT features [5] as the monocular cue. As the other cue we use visual words computed from viewpoint invariant patches (VIP) [13]. For the extraction of VIP features the local geometry of the scene needs to be known. In our case we compute dense stereo from the stereo system and use the depth map for VIP feature extraction. Both sets of features are combined and used as visual description of the location. The approach has already been evaluated and the scores achieved in the competition are given in this paper. Additionally, recognition rates on the *validation* competition datasets (which was used to prepare for the competition) are given, which demonstrate the excellent performance of our method by achieving 98% correct localization.

---

[1] This approach was ranked 1st in the ImageCLEF@ICPR 2010 RobotVision competition.

D. Ünay, Z. Çataltepe, and S. Aksoy (Eds.): ICPR 2010, LNCS 6388, pp. 180–189, 2010.

2   Related Work

Our visual word based place recognition system is related to [7,3,12,6] where a similar technique was used for image retrieval. It is also related to FABMAP [2], a visual word based approach to robot localization. However, FABMAP focuses on a probabilistic framework to identify matching locations, whereas we do a two-stage approach of visual ranking and geometric verification. The proposed geometric verification takes the planar motion constraints of a mobile robot into account. It is also related to [1], however they use a different technique of quantizing local features into visual words.

VIP features were firstly described in [13] and used for registration of 3D models. The local geometry was computed using a monocular structure-from-motion algorithm. In our approach we compute VIP features from a stereo system and use them for mobile robot localization. First results of this approach were already published in our previous work [4] and this work now explains the method in detail.

3   SIFT and VIP Features for Place Recognition

The availability of depth maps for every image pair of a stereo video sequence makes it possible to use viewpoint invariant patches (VIP). VIP's are image features extracted from images that are rectified with respect to the local geometry of the scene. The rectified texture can be seen as an ortho-texture of the 3D model which is viewpoint independent. This ortho-texture is computed using the depth map from the stereo system. This rectification step, which is the essential part of this concept delivers robustness to changes of viewpoint. We then determine the salient feature points of the ortho-textures and extract the feature description. For this the well known SIFT-features and their associated descriptor [5] is used. The SIFT-features are then transformed to a set of VIPs, made up of the feature's 3D position, patch scale, surface normal, local gradient orientation in the patch plane, in addition to the SIFT descriptor. Fig. 1 illustrates this concept. The original feature patches are the lower left and right patches, as seen from the gray and green camera. Rectification is performed by changing the viewpoints to the red cameras, so that the camera's image plane is parallel to the features scene plane. This results in the rectified VIP patches which are the center patches. It can be seen, that because of the rectification the VIP patches overlap perfectly.

Because of their viewpoint invariance, VIP features are a perfect choice for place recognition. For place recognition VIP features can be used instead of SIFT features from the original images or in addition to SIFT features (this is beneficial if the local geometry cannot be computed for the whole sequence). With view point invariant features place recognition will be possible with even large view point changes.

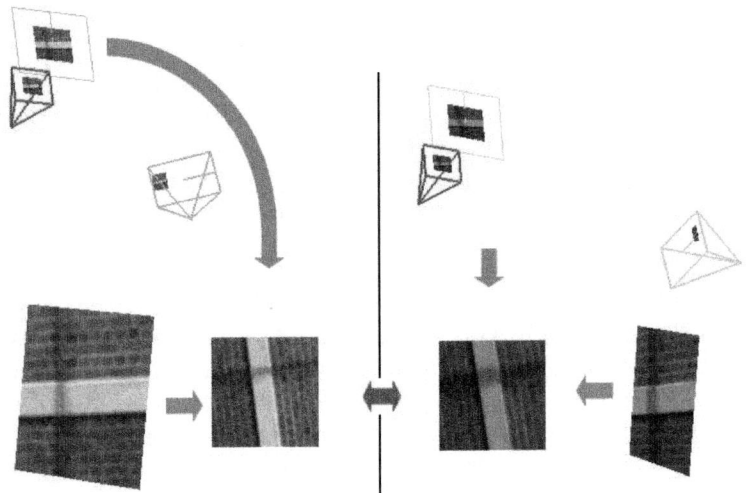

Fig. 1. Two corresponding viewpoint invariant patches (VIPs). The lower left and right patches are the original feature patches, while the center patches are the rectified VIP patches (see text for details).

4   Place Recognition and Verification

Robot localization can be phrased as a place recognition problem as described in [11]. The camera path is split up into distinct locations and the visual appearance of each location is described by visual features. A database of the environment is created holding the visual appearance of each location together with the actual coordinates of the location, and a label is assigned to each location. On performing global localization the current view of the robot is compared to all views in the database. The location with the most similar appearance is returned and the robot now knows its location up to the accuracy of the stored locations. For an efficient database search a visual word based approach is used. The approach quantizes a high-dimensional feature vector (in our case SIFT and VIP) by means of hierarchical k-means clustering, resulting in a so called hierarchical vocabulary tree. The quantization assigns a single integer value, called a visual word, to the originally high-dimensional feature vector. This results in a very compact image representation, where each location is represented by a list of visual words, each only of integer size. The list of visual words from one location forms a document vector which is a v-dimensional vector where v is the number of possible visual words (a typical choice would be $v = 10^6$). The document vector is a histogram of visual words normalized to 1. To compute the similarity matrix the L_2 distance between all document vectors is calculated. The document vectors are naturally very sparse and the organization of the database as an inverted file structure makes this very efficient. This scheme is illustrated in Fig. 2.

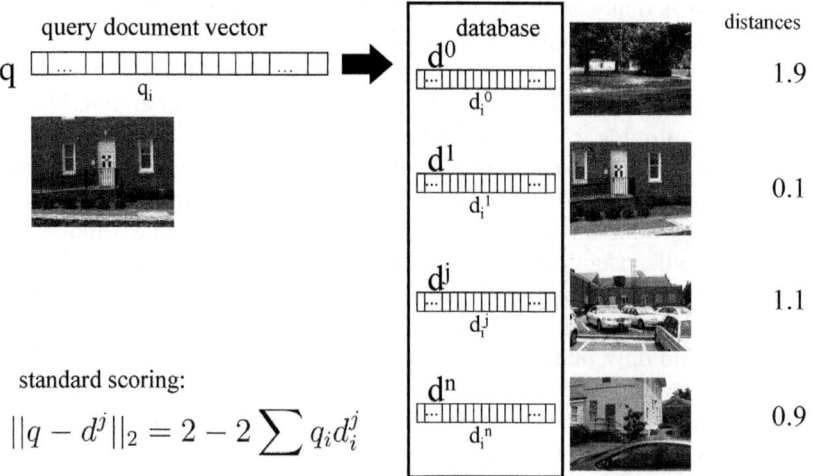

Fig. 2. Illustration of the visual place recognition system. A query image is represented by a query document vector (a set of visual words). A query works by computing the L_2-distance between the query document vector and all the document vectors in a database which represent the stored images, i.e. places. The L_2-distance is used as similarity score and is in the range of $[0, 2]$. A similarity score close to zero stands for a very similar image, while a similarity score close to 2 stands for a different image. The computation of the similarity score is using an inverted file structure for efficiency.

In our case robot localization is a 2-stage approach. First a similarity ranking is performed using the visual words, afterwards a geometric verification step tests the top-n results. Geometric verification is a very powerful cue. For each visual word the 2D image coordinates are stored, too. This makes it possible to compute the epipolar geometry between each database image and the query image. Only results that fulfill this epipolar constraint are considered.

5   Algorithm

This section describes the different steps of the localization algorithm.

5.1   Feature Extraction

Our approach uses SIFT features and VIP features. For the SIFT features only one image of the stereo pair is used. To extract SIFT features we use the implementation of Vedaldi[2]. This implementation is not real-time, but if real-time is required one could use GPU-Sift[3]. For every extracted feature we store the x, y position and the descriptor. Rotation and scale are not used in further processing.

---

[2] http://www.vlfeat.org/ vedaldi/code/sift.html
[3] http://www.cs.unc.edu/ ccwu/siftgpu/

The descriptor is quantized into a 128-dimensional byte-vector. The influence of this quantization for matching is basically negligible.

To extract VIP features a dense depth map is created first for each image pair by stereo matching. This is done by scan-line based stereo matching using dynamic programming [8]. As similarity measure the sum-of-absolute-differences (SAD) within a 9×9 pixel window is used. For efficient matching a maximum disparity range of 50 pixel is defined. To deal with textureless regions in the image the gray value variance is computed for each matching window. If the variance is smaller than a threshold the depth value is not used. This means that in textureless regions no VIP features are extracted. The next step is the detection of scene planes from the depth map. A RANSAC based plane detection is performed and only planes larger than a certain threshold are retained. Next, each detected scene plane is transformed into an orthographic view by using the GPU's texture warping function. For these rectified image regions SIFT features are extracted. VIP features are now composed of the location, rotation and descriptor of the SIFT features and of the 3D plane parameters.

For further processing both feature sets are combined. For this we use only x, y position and the descriptor of SIFT and VIP. This can be done as the descriptor part of VIP is a basic SIFT descriptor.

5.2 Quantization into Visual Words

For efficient localization the extracted SIFT descriptors are quantized into visual words. This is done by using a hierarchical vocabulary tree. Our vocabulary tree was created by k-means clustering of 100000 images from the web. We intentionally did not train on the images of the localization task. This means, we can do localization without the need of getting specific data of the environment first. The experiments showed that such a generically trained vocabulary tree will work sufficiently well. The vocabulary tree had a branching factor of 10 and 6 levels, which leads to 10^6 distinct visual words.

5.3 Visual Word Scoring

The visual words from a single image form the document vector. The document vector is a histogram of the visual words of the image. Each entry of the document vector is weighted by its importance, using the inverse document frequency (IDF) term. Finally the document vector is normalized to an L_2 norm of 1. The IDF weight for a visual word is computed by $w_i = \frac{|D|}{|D_i|}$, where $|D_i|$ is the number of documents in the database in which visual word i occurs and $|D|$ is the total number of documents in the database. Localization is done by searching for the most similar document vector d^j in the database to the query vector q. For this the L_2 distances between all the document vectors in the database and the query vector q are computed. The document vectors have the dimension of 10^6 (the number of visual words) but are very sparse. Each only contains as much non-zero entries as features in the image. Thus an inverted file structure can be used for efficient computation of the distances. In the inverted file structure for every visual word a list of images is stored in which this visual word appeared. A scoring vector s is initialized (to 0) that has as many entries as images in the database. A visual word from the query image is taken and the list of images from the inverted file is processed. For each image in the list a vote is cast by adding $q_i d_i^j$ to the corresponding position in the scoring array. The L_2 distance between q and d^j is then $||q - d^j||_2 = 2 - 2s_j$, if all the document vectors are normalized to 1. After sorting the scoring vector we get a similarity ranking of all the images in the database. Using the most similar image is denoted as the standard scoring method in the later experiments.

5.4   Geometric Verification

For geometric verification we compute a re-ranking of the top-n images based on a geometric criteria. The query image and the retrieved image need to fulfill the epipolar constraint to be accepted and for localization we select the one match that gives the most point correspondences. For this we perform SIFT feature matching between the query image and the top-n retrieved images. The SIFT features are already extracted in the previous step and just have to be read back in from hard disk. Next we compute the essential matrix from the point matches using a RANSAC scheme and use the number of inliers to RANSAC as the geometry score. The images are then re-ranked according to the geometry score. As the robot is performing planar motion only we can use this constraint for the essential matrix estimation by using the linear 3-pt algorithm [9]. If the geometry score is below a certain threshold we assume that the query image has no match in the database. This is used to identify the unknown areas in the localization task.

6   Evaluation

The evaluation of the algorithm was done within the framework of the Image-CLEF@ICPR 2010 Robotvision competition. For this competition a previously unpublished set of the COLD database [10] was used. The Robotvision competition consisted of two independent tasks for place recognition, *task1* and *task2*. *Task1* was conducted in a large office environment, which was divided into 13 distinct areas, which got different labels assigned. The goal for the robot was, to answer the question in which area it is given an input image, by assigning the correct label to the input image. For the competition, labeled image data was provided for training and validation and an unlabeled data set (*testing*) on which the competition was carried out. After the competition the ground truth labels for the *testing* set were released to the participants. The data sets were acquired with a stereo set consisting of two Prosilica GC1380C cameras mounted on a MobileRobots PowerBot robot platform. For the two training sets *training_easy* and *training_hard* the robot was driven through the environment and the captured images were labeled with the area code. The *training_easy* set consists of more images and more viewpoints than the *training_hard* set. The *validation* set comes from a run through the environment at a different time and was also labeled. Both training sets and also the validation set include only 9 of the 13 areas. With the use of a stereo system an image pair is available for each location which allows the use of depth information. However this was not required in the competition. The goal of *task1* was to label each image pair of the *testing* set with the correct area code. The *testing* set consists of 2551 image pairs taken at a different time and includes all 13 areas. This means that the 4 areas not included in the training sets needed to be labeled as "Unknown area". For *task1* each image pair had to be labeled independently without using knowledge from the labeling of the previous image pair. *Task2*, an optional task, was very similar to *task1*, however here it was possible to include sequential information into the labeling. Thus this *task2* is considerably easier than *task1*.

In the following we will present and discuss the recognition rates on the *validation* set and compare it to the ground truth and give the score achieved on the *testing* set in the competition. Table 1 shows the results on the *validation* set (2392 images) where ground truth data as labels is available. We measured recognition scores using standard scoring and standard scoring with geometric verification on both training sets. For the recognition score we compare the label of the top-ranked image (denoted as standard scoring) with the ground truth label and compute the number of correctly labeled images. For geometric verification the top-50 images from standard scoring get re-ranked according to the number of inliers to the epipolar constraint. Using the *training_easy* set the recognition rate with standard scoring was 96% and this number increased to 98% with geometric scoring. For the *training_hard* set the recognition rate with standard scoring was 87%. This number increased to 92% with subsequent geometric verification. Interestingly the recognition rates with standard scoring are already very high, geometric verification seems to give only a small improvement. However, only after geometric verification can one be sure that the database image really matches the query image. Standard scoring provides a ranking but the similarity measure does not guarantee that the top ranked image is really the matching one, e.g. if the query image is not in the database at all.

Table 2 shows the competition scores achieved on the *testing* set. The score is computed as follows:

- +1.0 points for a correctly classified image (includes the correct detection of an unknown location).
- -0.5 points for a misclassified image
- 0 points for an image that was not classified

The maximal achievable score would be 5102 points as the sum of the two 2551 points for each of the individual training sets. The top-10 ranked images are used for geometric verification. Image matches with less than 50 inlier matches were classified as "Unknown" location. We denote this parameterization as the "Competition method". Every image of the database was classified to either an area or the "Unknown" class, the option of refraining from a decision was not used. Table 3 shows the recognition rates for each individual class of the *testing* set. The table shows that some classes seem to be easier (100% recognition rate for the "Lab" class) while others seem to be harder. The table also confirms that

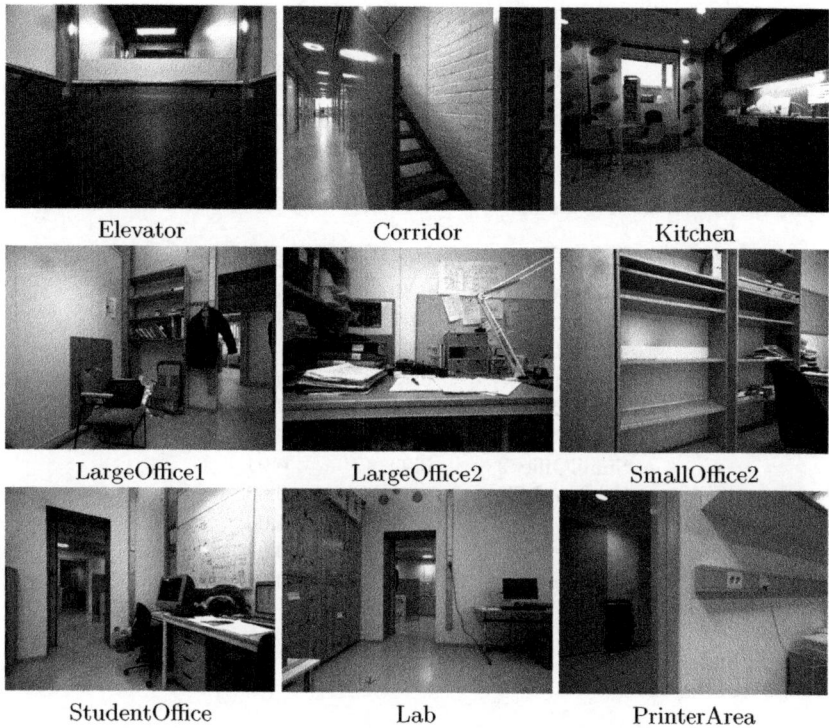

Fig. 3. Example images of the 9 classes in the training sets

the "Unknown" class is troublesome for our method, which has low recognition rates.

Finally we would like to give some runtime measurements from a 2.4GHz Intel Quadcore. An individual localization using standard scoring will take 26.4ms (including feature quantization into visual words). "Competition scoring" with geometric verification is currently taking 1.5s. Here the feature matching is not optimized and takes most of the time. Excluding the runtime for feature matching leaves 53.5ms for localization with epipolar geometry verification. Feature matching can easily be speeded up by using proper data structures, e.g. a kd-tree [5], so that real-time speed can be achieved with this approach.

Table 1. Recognition scores for the validation set with the different training sets and methods

Method	training_easy	training_hard
standard scoring	0.96	0.87
geometric verification	0.98	0.92

Table 2. Competition scores (with "Competition method") for task1

	training_easy	training_hard	combined score
task1	2047	1777	3824

Table 3. Recognition scores (with "Competition method") for the individual classes and the full set on the testing set

Class	training_easy	training_hard
Full set	0.80	0.70
Elevator	0.97	0.98
Corridor	0.96	0.86
Kitchen	0.99	0.96
LargeOffice1	0.93	0.63
LargeOffice2	0.96	0.83
SmallOffice2	0.99	0.94
StudentOffice	0.73	0.62
Lab	1.00	1.00
PrinterArea	0.99	0.64
Unknown	0.56	0.65

7   Conclusion

The ImageCLEF@ICPR 2010 competition provides a challenging dataset to evaluate different methods for robot localization. The use of a stereo camera as imaging system for the robot allowed us to combine monocular and stereo cues for robot localization.

Our approach achieves high recognition rates (e.g. 98% on the *validation*), which signals that the proposed approach is reliable enough to be used in practice. To foster the use of this approach we made the code for the visual word based similarity ranking publicly available[4].

References

1. Angeli, A., Filliat, D., Doncieux, S., Meyer, J.A.: Fast and incremental method for loop-closure detection using bags of visual words. IEEE Transactions on Robotics 24(5), 1027–1037 (2008)
2. Cummins, M., Newman, P.: FAB-MAP: Probabilistic Localization and Mapping in the Space of Appearance. The International Journal of Robotics Research 27(6), 647–665 (2008), http://ijr.sagepub.com/cgi/content/abstract/27/6/647
3. Fraundorfer, F., Wu, C., Frahm, J.M., Pollefeys, M.: Visual word based location recognition in 3d models using distance augmented weighting. In: Fourth International Symposium on 3D Data Processing, Visualization and Transmission (2008)

---

[4] http://www.cvg.ethz.ch/people/postgraduates/fraundof/vocsearch

4. Fraundorfer, F., Wu, C., Pollefeys, M.: Combining monocular and stereo cues for mobile robot localization using visual words. In: Ünay, D., Çataltepe, Z., Aksoy, S. (eds.) ICPR 2010. LNCS, vol. 6388, pp. 182–191. Springer, Heidelberg (2010)
5. Lowe, D.: Distinctive image features from scale-invariant keypoints. International Journal of Computer Vision 60(2), 91–110 (2004)
6. Murillo, A.C., Kosecka, J.: Experiments in place recognition using gist panoramas. In: IEEE Workshop on Omnidirectional Vision, Camera Netwoks and Non-Classical Cameras, ICCV 2009, pp. 1–8 (2009)
7. Nistér, D., Stewénius, H.: Scalable recognition with a vocabulary tree. In: Proc. IEEE Conference on Computer Vision and Pattern Recognition, New York City, New York, pp. 2161–2168 (2006)
8. Ohta, Y., Kanade, T.: Stereo by intra- and inter-scanline search using dynamic programming. PAMI 7(1), 139–154 (1985)
9. Ortín, D., Montiel, J.M.M.: Indoor robot motion based on monocular images. Robotica 19(3), 331–342 (2001)
10. Pronobis, A., Caputo, B.: Cold: Cosy localization database. International Journal of Robotics Research (IJRR) 28(5), 588–594 (2009)
11. Scaramuzza, D., Fraundorfer, F., Pollefeys, M., Siegwart, R.: Closing the loop in appearance-guided structure-from-motion for omnidirectional cameras. In: The Eight Workshop on Omnidirectional Vision, ECCV 2008, pp. 1–14 (2008)
12. Schindler, G., Brown, M., Szeliski, R.: City-scale location recognition. In: Proc. IEEE Conference on Computer Vision and Pattern Recognition, Minneapolis, Minnesota, pp. 1–7 (2007)
13. Wu, C., Clipp, B., Li, X., Frahm, J.M., Pollefeys, M.: 3d model matching with viewpoint invariant patches (vips). In: Proc. IEEE Conference on Computer Vision and Pattern Recognition (2008)

PicSOM Experiments in ImageCLEF RobotVision*

Mats Sjöberg, Markus Koskela, Ville Viitaniemi, and Jorma Laaksonen

Adaptive Informatics Research Centre
Aalto University School of Science and Technology
P.O. Box 15400, FI-00076 Aalto, Finland
`firstname.lastname@tkk.fi`

Abstract. The PicSOM multimedia analysis and retrieval system has previously been successfully applied to supervised concept detection in image and video databases. Such concepts include locations and events and objects of a particular type. In this paper we apply the general-purpose visual category recognition algorithm in PicSOM to the recognition of indoor locations in the ImageCLEF/ICPR RobotVision 2010 contest. The algorithm uses bag-of-visual-words and other visual features with fusion of SVM classifiers. The results show that given a large enough training set, a purely appearance-based method can perform very well – ranked first for one of the contest's training sets.

1 Introduction

In this paper we describe the application of our general-purpose content-based image and video retrieval system PicSOM [4] to the ImageCLEF/ICPR 2010 RobotVision contest task. Among other things, the PicSOM system implements a general visual category recognition algorithm using bag-of-visual-words and other low-level features together with fusion of SVM classifiers. This setup has been used successfully previously, for example in the NIST TRECVID 2008 and 2009 [10] high-level feature detection tasks [11], where events, locations and objects are detected in television broadcast videos. Our goal in the experiments described in this paper is to evaluate the suitability of this general-purpose visual category detection method to a more narrow domain in the indoor location detection setup of the RobotVision contest. We have not included any domain specific features in these experiments, such as depth information from stereo imaging. Thus, the only modality we consider is the current view from one or more forward-pointing cameras.

In addition to autonomous robots [8], a RobotVision-style setup arises, for example, in many applications of mobile augmented reality [2]. In fact, indoor localisation constitutes also one of the sub-tasks of our research platform for accessing abstract information in real-world pervasive computing environments through augmented reality displays [1]. In that context, objects, people, and

---

* This work has been supported by the Aalto University MIDE project UI-ART.

D. Ünay, Z. Çataltepe, and S. Aksoy (Eds.): ICPR 2010, LNCS 6388, pp. 190–199, 2010.

the environment serve as contextual channels to more information, and adaptive models are used to infer from implicit and explicit feedback signals the interests of users with respect to the environment. Results of proactive context-sensitive information retrieval are augmented onto the view of data glasses or other see-through displays.

Fig. 1 illustrates our visual category recognition approach. Given training images with location labels, we first train a separate detector for each location L_i. Section 2 describes these single-location detectors that employ fusion of several SVM detectors, each based on a different visual feature. The probabilistic outcomes of the detectors are then used as inputs to the multi-class classification step that determines the final location label $\hat{L}$ for each test image. This step is described in Section 3. The predicted $\hat{L}$ is either one of the known locations L_i, or alternatively, the system can predict that the image is taken in a novel unknown location, or declare the location to be uncertain. In Section 4 we describe our experiments in RobotVision 2010 and summarise the results. Finally, conclusions are drawn in Section 5.

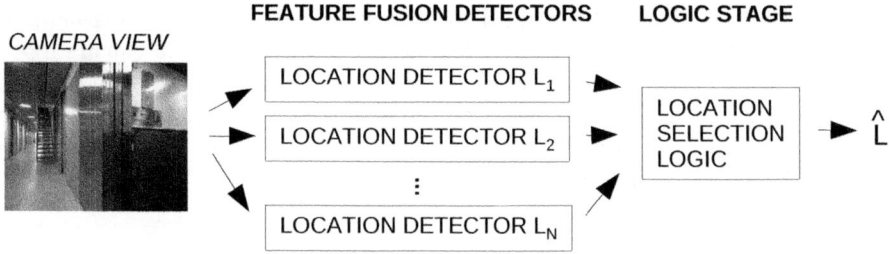

Fig. 1. General architecture for predicting location $\hat{L}$ based on a camera view

2   Single-Location Detectors

For detecting a single location L_i, our system employs the architecture illustrated in Fig. 2. The training phase begins with the extraction of a large set of low-level visual features. The features and binary location labels of the training images (L_i or non-L_i) are then used to train a set of probabilistic two-class SVM classifiers. A separate SVM is trained for each visual feature.

After training, the detector can estimate the probability of a novel test image depicting location L_i. This is achieved by first extracting the same set of visual features from the test image that was extracted from the training images. The trained feature-wise SVM detectors produce a set of probability estimates that are combined to a final probability estimate in a fusion stage. The location-wise estimates are then combined in a multi-class classification stage to determine the location of the test image.

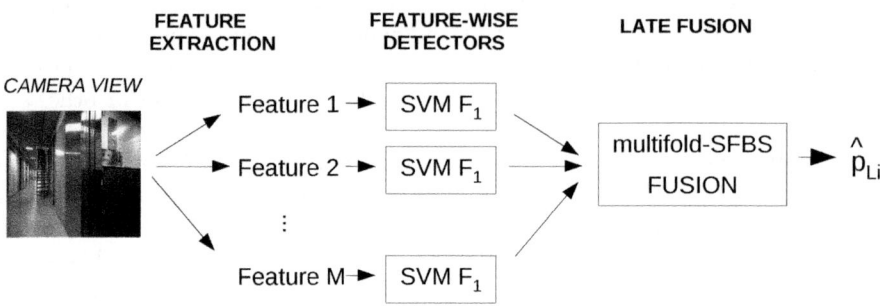

Fig. 2. Architecture for estimating the probability $\hat{p}_{Li}$ that the given camera view is from location L_i

2.1   Feature Extraction

From each image, a set of low-level visual features is extracted. We use our own implementations of the following MPEG-7 descriptors: *Color Layout, Dominant Color, Scalable Color*, and *Edge Histogram* [3]. Additionally, we calculate several non-standard low-level appearance features: *Average Color, Color Moments, Texture Neighbourhood, Edge Histogram, Edge Co-occurrence* and *Edge Fourier*. The non-standard features are calculated for five spatial zones of each image (Figure 3) and the values concatenated to one image-wise vector.

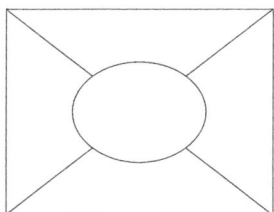

Fig. 3. The five-part center-surround zoning mask for image feature extraction

Of the non-standard features, the Average Color feature is a three-element vector that contains the average RGB values of all the pixels within the zone. The Color Moments feature treats the HSV colour channels from the zone as probability distributions, and calculates the first three central moments (mean, variance and skewness) of each distribution. For the Texture Neighbourhood feature, relative values of the Y (luminance) component of the YIQ colour representation in all 8-neighbourhoods within the zone are characterised. The probabilities for neighbouring pixels being more luminous than the central pixel are estimated separately for all the eight surrounding relative pixel positions, and collected as a feature vector. Edge Histogram is the histogram of four Sobel edge directions. The feature differs in details from the similarly named MPEG-7 descriptor. Edge Co-occurrence gives the co-occurrence matrix of four Sobel edge directions.

Furthermore, eight different bag-of-visual-words (BoV) features are also extracted. In the BoV model images are represented by histograms of local image descriptors. The eight features result from combining number of independent design choices. First, we use either the *SIFT* [6] or the opponent colour space version of the *Color SIFT* descriptor [9]. Second, we employ either the Harris-Laplace detector [7] or use dense sampling of image locations as the interest point detector. Third, we have the option to use the soft-histogram refinement of the BoV codebooks [9]. Finally, for some of the features, we have used the spatial pyramid extension of the BoV model [5].

2.2   Feature-Wise Detectors

In our location recognition system, the association between an image's visual features and its location is learned using the SVM supervised learning algorithm. The SVM implementation we use in our system is an adaptation of the C-SVC classifier of the LIBSVM[1] software library. For all histogram-like visual features we employ the χ^2 kernel

$$g_{\chi^2}(\mathbf{x}, \mathbf{x}') = \exp\left(-\gamma \sum_{i=1}^{d} \frac{(x_i - x_i')^2}{x_i + x_i'}\right). \tag{1}$$

The radial basis function (RBF) SVM kernel

$$g_{\mathrm{RBF}}(\mathbf{x}, \mathbf{x}') = \exp\left(-\gamma \|\mathbf{x} - \mathbf{x}'\|^2\right) \tag{2}$$

is used for all the other features. The motivation for this is the well-known empirical observation that χ^2 distance is well-suited for comparing histograms.

The free parameters of the SVMs are selected with an approximate 10-fold cross-validation search procedure that consists of a heuristic line search to identify a promising parameter region, followed by a grid search in that region. To speed up the computation, the data set is radically downsampled for the parameter search phase. Further speed-up is gained by optimising the C-SVC cost function only very approximately during the search.

For the final detectors we also downsample the data set, but less radically than in the parameter search phase. Usually there are much fewer annotated example shots of a location (positive examples) than there are example shots not exhibiting that location (negative examples). Consequently, for most of the locations, the sampling is able to retain all the positive examples and just limit the number of negative examples. The exact amount of applied sampling varies according to the computation resources available and the required accuracy of the outputs. Generally we have observed the downsampling to degrade detection accuracy.

---

[1] http://www.csie.ntu.edu.tw/~cjlin/libsvm

2.3 Fusion

The supervised fusion stage of our location recognition system is based on the geometric mean of feature-wise detector outcomes. However, instead of calculating the mean of all feature-wise detectors we select the set using sequential forward-backward search (SFBS). This supervised variable selection technique requires detector outcomes also for training images. These outcomes are obtained via 10-fold cross-validation.

Our search technique refines the basic SFBS approach by partitioning the training set into multiple folds. In our implementation we have used a fixed number of six folds. The SFBS algorithm is run several times, each time leaving one fold outside the training set. The final fusion outcome is the geometric mean of the fold-wise geometric means.

3 Multi-class Classification

The fusion of the feature-wise detector scores described in the previous section provides probability estimates for each location given a particular image. The final classification step is a traditional multi-class classification, where we combine several one-versus-the-rest SVM classifiers. The straightforward solution is to classify the image to the class with the highest probability estimate. However, in the current scenario, we must also be able to detect *unknown* categories, i.e. images of new locations that have not been seen before. We have implemented this by a heuristic method with two thresholds. First, if there are detector scores above a high threshold T_1, then we deem the system to be confident enough to simply pick the class with highest score. Second, if all scores are below a low threshold T_2, this is interpreted to mean that we are seeing an unknown class for which we have not trained a detector. Finally, if none of the above conditions apply, there must be one or more scores that are above T_2, which can be seen as potential detections.

These scores of such potential detections are all smaller than T_1 (since the first condition was not true), and can be seen as potential, but not strong detections. Trivially, if the number of such scores equals one, this one is selected as the detected class. If the number of such detection scores is higher than two we deem the situation to be too uncertain and decline to classify it (i.e. it is the *reject* class). If the number equals two we select the highest one. This is due to the particular performance measure used in RobotVision, which rewards a correct choice with $+1.0$ and an incorrect choice with -0.5. This means that if the correct class is either of the two potential candidates the expectation value of the performance measure score is still positive even by selecting either at random.

4 Experiments

4.1 Recognition with Stereo Images

In the RobotVision setup, the presence of a stereo image pair demands some additional considerations. For example, we might learn a separate model for each camera, i.e. independent models for the images of left camera and right camera. On the other hand, since the two cameras show the same scene from somewhat different angles, they are certainly not independent. In fact, if we consider the set of images taken from each point in every possible angle, the left and right images are just two samples from the same distribution. I.e. the image seen in the left camera might be seen in the right camera at some other point in time if the robot happens to be at the same point in space but at a slightly different angle. This view would then support the approach of simply using all images as training data discarding the left/right distinction. In the end we made two models, one using all images, and one using only the images from the left camera (i.e. only half of the data). After the competition we also made a model from the right camera images for comparison.

In the final stage, when categorising stereo images at particular times, one must have a strategy for combining the detections scores for the two cameras. We tried taking their average, maximum, minimum, or only taking the left camera, or only right camera result. The stereo image pair could also be utilised for stereo imaging and the contest organisers provided the camera calibration data for the image sequences. Depth information would undoubtedly be an useful feature for location recognition. We have, however, not utilised such domain specific features in the present work.

4.2   Parameter Selection

The combining logic described in Section 3 was used, and the two thresholds, T_1 and T_2 were determined by simple grid search maximising the performance score in the validation set. Because the testing set had four unseen rooms, we tried to simulate this situation in the training by leaving out three rooms (roughly the same ratio of unseen to seen) and use this setup when determining the thresholds. We did this both with the regular detectors trained on the full set of rooms, and with detectors trained on the reduced set. Those trained on the full set we thought might be unrealistic since they had used the three removed rooms as negative examples in their training. Using detectors trained on the reduced set tended to increase the lower threshold T_2 from the level it had when using detectors trained on all known rooms. It turned out however that the lower thresholds worked better in the testing set.

The threshold T_1, which gives a limit for when to decline from classification (*reject*) did not affect the results significantly, and is not included in the results presented in the following.

4.3   Results

Our best submitted result for the easy set received a score of 2176.0, which is 85% of the best possible score. This result was based on detectors trained on the left camera data only, and it obtained the overall highest score in the competition for the easy set. The same setup achieved our best result (1117.0) for the hard set as well. Fig. 4 visualises this run compared to the groundtruth. For the hard set, our result was slightly above the median of the submitted results using the hard training data. The overall best submitted run to the hard set was 1777.0.

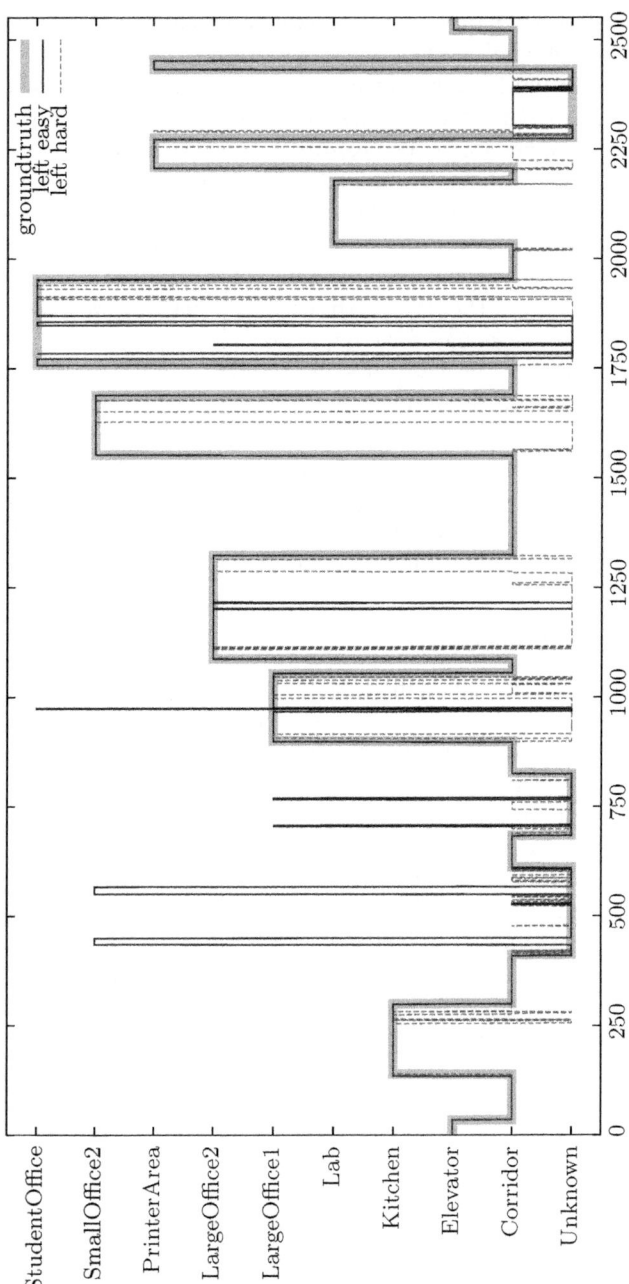

Fig. 4. Recognition results over time (frame indices) based on left camera images, trained on easy (blue line) and hard (red dashed) sets. The groundtruth is shown as a thick grey line.

These and some additional runs are summarised in Table 1, with "•" denoting that the run was submitted to the competition. The first column in the table specifies how the training data was selected with regard to the cameras. The word "separate" indicates that separate models were trained for each camera and then averaged, while "both" uses all images to train a single model. The camera-wise scores were combined by taking the average value when using datasets with images from both cameras. The second column states whether fusion of single-feature classifiers (Section 2.3) or just the single best performing feature (ColorSIFT with dense sampling over a spatial pyramid and soft clustering) is used.

Somewhat surprisingly, using information from both cameras does not improve the results, in fact using a single camera works better than using a single model trained on all images. This difference is especially notable on the hard training set. Also, using the separate left and right models together gives no improvement over using just one of them. Finally, in Table 1 we can also see that the feature fusion is highly beneficial: with a single feature the results are significantly weaker.

Table 1. RobotVision recognition scores

cameras	features	easy	hard	total
• left only	fusion	2176.0	1117.0	3293.0
right only	fusion	2210.5	1072.0	3282.5
separate	fusion	2207.5	1057.0	3264.5
• both	fusion	2065.0	665.5	2730.5
• both	single	964.0	554.5	1518.5

After the competition, each participant was given access to the labels for the testing dataset, and we were able to perform some more tests, and determine optimal parameters T_1 and T_2 in the testing set. These additional tests are summarised in Table 2 using fusion of single-feature classifiers. The first column specifies how the training data was selected with regards to the cameras: using images from both together, or just using the images from the left camera. The second column shows how the detection scores from the two cameras were combined to form the final detection score: by taking the average of the left and the right, by taking the maximum or minimum, or by simply taking the left or the right camera scores directly. It can clearly be seen that using different ways of combining the stereo-vision scores makes very little difference, the choice of training data is much more important. Even using a model trained on the left camera images for the right camera is better than using the dataset with images from both cameras with any score combination method.

Note that the results shown in Table 2 are not comparable with other competition submissions since they have been optimised against the testing set, and are thus "oracle" results. They are however interesting for a comparison between

Table 2. "Oracle" detection scores

cameras	selection	easy	hard	total
both	average	2126.5	910.0	3036.5
left	average	2176.0	1144.5	3320.5
both	left	2090.5	908.5	2999.0
left	left	2174.5	1160.5	3335.0
both	max	2114.5	912.5	3027.0
left	max	2179.0	1148.5	3327.5
both	min	2099.5	901.5	3001.0
left	min	2164.0	1157.5	3321.5
both	right	2116.0	905.0	3021.0
left	right	2152.0	1135.0	3287.0

different alternations of our method. Furthermore, it can be observed that the oracle results are only slightly better than the submitted ones, indicating that the system performance is not very sensitive to the threshold parameters.

5   Conclusions

Our results indicate that a general-purpose algorithm for visual category recognition can perform well in indoor location recognition, given that enough training data is available. The generality of our approach is illustrated e.g. by its successful application to image and video retrieval [10]. With limited training data, however, the performance of our purely appearance-based method is less competitive. There are several possible explanations for this. It might be that the generic scene appearance features utilise the limited training data uneconomically and other domain-specific modalities would be needed to take best use of the scarce training examples. For location recognition, these could include the depth information, the temporal continuity of the frame sequence and information based on pair-wise matching of images.

Yet, it is also possible that better performance could be achieved on basis of the generic appearance features by better system design. In particular, there might be some overlearning issues. With the larger training set, just memorising all the camera views appearing in the training material might be a viable strategy, whereas the smaller training set calls for generalising between views. A naive use (such as here) of a rich and distinctive scene representation might actually lead to worse performance than a feature extraction scheme with more limited distinguishing power if the inter-view generalisation issue is not properly taken care of. Our experiments reported here are insufficient to confirm either one of these hypotheses.

Our experiments back up our earlier findings that fusion of a large set of features consistently results in a much better visual category recognition accuracy than the use of any single feature alone.

References

1. Ajanki, A., Billinghurst, M., Kandemir, M., Kaski, S., Koskela, M., Kurimo, M., Laaksonen, J., Puolamäki, K., Tossavainen, T.: Ubiquitous contextual information access with proactive retrieval and augmentation. In: Proceedings of 4th International Workshop on Ubiquitous Virtual Reality 2010 at Pervasive 2010, Helsinki, Finland (May 2010)
2. Feiner, S., MacIntyre, B., Höllerer, T., Webster, A.: A touring machine: Prototyping 3D mobile augmented reality systems for exploring the urban environment. Personal and Ubiquitous Computing 1(4), 208–217 (1997)
3. ISO/IEC: Information technology - Multimedia content description interface - Part 3: Visual, 15938-3:2002(E) (2002)
4. Laaksonen, J., Koskela, M., Oja, E.: PicSOM—Self-organizing image retrieval with MPEG-7 content descriptions. IEEE Transactions on Neural Networks, Special Issue on Intelligent Multimedia Processing 13(4), 841–853 (2002)
5. Lazebnik, S., Schmid, C., Ponce, J.: Beyond bags of features: Spatial pyramid matching for recognizing natural scene categories. In: Proc. of IEEE CVPR, vol. 2, pp. 2169–2178 (2006)
6. Lowe, D.G.: Distinctive image features from scale-invariant keypoints. International Journal of Computer Vision 60(2), 91–110 (2004)
7. Mikolajcyk, K., Schmid, C.: Scale and affine point invariant interest point detectors. International Journal of Computer Vision 60(1), 68–86 (2004)
8. Pronobis, A., Caputo, B.: COLD: COsy Localization Database. The International Journal of Robotics Research (IJRR) 28(5) (May 2009)
9. van de Sande, K.E.A., Gevers, T., Snoek, C.G.M.: Evaluating color descriptors for object and scene recognition. IEEE Transactions on Pattern Analysis and Machine Intelligence (in press, 2010)
10. Sjöberg, M., Viitaniemi, V., Koskela, M., Laaksonen, J.: PicSOM experiments in TRECVID 2009. In: Proceedings of the TRECVID 2009 Workshop, Gaithersburg, MD, USA (November 2009)
11. Smeaton, A.F., Over, P., Kraaij, W.: High-Level Feature Detection from Video in TRECVid: a 5-Year Retrospective of Achievements. In: Divakaran, A. (ed.) Multimedia Content Analysis, Theory and Applications, pp. 151–174. Springer, Berlin (2009)

Combining Image Invariant Features and Clustering Techniques for Visual Place Classification

Jesús Martínez-Gómez, Alejandro Jiménez-Picazo, José A. Gámez,
and Ismael García-Varea

Computing Systems Department, SIMD i^3A
University of Castilla-la Mancha, Albacete, Spain
{jesus_martinez,ajimenez,jgamez,ivarea}@dsi.uclm.es

Abstract. This paper presents the techniques developed by the SIMD group and the results obtained for the 2010 RobotVision task in the ImageCLEF competition. The approach presented tries to solve the problem of robot localization using only visual information. The proposed system presents a classification method using training sequences acquired under different lighting conditions. Well-known SIFT and RANSAC techniques are used to extract invariant points from the images used as training information. Results obtained in the RobotVision@ImageCLEF competition proved the goodness of the proposal.

1 Introduction

The Cross Language Evaluation Forum (CLEF[1]) promotes the research in several research lines, related to information retrieval. This forum was created in 2000 and the number of proposed tracks has been increasing notoriously until 2010. CLEF started with three evaluation tracks in 2000 (Multilingual, Bilingual and Monolingual information retrieval) but the 2009 CLEF edition offered 8 main tracks. In 2003, a new track related to cross language retrieval of images via their associated textual captions was introduced. This new track aims to explore the relationship between images during the retrieval process.

A new subtrack inside the ImageCLEF track appeared in the 2009 CLEF edition. This new subtrack, called RobotVision, addresses the problem of topological localization of mobile robots using visual information. Our group started working inside the CLEF challenge with the release of this subtrack, due to the high relationship between the aim of the RobotVision task and the background of the group (robotics and information retrieval).

For mobile robotics, image processing has become a keystone. Visual cameras are the most common robot sensor, providing a huge amount of data with low cost, but all these data should be real-time processed to retrieve relevant information. Most of the robot decisions are based on the information sensed from the environment (mainly through visual sensors). Decision making is based on the strategy of the robot and the own robot position, which is estimated with a localization algorithm. Robot localization is usually performed by using odometry and the information obtained from sensors.

---

[1] http://www.clef-campaign.org/

D. Ünay, Z. Çataltepe, and S. Aksoy (Eds.): ICPR 2010, LNCS 6388, pp. 200–209, 2010.

For the RobotVision task, only visual information is available. The 2009 RobotVision task was focused on robot localization, but the new 2010 edition presents the problem of visual place classification: participants are asked to classify rooms on the basis of image sequences, captured by a mobile robot within a controlled environment. Image classification has become one of the most difficult problems in computer vision research. This problem becomes highly complex when images are captured by a robot's camera in dynamic environments. One of the main applications of visual classification is robot localization, but this adds some extra constraints to the process. The most important one is the processing time, because images need to be handled in real-time.

The 2009 RobotVision challenge provided information related to the real robot's pose embedded with the labels of the images. SIMD proposal combined all this information developing a SIFT-Montecarlo localization algorithm solving the task proposed as a pure robot localization problem, with a well defined iterative process with two main steps: an odometry and a visual phase.

RobotVision@ICPR challenge provides training images labelled with the room from where they were taken but not with the real robot's pose. Moreover, the number of frames of the training sequence that is mandatory to use for the final experiment is greater. Based on these new premises, current SIMD proposal is focused on increasing the accuracy and reducing the processing time of the classification methods (based on features similarity) developed for the 2009 edition.

RobotVision@ICPR task deals with visual place classification, where the visual information provided is restricted to images taken by a robot in an indoor work environment. The approach presented here carries out classification by using the Scale-Invariant Feature Transform[5] combined with RANSAC[2]. SIFT is used to extract invariant features from images and to perform an initial matching. RANSAC improves this matching by discarding invalid correspondences.

The experiments were carried out following the proposed procedure, using two training sequences (easy and hard) and a final test sequence. Our proposal was evaluated for the two proposed tasks: obligatory (classification must be performed separately for each test image) and optional (the algorithm can exploit the continuity of the sequences), which is a more realistic localization task.

The rest of the paper is organized as follows: a description of the data used for the task can be observed in Section 2. Invariant features and matching techniques are outlined in Section 3. In Section 4 a description of the system training process in presented. Next, in Section 5 the SIMD approach to the task is proposed. Section 6 describes the experiments performed and the results obtained. Finally, the conclusions and areas for future work are given in Section 7.

2 RobotVision Data Description

RobotVision participants are asked to classify unlabeled frames (final test frame sequence) by using training frame sequences from the *COLD-Stockholm* database[8]. Test and training frames were acquired in the same environment, but test sequence includes additional rooms not previously imaged. Different lighting conditions were used of for the different sequences, and preliminary proposals could be evaluated by using a

Table 1. RobotVision@ImageCLEF data set information

Sequence	Frames	Rooms	Lighting conditions
Training Easy	8149	9	Cloudy
Training Hard	4535	9	Cloudy
Validation	4783	9	Night
Test	5102	12	Night

validation data set released before the final test data set. A complete description of all these data sets can be observed in Table 1.

3 Invariant Features and Matching

In order to perform a correct comparison between two (or more) images representing the same scenario, appropriate features should be used. These features have to be invariant to changes in the viewpoint where the frames were captured from. Different frames to be compared can be acquired over different lighting conditions but such changes should not affect the performance of the comparison. Small variations could happen within the environment (due to the environments are dynamic) and important elements used to train preliminary classifying algorithms can be removed or replaced.

Taking into account all these factors, it makes no sense to use classical image processing techniques based on edge detection to solve the problem of visual place classification. These techniques rely on specific elements of the environment liable to be removed or replaced and moreover, they are dependent on rotation and scale.

One of the most popular technique for extracting relevant invariant features from images is the Scale-Invariant Feature Transform algorithm (SIFT) [6]. The main idea of the algorithm is to apply different transformations and study the points of the image which remain invariant under these transformations.

Some authors as B. Caputo and A. Pronobis [9] propose the developing of feature extraction techniques based on the information extracted from the available training frames instead of using standard techniques as SIFT. All these features, which can be extracted using high dimensional Composed Receptive Field Histograms (CRFH) [4], will be used as the input to train a classifier. The classifier will separate training data by a hyper plane in a high dimensional feature space. The classifier proposed by the authors is a Support Vector Machine (SVM) [10].

Features extracted with SIFT are robust to noise and changes in viewpoint and also invariant to image scale and rotation. An important characteristic of systems developed to perform object recognition using SIFT is that they are robust to partial object occlusions. In order to deal with the considerable processing time of the algorithm, an implementation of the algorithm over the graphics processor unit (GPU) was considered. The selected implementation (named "SiftGPU"[2]) speeds up the process, reducing the processing time to less than 0.025 seconds for extraction and matching.

---

[2] http://www.cs.unc.edu/ ccwu/siftgpu/

3.1 RANSAC

Random Sample Consensus[2] (RANSAC) is an iterative method for estimating a mathematical model from a set of data which contains outliers. RANSAC was developed as a non-deterministic iterative algorithm where different models are estimated using a random subset of the original data. All these models are evaluated using the complete data set and the algorithm finishes when some constraints (related to the fitness of the model to the data) are overcome or after a certain number of iterations.

Fig. 1. Matching between two images where outliers (red lines) are discarded

Our proposal uses RANSAC to improve the preliminary matching obtained with SIFT techniques. Such initial matching obtains a high number of outliers that do not fit the real correspondence between two candidate images. The data set includes all the a priori correspondences and the models we want to estimate are those capable of representing real matching between two images.

Fig. 1 illustrates the result of a matching between features extracted from two images and how the outliers (red lines) are discarded using RANSAC.

4   System Training

Our system was trained using the two available training data sets: easy and hard. These data sets contain frames taken by a robot's camera while the robot was driven through the environment. The main difference between the two sequences is the similarity between that sequence and the final test sequence. The amount of information provided by the "easy data set" is greater than that provided by the "hard one", because the environment is observed from a higher number of viewpoints. There are too many training frames to perform a complete real-time comparison and so the number of training frames to work with should be reduced. Our proposal consists of: first, discarding redundant frames and second, selecting from those frames the ones that are the most representative. These steps are described below in detail.

4.1   Training Sequence Pre-processing

There are many training frames that share the viewpoint from they were taken and containing redundant information. These different sets of frames with similar information should be reduced to a single frame set. We applied the following process. Firstly, all the frames were converted to greyscale. After that, we computed the difference between two images as the absolute difference for the grey value pixel by pixel. A frame will be removed from the training sequence when the difference between it and the last non-removed frame is lower than a certain threshold, which was obtained empirically.

Fig. 2 illustrates how the difference between two frames is computed. Upper images are the original frames to compute its difference and bottom image shows the visual result of this difference.

Fig. 2. Absolute difference computed between two frames

4.2   Extracting Most Representative Training Frames - Clustering Process

Once we had reduced the number of redundant training frames, the next step was to select a subset of the most representative frames. This step was applied separately for each room from the training sequence. Each most representative frame should have a high similarity with some of the discarded frames and a high dissimilarity with the other most representative frames.

These characteristics are ideal for a clustering process and therefore we applied a k-medoids algorithm[3]. The similarity between frames was computed using a SIFT matching (without RANSAC) between the features extracted from them. The value of k was selected as a percentage of training frames for the room that was processed, so its value was different for the different rooms.

An example of the complete training process is illustrated in Fig. 3. First row shows all the original training frames that are available. Thanks to the use of the difference between frames, redundant ones are discarded in the second row. Finally, third row shows the key training frames selected to compare with future test frames.

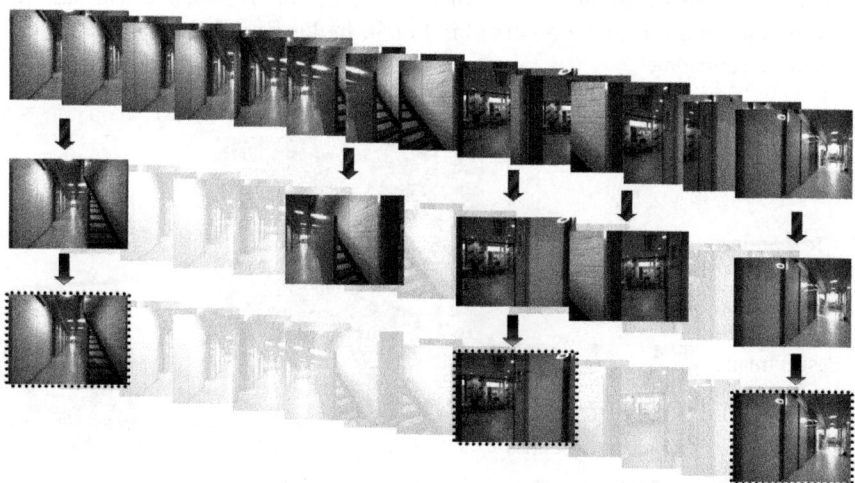

Fig. 3. An example of sequence pre-processing where redundant frames are discarded and best candidates are selected

5   Complete System

The complete process for classifying a test frame consists of three steps. First, SIFT points are extracted from the current frame. Second, we compute the similarity between the current frame and the training frames by means of the percentage of matching points (SIFT + RANSAC). Finally, these similarities are used to classify the test frame. This process is explained in detail below.

5.1   Matching Test Frames with Most Representative Training Frames

The similarity value between a test and a training frame is obtained using SIFT matching and RANSAC. After all the SIFT matching points are obtained, RANSAC is applied to discard the outliers. The percentage of common points between both frames is stored as the similarity value.

5.2   Frame Classification

Each test frame can be classified as a specific room (the class), marked as unknown or not classified. A ranking with the n-best values of similarity and its associated room is obtained. We compute the sum of the similarity values separately for the different rooms in the ranking. The test frame will be classified as the room with the highest value when this value clearly exceeds all the other ranking rooms, otherwise it will not be classified. Unknown class is used to denote a test frame acquired in a room not included in the training rooms and will be used when the maximum similarity value is below a certain threshold.

A complete classification process where the test frame is matched with all the selected training frames can be observed in Fig. 4. In this case the test frame should be labelled as a corridor.

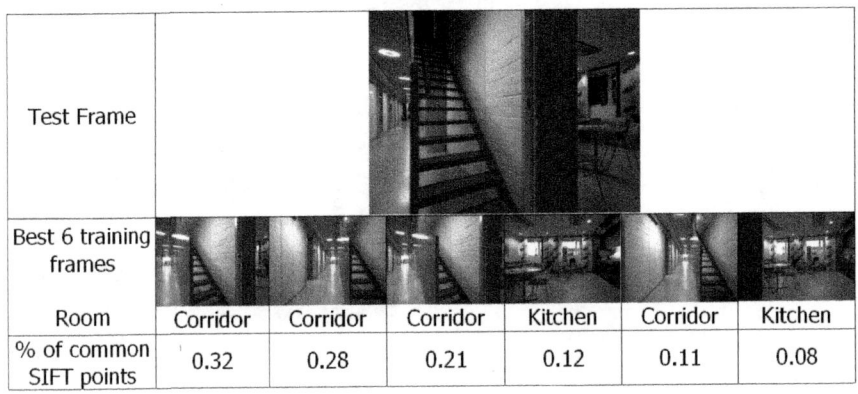

Test Frame						
Best 6 training frames						
Room	Corridor	Corridor	Corridor	Kitchen	Corridor	Kitchen
% of common SIFT points	0.32	0.28	0.21	0.12	0.11	0.08

Fig. 4. Test frame classification using the 6-best training frames

6   Experimental Results

Our proposal was evaluated using the procedure proposed by RobotVision task at ImageCLEF@ICPR. The test sequence contains 2551 frames to be classified. The performance of the algorithm was evaluated using a score, which is computed as: for each correctly classified frame the score is updated by +1.0, for each misclassified one the score is reduced -0.5. A non-classified frame does not vary the score. Therefore, the maximum score obtainable is 2551, when all test frames are correctly classified.

Test frames were classified as unknown when the best matching obtained a percentage value below 2.8% and marked as not classified when the maximum sum of percentages for a room was not higher than 34% of the sum of all the weights, for the most similar 7 frames. These thresholds were obtained empirically using preliminary tests.

Each run consisted of submitting the results for the two training sets: easy and hard. There were two separate tasks, obligatory and optional. The difference between both tasks was that the classification had to be performed separately for each test frame for the obligatory task but, for the optional task, the algorithm was allowed to exploit the continuity of the sequences of frames.

6.1 Obligatory Task

The complete test sequence was processed twice, using the easy and the hard training sequences. Each one of the test frames can be classified separately and the order of the sequence is not important. Table 2 shows the final score and the score obtained using the easy and the hard training sequence.

Table 2. Results for the obligatory task

Training Sequence	Total	Easy	Hard
Score	**3372.5**	2000.0	1372.5
Correctly Classified	3886.0	2180.0	1706.0
Missclasified	1027.0	360.0	667.0
Not Classified	189.0	11.0	178.0

Our run achieved the 3rd place for this task, for which 8 different research groups submitted results. The winner for the obligatory track was the Computer Vision and Geometry Lab (CVG) group, from ETH Zürich.

6.2 Optional Task

For the optional task, we took into account the test frame we were going to classify and the last 4 test frames already classified. Test frames initially labelled as not classified were labelled as the room used to classify the last 4 frames, when this room was the same. Additional verifications were performed to avoid passing from one room to another without using the corridor. The final score was 3881.0 and the complete results can be observed in Table 3.

Table 3. Results for the optional task

Training Sequence	Total	Easy	Hard
Score	**3881.0**	2230.5	1650.5
Correctly Classified	4224	2332	1892
Missclasified	686	203	483
Not Classified	192	16	176

Our group achieved the 1st place for this task, for which 4 different groups submitted results.

7 Conclusions and Future Work

According to the results and the processing time obtained, our proposal constitutes a real-time image classification method applicable to robot localization. The method can be used for the visual step of formal robot localization methods, such as Monte Carlo[1] or (Extended) Kalman filters [7]. The set of training frames used to classify a test frame can be considerably reduced by using the information obtained from the localization method.

The percentage of common points between the test frame and the best training frame can be used to ponder the classification. This information can be highly useful for estimating the quality or the performance of the classification.

As future work, we have in mind the application of this method within a complete localization algorithm. The optimal value of the thresholds could be tuned using evolutionary computation techniques.

We consider also to use the stereo frames to extract additional information from the environment and to use other features extractors as the proposed in the literature.

Acknowledgements

Work partially supported by the EC (FEDER/ESF), the Spanish goverment under the MIPRCV "Consolider Ingenio 2010" program (CSD2007-00018), and the Spanish Junta de Comunidades de Castilla-La Mancha regional goverment under projects PBI08-02010-7127 and PCI08-0048-8577.

References

1. Dellaert, F., Fox, D., Burgard, W., Thrun, S.: Monte carlo localization for mobile robots. In: IEEE International Conference on Robotics and Automation, ICRA 1999 (May 1999)
2. Fischler, M.A., Bolles, R.C.: Random sample consensus: A paradigm for model fitting with applications to image analysis and automated cartography. Communications of the ACM 24(6), 381–395 (1981)
3. Kaufman, L., Rousseeuw, P.: Clustering by means of medoids. In: Dodge, Y. (ed.) Statistical Data Analysis Based on the L_1-Norm and Related Methods. North-Holland, Amsterdam (1987)
4. Linde, O., Lindeberg, T.: Object recognition using composed receptive field histograms of higher dimensionality. In: International Conference on Pattern Recognition, vol. 4, pp. 1–6 (2004)
5. Lowe, D.: Object recognition from local scale-invariant features. In: 17th International Conference on Computer Vision, Corfu, Greece, vol. 2, pp. 1150–1157 (1999)

6. Lowe, D.: Distinctive image features from scale-invariant keypoints. International Journal of Computer Vision 60(2), 91–110 (2004)
7. Negenborn, R.: Robot Localization and Kalman Filters. Ph.D. thesis, Institute of Information and Computer Science, Copenhagen University (September 2003)
8. Pronobis, A., Caputo, B.: COLD: The CoSy Localization Database. International Journal of Robotics Research 28(5), 588 (2009)
9. Pronobis, A., Caputo, B., Jensfelt, P., Christensen, H.: A realistic benchmark for visual indoor place recognition. Robotics and Autonomous Systems (RAS) 58(1), 81–96 (2010)
10. Vapnik, V.: Statistical learning theory. Wiley, New York (1998)

On the Results of the First Mobile Biometry (MOBIO) Face and Speaker Verification Evaluation

Sébastien Marcel[1], Chris McCool[1], Pavel Matějka[3], Timo Ahonen[2], Jan Černocký[3],
Shayok Chakraborty[4], Vineeth Balasubramanian[4], Sethuraman Panchanathan[4],
Chi Ho Chan[5], Josef Kittler[5], Norman Poh[5], Benoît Fauve[6], Ondřej Glembek[3],
Oldřich Plchot[3], Zdeněk Jančík[3], Anthony Larcher[7], Christophe Lévy[7],
Driss Matrouf[7], Jean-François Bonastre[7], Ping-Han Lee[8], Jui-Yu Hung[8], Si-Wei Wu[8],
Yi-Ping Hung[8], Lukáš Machlica[9], John Mason[10], Sandra Mau[11], Conrad Sanderson[11],
David Monzo[12], Antonio Albiol[12], Hieu V. Nguyen[13], Li Bai[13], Yan Wang[13],
Matti Niskanen[14], Markus Turtinen[14], Juan Arturo Nolazco-Flores[15],
Leibny Paola Garcia-Perera[15], Roberto Aceves-Lopez[15],
Mauricio Villegas[16], and Roberto Paredes[16]

[1] Idiap Research Institute, CH
[2] University of Oulu, FI
[3] Brno University of Technology, CZ
[4] Center for Cognitive Ubiquitous Computing, Arizona State University, USA
[5] Centre for Vision, Speech and Signal Processing, University of Surrey, UK
[6] Validsoft Ltd., UK
[7] University of Avignon, LIA, FR
[8] National Taiwan University, TW
[9] University of West Bohemia, CZ
[10] Swansea University, UK
[11] NICTA, AU
[12] iTEAM, Universidad Politecnica de Valencia, ES
[13] University of Nottingham, UK
[14] Visidon Ltd, FI
[15] Tecnologico de Monterrey, MX
[16] Instituto Tecnológico de Informática, Universidad Politécnica de Valencia, ES
http://www.mobioproject.org/icpr-2010

Abstract. This paper evaluates the performance of face and speaker verification techniques in the context of a mobile environment. The mobile environment was chosen as it provides a realistic and challenging test-bed for biometric person verification techniques to operate. For instance the audio environment is quite noisy and there is limited control over the illumination conditions and the pose of the subject for the video. To conduct this evaluation, a part of a database captured during the "Mobile Biometry" (MOBIO) European Project was used. In total there were nine participants to the evaluation who submitted a face verification system and five participants who submitted speaker verification systems. The results have shown that the best performing face and speaker verification systems obtained the same level of performance, respectively 10.9% and 10.6% of HTER.

Keywords: mobile, biometric, face recognition, speaker recognition, evaluation.

D. Ünay, Z. Çataltepe, and S. Aksoy (Eds.): ICPR 2010, LNCS 6388, pp. 210–225, 2010.

1   Introduction

Face and speaker recognition are both mature fields of research. Face recognition has been explored since the mid 1960's [5]. Speaker recognition by humans has been done since the invention by the first recording devices, but automatic speaker recognition is a topic extensively investigated only since 1970 [6]. However, these two fields have often been considered in isolation to one another as very few joint databases exist.

For speaker recognition there is a regular evaluation organised by the National Institute of Standards and Technology (NIST)[1] called the NIST Speaker Recognition Evaluation (SRE). NIST has been coordinating SRE since 1996 and since then over 50 research sites have participated in the evaluations. The goal of this evaluation series is to contribute to the direction of research efforts and the calibration of technical capabilities of text independent speaker recognition. The overarching objective of the evaluations has always been to drive the technology forward, to measure the state-of-the-art, and to find the most promising algorithmic approaches.

Although there is no regular face recognition competition, there have been several competitions and evaluations for face recognition. These include those led by academic institutions, such as the 2004 ICPR Face Verification Competition [25], in addition to other major evaluations such as the Face Recognition Grand Challenge [27] organised by NIST.

The MOBIO Face and Speaker Verification Evaluation provides the unique opportunity to analyse two mature biometrics side by side in a mobile environment. The mobile environment offers challenging recording conditions including adverse illumination, noisy background and noisy audio data. This evaluation is the first planned of a series of evaluations and so only examines uni-modal face and speaker verification techniques.

In the next section, we briefly present the state-of-the-art in face and speaker verification. Then, we introduce in section 3 the MOBIO database and its evaluation protocol. In sections 4 and 5, we shortly describe the individual face and speaker verification systems involved in this evaluation. The reader can be referred to [24] for a more detailed description of these systems. Finally in section 6, we present the results obtained and discuss them.

2   Face and Speaker Verification

2.1   Face Verification

The face is a very natural biometric as it is one that humans use everyday in passports, drivers licences and other identity cards. It is also relatively easy to capture the 2D face image as no special sensors, apart from a camera that already exist on many mobile devices, are needed.

Despite the ease with which humans perform face recognition the task of automatic face recognition (for a computer) remains very challenging. Some of the key challenges

---

[1] http://www.nist.gov

include coping with changes in the facial appearance due to facial expression, pose, lighting and aging of the subjects.

There have been surveys of both face recognition [40] [34] and video based analysis [35]. From all of these it can be seen that there are many different ways to address the problem of face recognition in general, and more particularly of face verification in theis paper. Some of the solutions can include (but are not limited to) steps such as image preprocessing, face detection, facial feature point detection, face preprocessing for illumination and 2D or 3D geometric normalisation, quality assessment feature extraction, score computation based on client-specific and world models, score normalisation and finally decision making. However, the actual steps taken vary drastically from one system to another.

2.2 Speaker Verification

The most prevalent technique for speaker verification is the Gaussian Mixture Model (GMM) paradigm that uses a Universal Background Model (UBM). In this paradigm a UBM is trained on a set of independent speakers. Then a client is enroled by adapting from this UBM using the speaker specific data. When testing two likelihoods are produced, one for the UBM and one for the client specific model, and these two scores are combined using the log-likelihood ratio and compared to a threshold to produce a "client/imposter" decision [29].

Many other techniques for speaker verification have been proposed. These techniques range from Support Vector Machines [9], Joint Factor Analysis [20] and other group based on Large Vocabulary Continuous Speech Recognition systems [33] through to prosodic and other high level based features for speaker verification [32]. One common thread with the speaker verification techniques proposed nowadays is the ability to cope with inter-session variability which can come from the: communication channel, acoustic environment, state of the speaker (mood/health/stress), and language.

3 MOBIO Database and Evaluation Protocol

3.1 The MOBIO Database

The MOBIO database was captured to address several issues in the field of face and speaker recognition. These issues include: (1) having consistent data over a period of time to study the problem of model adaptation, (2) having video captured in realistic settings with people answering questions or talking with variable illumination and poses, (3) having audio captured on a mobile platform with varying degrees of noise.

The MOBIO database consists of two phases, only one of which was used for this competition. The first phase (Phase I) of the MOBIO database was captured at six separate sites in five different countries. These sites are at the: University of Manchester (UMAN), University of Surrey (UNIS), Idiap Research Institute (IDIAP), Brno University of Technology (BUT), University of Avignon (LIA) and University of Oulu (UOULU). It includes both native and non-native English speakers (speaking only English).

The database was acquired primarily on a mobile phone. The Phase I of the database contains 160 participants who completed six sessions. In each session the participants were asked to answer a set of questions which were classified as: i) set responses, ii) read speech from a paper, and iii) free speech. Each session consisted of 21 questions: 5 set response questions, 1 read speech question and 15 free speech questions. In total there were five **Set responses** to questions and **fake responses** were supplied to each user. **Read speech** was obtained from each user by supplying all users with the same text to read. **Free speech** was obtained from each user by prompting the user with a random question. For five of these questions the user was asked to speak for five seconds (short free speech) and for ten questions the user was asked to speak for ten seconds (long free speech), this gives a total of fifteen such questions.

3.2   The MOBIO Evaluation Protocol

The database is split into three distinct sets: one for training, one for development and one for testing. The data is split so that two sites are used in totality for one set, this means that the three sets are completely separate with no information regarding individuals or the conditions being shared between any of the three sets.

The training data set could be used in any way deemed appropriate and all of the data was available for use. Normally the training set would be used to derive background models, for instance training a world background model or an LDA sub-space. The development data set had to be used to derive a threshold that is then applied to the test data. However, for this competition it was also allowed to derive fusion parameters if the participants chose to do so. To facilitate the use of the development set, the same protocol for enrolling and testing clients was used in the development and test splits. The test split was used to derive the final set of scores. No parameters could be derived from this set, with only the enrolment data for each client available for use; no knowledge about the other clients was to be used. To help ensure that this was the case the data was encoded so that the filename gave no clue as to the identity of the user.

The protocol for enrolling and testing were the same for the development split and the test split. The first session is used to enrol the user but only the five set response questions can be used for enrolment. Testing is then conducted on each individual file for sessions two to six (there are five sessions used for development/testing) and only the free speech questions are used for testing. This leads to five enrolment videos for each user and 75 test client (positive sample) videos for each user (15 from each session). When producing imposter scores all the other clients are used, for instance if in total there were 50 clients then the other 49 clients would perform an imposter attack.

3.3   Performance Evaluation

Person verification (either based on the face, the speech or any other modality) is subject to two type of errors, either the true client is rejected (false rejection) or an imposter is accepted (false acceptance). In order to measure the performance of verification systems, we use the Half Total Error Rate (HTER), which combines the False Rejection Rate (FRR) and the False Acceptance Rate (FAR) and is defined as:

$$HTER(\tau, \mathcal{D}) = \frac{FAR(\tau, \mathcal{D}) + FRR(\tau, \mathcal{D})}{2} \quad [\%] \quad (1)$$

where $\mathcal{D}$ denotes the used dataset. Since both the FAR and the FRR depends on the threshold τ, they are strongly related to each other: increasing the FAR will reduce the FRR and vice-versa. For this reason, verification results are often presented using either Receiver Operating Characteristic (ROC) or Detection-Error Tradeoff (DET) curves, which basically plots the FAR versus the FRR for different values of the threshold. Another widely used measure to summarise the performance of a system is the Equal Error Rate (EER), defined as the point along the ROC or DET curve where the FAR equals the FRR.

However, it was noted in [4] that ROC and DET curves may be misleading when comparing systems. Hence, the so-called Expected Performance Curve (EPC) was proposed, and consists in an unbiased estimate of the reachable performance of a system at various operating points. Indeed, in real-world scenario, the threshold τ has to be set a priori: this is typically done using a development set (also called validation set). Nevertheless, the optimal threshold can be different depending on the relative importance given to the FAR and the FRR. Hence, in the EPC framework, $\beta \in [0; 1]$ is defined as the tradeoff between FAR and FRR. The optimal threshold τ^* is then computed using different values of β, corresponding to different operating points:

$$\tau^* = \underset{\tau}{\operatorname{argmin}} \quad \beta \cdot \text{FAR}(\tau, \mathcal{D}_d) + (1 - \beta) \cdot \text{FRR}(\tau, \mathcal{D}_d) \tag{2}$$

where $\mathcal{D}_d$ denotes the development set.

Performance for different values of β is then computed on the test set $\mathcal{D}_t$ using the previously found threshold. Note that setting β to 0.5 yields to the Half Total Error Rate (HTER) as defined in Equation (1). It should be also noted that for fair evaluations this threshold is not estimated by the participants but by the organizers.

4   Face Verification Systems

4.1   Idiap Research Institute (IDIAP)

The Idiap Research Institute submitted two face (video) recognition systems. The two used exactly the same verification method using a mixture of Gaussians to model a parts-based topology, as described in [10], and so differed only in the way in which the faces were found in the video sequence (the face detection method). The systems submitted by the Idiap Research Institute served as baseline systems for the face (video) portion of the competition.

System 1 is referred to as a frontal face detector as it uses only a frontal face detector. **System 2** is referred to as a multi-view face detector as it uses a set of face detectors for different poses. Both frontal and multi-view face detection systems are taken from [30].

4.2   Instituto Tecnológico de Informática (ITI)

Two face recognition systems were submitted by the Instituto Tecnológico de Informática. Both systems, first detect faces every 0.1 seconds up to a maximum of 2.4 seconds of video. For enrolment or verification, only a few of the detected faces are selected based on a quality measure. The face verification approach was based on [37]. Each face is cropped to 64×64 pixels and 9×9 pixel patches are extracted at overlapping positions every 2 pixels, 784 features in total. The verification score is obtained using a Nearest-Neighbor classifier and a voting scheme. For further details refer to [37].

System 1 used the *haarcascade_frontalface_alt2* detection model that is included with the OpenCV library, and as quality measure used the confidence of a face-not-face classifier learnt using [36]. For verification, 10 face images are used. **System 2** used the face detector from the commercial OmniPerception's SDK and as quality the average of the confidences of the detector and the face-not-face classifier. For verification, 5 face images are used.

4.3   NICTA

NICTA submitted two video face recognition systems. Both systems used OpenCV for face detection in conjunction with a modified version of the Multi-Region Histogram (MRH) face comparison method [31]. To extend MRH from still-to-still to video-to-video comparison, a single MRH signature was generated for each video sequence by averaging the histograms for each region over the available frames. Two signatures are then compared through an L_1-norm based distance. If a person has several video sequences for enrolment, multiple signatures are associated with their gallery profile, and the minimum distance of those to the probe video signature is taken as the final result. For normalisation, each raw measurement is divided by the average similarity of each probe-gallery pair to a set of cohort signatures from the training set [31].

System 1 used only closely cropped faces (of size 64×64 pixels) which excluded image areas susceptible to disguises, such as hair and chin. **System 2** used information from those surrounding regions as well, resulting 96×96 pixel sized faces. The results show that the use of the surrounding regions considerably improved the recognition performance for the female set.

4.4   Tecnologico de Monterrey, Mexico and Arizona State University, USA (TEC-ASU)

The CUbiC-FVS (CUbiC-Face Verification System) was based on distance computations using a nearest neighbor classifier [14]. Each video stream was sliced into images and a face detection algorithm based on the mean-shift algorithm [13] was used to localize a face in a given frame. The block based discrete cosine transform (DCT) was used to derive facial features [16], since this feature is known to be robust to illumination changes.

For each user U_i, all the respective feature vectors were assembled into a training matrix M_i. A distance measure, D_{true}, was computed as the minimum distance of T (the test data) from the feature vectors of matrix M_k of the claimed identity k. Similarly, D_{imp} was computed as the minimum distance of T from the feature vectors of all matrices other than M_k. The ratio of D_{true} to D_{imp} was used to decide whether the claim has to be accepted or not. The scores were scaled so that clients have a positive score and imposters have a negative score.

4.5   University of Surrey (UNIS)

In total, UNIS submitted 4 systems which can be divided into two categories: *fusion* systems (FS) as well as *single descriptor* systems (SDS). FS is composed of two subsystems which differ mainly in the feature representation, one based on Multiscale Local Binary Pattern Histogram (MLBPH) [12] and the other based on Multiscale Local Phase Quantisation Histogram (MLPQH) [11]. SDS above refers to MLBPH. In each category, we have *basic* and *updated* versions. Hence, the 4 systems are: **System 1** (Basic+SDS), **System 2** (Updated+SDS), **System 3** (Basic+FS), **System 4** (Updated+FS). The *basic* and *updated* systems differ in terms of image selection strategies and data sets for the LDA matrix training. Regarding the image selection strategy, a basic system chooses a single face image, while an updated system selects 15 images from the video sequence. For training the LDA matrix, the training set of the MOBIO database is used in the basic system, while the updated system uses an external database. In each version, we measure the difference between the results of those 4 systems (without score normalisation) and the results of these systems with test-normalisation, using the training set of the MOBIO database.

4.6   Visidon Ltd (VISIDON)

Visidon face identification and verification system is originally designed for embedded usage, in order to quickly recognize persons in still images using a mobile phone, for example [1]. Thanks to a real-time frame performance, additional information provided by video can be easily utilized to improve the accuracy.

Both object detector (used for face and facial feature detection) and person recognition modules are based on our patented technology.

4.7   University of Nottingham (UON)

We implemented two methods: video-based (**System 1**) and image-based (**System 2**). System 1 makes use of all frames in a video and bases on the idea of Locally Linear Embedding [18]. System 2 uses only a couple of frames in a video and bases on 4 facial descriptors: Raw Image Intensity, Local Binary Patterns [2], Gabor Filters, Local Gabor Binary Patterns [39,19]; 2 subspace learning methods: Whitened PCA, One-shot LDA [38]; and Radial Basis Function SVM for verification.

In our experiments, system 2 performs much better than system 1. However, system 2 didn't perform as well as it could be because we made a mistake in the training process which makes the final SVM over-fitted. Another observation is that face detection is very important to get high accuracy.

4.8   National Taiwan University (NTU)

In each frame, we detected and aligned faces according to their eye and mouth positions. We also corrected the in-plane and out-of-plane rotations of the faces. We further rejected false face detections using a face-non face SVM classifier.

We proposed two systems: **System 1** applied the Facial Trait Code (FTC) [21]. FTC is a component based approach. It defines the N most discriminative local facial features on human faces. For each local feature, some prominent patterns are defined and symbolized for facial coding. The original version of FTC encodes a facial image into a codeword composed of N integers. Each integer represents a pattern for a local feature. In this competition, we used 100 local facial features, each had exactly 100 patterns, and it made up a feature vector of 100 integer numbers for each face. **System 2** applied the Probabilistic Facial Trait Code (PFTC), which is an extension of FTC. PFTC encodes a facial image into a codeword composed of N probability distributions. These distributions gives more information on similarity and dissimilarity between a local facial image patch and prominent patch patterns, and the PFTC is argued to outperform the original FTC. The associating study is currently under review. In this competition, we used 100 local facial features, each had exactly 100 patterns, and it made up a feature vector of 10000 real numbers for each face.

We collected at most 10 faces (in 10 frames) from an enrollment video. Each collected face was encoded into a gallery codeword. We collected at most 5 faces from a testing video. Each collected face was encoded into a probe codeword. Then, this probe codeword was matched against known gallery codewords.

4.9   iTEAM, Universidad Politecnica Valencia (UPV)

The UPV submitted two face recognition systems. Both systems use the same method for feature extraction and dimensionality reduction which are based on [3] and [23] respectively. KFA was trained using face images from the FERET database [28] and ten face images of each person of the MOBIO training set. Similarity measurements are computed using the cosine distance. Our systems differed only in the way in which the faces were extracted from the video sequence. **System 1** extracts faces from each frame independently using the OpenCV AdaBoost implementation [22] . **System 2** uses a commercial closed solution [26] for face detection and also introduces a Kalman filter to track the eyes and reduce the eye detection noise.

5   Speaker Verification Systems

5.1   Brno University of Technology (BUT)

Brno University of Technology submitted two audio speaker verification systems and one fusion of these two systems. The first system is Joint Factor Analysis and the second one is I-vector system. Both systems used for training the MOBIO data but also other data mainly from NIST SRE evaluations. Both system use 2048 Gaussians in UBM.

System 1 – Joint factor analysis (JFA) system closely follows the description of "Large Factor Analysis model" in Patrick Kenny's paper [20]. **System 2** – I-vector system was published in [15] and is closely related to the JFA framework. While JFA effectively splits model parameter space into wanted and unwanted variability subspaces, i-vector system aims at describing the subspace with the highest overall variability.

5.2   University of Avignon (LIA)

The LIA submitted two speakers recognition systems. Both are based on the UBM/GMM (Universal Background Model / Gaussian Mixture Model) paradigm without factor analysis. During this evaluation, development and training (even UBM training) were processed by using only MOBIO corpus.

The two systems, **LIA system 1** and **LIA system 2** differ by the acoustic parametrisation and the number of Gaussian components into the UBM. For the LIA system 1, the acoustic vectors are composed of 70 coefficients and the UBM has 512 components while LIA system 2 has only 50 coefficients, a bandwidth limited to the 300-3400Hz range and a UBM with 256 Gaussian components.

5.3 Tecnologico de Monterrey, Mexico and Arizona State University, USA (TEC-ASU)

Our speaker verification system, named TECHila, is based on a Gaussian Mixture Model (GMM) framework. The speech signal was downsampled to 8 KHz and a short-time 256-pt Fourier analysis is performed on a 25ms Hamming window (10ms frame rate). Every frame log-energy was tagged as high, medium and low (low and 80% of the medium log-energy frames were discarded). The magnitude spectrum was transformed to a vector of Mel-Frequency Cepstral Coefficients (MFCCs). Further, a feature warping algorithm is applied on the obtained features. Afterwards, a gender-dependent 512-mixture GMM UBM was initialised using k-means algorithm and then trained by estimating the GMM parameters via the EM (expectation maximization) algorithm. Target-dependent models were then obtained with MAP (maximum a posteriori) speaker adaptation. Finally, the score computation followed a hypothesis test framework.

Two approaches were used: a) *System 1* composed of 16 static Cepstral, 1 log Energy, and 16 delta Ceptral coefficient and single file adaptation (7 seconds of speech). b) *System 2* composed of 16 static Cepstral, 1 log Energy, 16 delta Ceptral coefficient, 16 double delta coefficient and all file adaptation (using the set of all target files).

5.4 University of West Bohemia (UWB)

Systems proposed by UWB made use of Gaussian Mixture Models (GMMs) and Support Vector Machines (SVMs), 4 systems were submitted. In the feature extraction process the speech signal was downsampled to 16kHz and voice activity detector was applied to discard non-speech frames. Subsystems exploited MFCCs extracted each 10 ms utilizing a 25 ms hamming window, delta's were added, simple mean and variance normalization was applied. GMMs were adapted from Universal Background Model (UBM) according to MAP adaptation with relevance factor 14. UBM consisted of 510 mixtures. UBM and impostors for SVM modeling were chosen from the world-set supplied by MOBIO in a gender specific manner. Score normalization was not utilized.

The specific systems were **System 1**: GMM-UBM [29], **System 2**: SVM-GLDS [7], **System 3**: SVM-GSV [8], and **System 4** was their combination. Regarding low amount of impostor data, the best performing system turned out to be **System 1** followed by **System 4**. However, for females **System 4** slightly outperformed **System 3**.

5.5 Swansea University and Validsoft (SUV)

The speaker verification systems submitted by Swansea University and Validsoft are based on standard GMM-MAP systems [29], whose originality lies in the use of wide band (0 to 24 kHz) mel frequency cepstral coefficients (MFCCs) features, an idea already explored by Swansea University during the Biosecure evaluation campaign [17].

System 1 is a GMM-MAP system with a large number filter bands (50) and cepstral coefficients (29). **System 2** is a GMM-MAP system based on a standard number of filter bands (24) and cepstral coefficients (16). **System 3** is a score level fusion of System 1 and System 2 after T-normalisation.

6   Discussion

In this section, we summarize and discuss the results of this evaluation. To facilitate the comparison between participants, we selected the best performing face or speaker verification system for each participant[2].

Table 1. Table presenting the results (HTER) of the best performing face verification systems for each participants on the Test set

	Male	Female	Average
IDIAP (Face System A)	25.45%	24.39%	24.92%
ITI (Face System B)	16.92%	17.85%	17.38%
NICTA (Face System C)	25.43%	20.83%	23.13%
NTU (Face System D)	20.50%	27.26%	23.88%
TEC (Face System E)	31.36%	29.08%	30.22%
UNIS (Face System F)	9.75%	12.07%	10.91%
UON (Face System G)	29.80%	23.89%	26.85%
UPV (Face System H)	21.86%	23.84%	22.85%
VISIDON (Face System I)	10.30%	14.95%	12.62%

Table 2. Table presenting the results (HTER) of the best performing speaker verification systems for each participants on the Test set

	Male	Female	Average
BUT (Speech System A)	10.47%	10.85%	10.66%
LIA (Speech System B)	14.49%	15.70%	15.10%
SUV (Speech System C)	13.57%	15.27%	14.42%
TEC (Speech System D)	15.45%	17.41%	16.43%
UWB (Speech System E)	11.18%	10.00%	10.59%

6.1   Face Verification

A summary of the results of the face verification systems can be found in Table 1. The results of the same systems are also presented in the DET plots in Figure 1 (male trials) and in Figure 2 (female trials).

From the plots, it can be observed mainly three groups of systems (more distinctly for female trials). The first group is composed by the two best performing systems. The best performance, with an HTER of 10.9%, is obtained by the UNIS System 4 (norm) which is fusing multiple cues and is post-processing the scores (score normalisation). This system without score normalisation, UNIS System 4, obtained an HTER of 12.9%. The second best performance is obtained by the VISIDON System 1 with an HTER of 12.6% and is using local filters but no score normalisation. Interestingly, it should be noticed that these systems use a proprietary software for the task of face detection. The second group is composed of two systems, ITI System 2 and NICTA System 2 (norm). ITI System 2 is also using a proprietary software for face detection (the same than UNIS System 4) while NICTA System 2 (norm) is using OpenCV for that task.

Interestingly, NICTA System 2 (with normalisation) performs better on the female test set than on the male test. This is the opposite trend to what occurs for most of the other systems (such as the UNIS, VISIDON and ITI systems) where better results are

---

[2] Please note that similarly to the previous sections the systems of each participant are numbered *1, 2, ...* In this section the best systems, one for each participants, are numbered *A, B, ...*

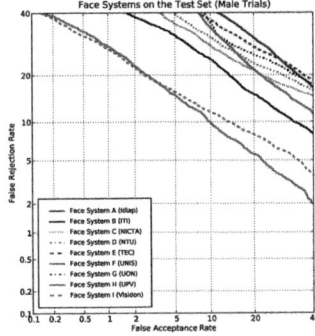
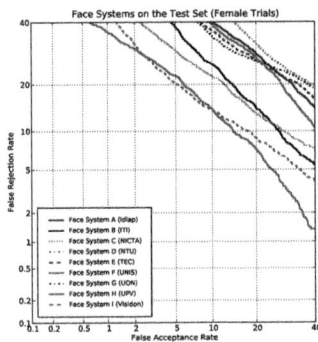

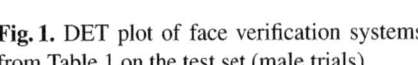

Fig. 1. DET plot of face verification systems from Table 1 on the test set (male trials)

Fig. 2. DET plot of face verification systems from Table 1 on the test set (female trials)

obtained on the male test set than on the female test set. The third group is composed mainly by all the remaining systems and obtained an HTER of more than 20%. The majority of these systems uses an OpenCV like face detection scheme and all seem to have similar performance.

From these results we can draw two conclusions: (1) the choice of the face detection system can have an important impact on the face verification performance, and (2) the role of score normalisation on the performance is difficult to establish clearly.

The impact of the face detection algorithm can be seen clearly when examining the two systems from ITI. The difference between these two systems from ITI comes only from the use of a different face detection technique: ITI System 1 uses the frontal OpenCV face detector and ITI System 2 uses the OmniPerception SDK. The difference in face detector alone leads to an absolute improvement of the average HTER of more than 4%. This leads us to conclude that one of the biggest challenges for video based face recognition is the problem of accurate face detection.

A second interesting conclusion is that score normalisation can be difficult to apply to face recognition. This can be seen by examining the performance of the systems from UNIS and NICTA. The NICTA results show that score normalisation provides a minor but noticeable improvement in performance. However, the UNIS systems provide conflicting results as score normalisation on Systems 1 and 2 degrades performance whereas score normalisation on Systems 3 and 4 improves performance. The only conclusion that can be brought from this is that more work is necessary to be able to successfully apply score normalisation to face verification.

6.2 Speaker Verification

A summary of the results for the speaker verification systems is presented in terms of HTER in Table 2 and also in DET plots in Figure 3 (male trials) and in Figure 4 (female trials). Generally, the audio systems exhibit smaller dispersion of HTER scores than their video counterparts, which can be attributed to lesser differences between individual audio systems than between those for videos.

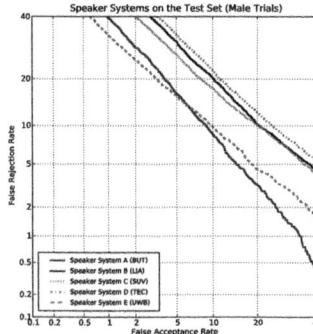
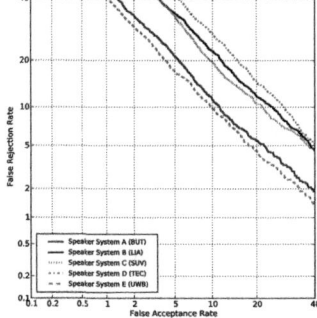

Fig. 3. DET plot of speaker verification systems from Table 2 on the test set (male trials)

Fig. 4. DET plot of speaker verification systems from Table 2 on the test set (female trials)

From the results it can be seen that voice activity detection (VAD) is crucial for all audio systems (just as face detection is crucial for face verification). The participants use approaches ranging from classical energy based (LIA, TEC-ASU) through to sub-band quality measures (UWB) and the use of phone recognizers (BUT). By contrast , the variability in feature extraction is much smaller with most participants using standard MFCC coefficients with some variants.

For the speaker verification part, two approaches were adopted: GMM-UBM and SVM-based. The former ones were generally weaker in performances, with the exception of UWB System1 - a pure GMM-UBM based system that was the best performing single system. This performance is probably due to UWB VAD, their system is also fully trained on MOBIO 16kHz data.

The latter approach (SVM) performed well both on standard GMM means (UWB) as well as on JFA-derived speaker factors (BUT System1). This supports the conclusion that SVMs provide superior performance on shorter segments of speech.

The importance of score normalisation was also confirmed, mainly for the systems not based on SVMs. However, it was hard to derive representative gender dependent ZT-norm cohorts, mainly because there were too few speakers in the world-set of the MOBIO database.

Another lesson learned was the importance of the target (MOBIO) data for training when compared to the hundreds hours of non-target (NIST) telephone data. It can be seen that the SVM-based techniques largely benefit from having this data in their imposter sets. On the other hand, JFA does not improve with this data as the utterances are too short and too few.

7   Conclusion

This paper presented the results of several uni-modal face and speaker verification techniques on the MOBIO database (Phase I). This database provides realistic and challenging conditions as it was captured on a mobile device and in uncontrolled environments.

The evaluation was organised in two stages. During the first stage, the training and development sets of the database was distributed among the participants (from December 1 2009 to January 27 2010). The deadline for the submission of the first results by the participants on the development set was February 1 2010. During the second stage, the test set was distributed only to the participants that met the first deadline. The deadline for the submission of the results on the test set was March 8 2010.

Out of the thirty teams that signed the End User License Agreement (EULA) of the database and downloaded it, finally, fourteen teams have participated to this evaluation. Eight teams participated to the face verification part of the evaluation, four teams participated to the speaker verification part of the evaluation and one team participated both to the face and the speaker part. Only one team dropped from the competition during the second stage. Each participant provided at least the results of one system but were allowed to submit the results of several systems.

This evaluation produced three interesting findings. First, it can be observed that face verification and speaker verification obtained the same level of performance. This is particularly interesting because it is generally observed that speaker verification performs much better than face verification in general. Second, it has been highlighted that segmentation (face detection and voice activity detection) was critical both for face and speaker verification. Finally, it has been shown that the two modalities are complementary as a clear gain in performance can be obtained simply by fusing the individual face and speaker verification scores.

Overall, it was shown that the MOBIO database provides a challenging test-bed both for face verification, for speaker verification but also for bi-modal verification. This evaluation would have established baseline performance for the MOBIO database.

The MOBIO consortium is planning to distribute the database (Phase I) in August 2010 together with the results and the annotations (face detection output) generated by the participants during this evaluation. It is foreseen as well to distribute the Phase II of the MOBIO database before the end of 2010.

Acknowledgements

This work has been performed by the MOBIO project 7th Framework Research Programme of the European Union (EU), grant agreement number: 214324. The authors would like to thank the EU for the financial support and the partners within the consortium for a fruitful collaboration. For more information about the MOBIO consortium please visit http://www.mobioproject.org.

The authors would also like to thank Phil Tresadern (University of Manchester), Bastien Crettol (Idiap Research Institute), Norman Poh (University of Surrey), Christophe Levy (University of Avignon), Driss Matrouf (University of Avignon), Timo Ahonen (University of Oulu), Honza Cernocky (Brno University of Technology) and Kamil Chalupnicek (Brno University of Technology) for their work in capturing this database and development of the protocol.

NICTA is funded by the Australian Government as represented by the *Department of Broadband, Communications and the Digital Economy* as well as the Australian Research Council through the *ICT Centre of Excellence* program.

References

1. Visidon ltd., http://www.visidon.fi
2. Ahonen, T., Hadid, A., Pietikainen, M.: Face Recognition with Local Binary Patterns. In: Pajdla, T., Matas, J(G.) (eds.) ECCV 2004. LNCS, vol. 3021, pp. 469–481. Springer, Heidelberg (2004)
3. Albiol, A., Monzo, D., Martin, A., Sastre, J., Albiol, A.: Face recognition using hog-ebgm. Pattern Recognition Letters 29(10), 1537–1543 (2008)
4. Bengio, S., Mariéthoz, J., Keller, M.: The Expected Performance Curve. In: Intl Conf. On Machine Learning, ICML (2005)
5. Bledsoe, W.W.: The model method in facial recognition. Tech. rep., Panoramic Research Inc (1966)
6. Campbell, J.P.: Speaker recognition: A tutorial. In: Proceedings of the IEEE 85(9) (September 1997)
7. Campbell, W.: Generalized linear discriminant sequence kernels for speaker recognition. In: IEEE International Conference on Acoustics, Speech and Signal Processing, ICASSP 2002, vol. 1, pp. I–161–I–164 (2002)
8. Campbell, W., Sturim, D., Reynolds, D.: Support vector machines using gmm supervectors for speaker verification. IEEE Signal Processing Letters 13(5), 308–311 (2006)
9. Campbell, W., Sturim, D., Reynolds, D., Solomonoff, A.: Svm based speaker verification using a gmm supervector kernel and nap variability compensation. In: Acoustics, Speech and Signal Processing ICASSP 2006 Proceedings, vol. 1, pp. I–I (2006)
10. Cardinaux, F., Sanderson, C., Marcel, S.: Comparison of mlp and gmm classifiers for face verification on xm2vts. In: International Conference on Audio- and Video-based Biometric Person Authentication, pp. 1058–1059 (2003)
11. Chan, C., Kittler, J., Poh, N., Ahonen, T., Pietikäinen, M.: (multiscale) local phase quantization histogram discriminant analysis with score normalisation for robust face recognition. In: VOEC, pp. 633–640 (2009)
12. Chan, C.H., Kittler, J., Messer, K.: Multi-scale local binary pattern histograms for face recognition. In: Lee, S.-W., Li, S.Z. (eds.) ICB 2007. LNCS, vol. 4642, pp. 809–818. Springer, Heidelberg (2007)
13. Comaniciu, D., Ramesh, V., Meer, P.: Real-time tracking of non-rigid objects using mean shift. In: Proceedings of IEEE Conference on Computer Vision and Pattern Recognition (CVPR 2000), pp. 142–149 (2000)
14. Das, A.: Audio visual person authentication by multiple nearest neighbor classifiers. In: SpringerLink (2007)
15. Dehak, N.., Dehak, R.., Kenny, P., Brümmer, N., Ouellet, P., Dumouchel, P.:.Support vector machines versus fast scoring in the low-dimensional total variability space for speaker verification. In: Proc. International Conferences on Spoken Language Processing (ICSLP). pp. 1559–1562 (September 2009)
16. Ekenel, H., Fischer, M., Jin, Q., Stiefelhagen, R.: Multi-modal person identification in a smart environment. In: IEEE CVPR (2007)
17. Fauve, B., Bredin, H., Karam, W., Verdet, F., Mayoue, A., Chollet, G., Hennebert, J., Lewis, R., Mason, J., Mokbel, C., Petrovska, D.: Some results from the biosecure talking face evaluation campaign. In: Proceedings of International Conference on Acoustics Speech and Signal Processing, ICASSP (2008)
18. Hadid, A., Pietikäinen, M.: Manifold learning for video-to-video face recognition. In: COST 2101/2102 Conference, pp. 9–16 (2009)
19. Hieu, N., Bai, L., Shen, L.: Local gabor binary pattern whitened pca: A novel approach for face recognition from single image per person. In: Proceedings of The 3rd IAPR/IEEE International Conference on Biometrics 2009 (2009)

20. Kenny, P., Ouellet, P., Dehak, N., Gupta, V., Dumouchel, P.: A study of inter-speaker variability in speaker verification. In: IEEE Transactions on Audio, Speech and Language Processing (July 2008)
21. Lee, P.H., Hsu, G.S., Hung, Y.P.: Face verification and identification using facial trait code. In: IEEE Conference on Computer Vision and Pattern Recognition, pp. 1613–1620 (2009)
22. Lienhart, R., Kuranov, A., Pisarevsky, V.: Empirical analysis of detection cascades of boosted classifiers for rapid object detection. In: Michaelis, B., Krell, G. (eds.) DAGM 2003. LNCS, vol. 2781, pp. 297–304. Springer, Heidelberg (2003)
23. Liu, C.: Capitalize on dimensionality increasing techniques for improving face recognition grand challenge performance. IEEE Transactions on Pattern Analysis and Machine Intelligence 28, 725–737 (2006)
24. Marcel, S., McCool, C., Matějka, P., Ahonen, T., Černocký, J., Chakraborty, S., Balasubramanian, V., Panchanathan, S., Chan, C.H., Kittler, J., Poh, N., Fauve, B., Glembek, O., Plchot, O., Jančík, Z., Larcher, A., Lévy, C., Matrouf, D., Bonastre, J.F., Lee, P.H., Hung, J.Y., Wu, S.W., Hung, Y.P., Machlica, L., Mason, J., Mau, S., Sanderson, C., Monzo, D., Albiol, A., Albiol, A., Nguyen, H., Li, B., Wang, Y., Niskanen, M., Turtinen, M., Nolazco-Flores, J.A., Garcia-Perera, L.P., Aceves-Lopez, R., Villegas, M., Paredes, R.: Mobile biometry (MOBIO) face and speaker verification evaluation. Idiap-RR Idiap-RR-09-2010. Idiap Research Institute (May 2010)
25. Messer, K., Kittler, J., Sadeghi, M., Hamouz, M., Kostin, A., Cardinaux, F., Marcel, S., Bengio, S., Sanderson, C., Poh, N., Rodriguez, Y., Czyz, J., Vandendorpe, L., McCool, C., Lowther, S., Sridharan, S., Chandran, V., Palacios, R.P., Vidal, E., Bai, L., Shen, L., Wang, Y., Yueh-Hsuan, C., Hsien-Chang, L., Yi-Ping, H., Heinrichs, A., Muller, M., Tewes, A., von der Malsburg, C., Wurtz, R., Wang, Z., Xue, F., Ma, Y., Yang, Q., Fang, C., Ding, X., Lucey, S., Goss, R., Schneiderman, H.: Face authentication test on the banca database. In: Proceedings of the 17th International Conference on Pattern Recognition, vol. 4, pp. 523–532 (2004)
26. Neurotechnologija: Verilook SDK, neurotechnologija Biometrical and Artificial Intelligence Technologies, http://www.neurotechnologija.com
27. Phillips, J., Flynn, P., Scruggs, T., Bowyer, K., Chang, J., Hoffman, K., Marques, J., Min, J., Worek, W.: Overview of the face recognition grand challenge. In: IEEE Conference of Computer Vision and Pattern Recognition, vol. 1, pp. 947–954 (2005)
28. Phillips, J.P., Moon, H., Rizv, S., Rauss, P.J.: The FERET evaluation methodology for face-recognition algorithms. IEEE Transactions on Pattern Analysis and Machine Intelligence 22(10), 1090–1104 (2000)
29. Reynolds, D.A., Quatieri, T.F., Dunn, R.B.: Speaker verification using adapted gaussian mixture models. Digital Signal Processing 10(1-3), 19–41 (2000)
30. Rodriguez, Y.: Face Detection and Verification using Local Binary Patterns. Ph.D. thesis, EPFL (2006)
31. Sanderson, C., Lovell, B.C.: Multi-region probabilistic histograms for robust and scalable identity inference. In: Tistarelli, M., Nixon, M.S. (eds.) ICB 2009. LNCS, vol. 5558, pp. 199–208. Springer, Heidelberg (2009)
32. Shriberg, E., Ferrer, L., Kajarekar, S.: Svm modeling of snerf-grams for speaker recognition. In: International Conference on Spoken Language Processing (ICSLP), Jeju Island, Korea (October 2004)
33. Stolcke, A., Ferrer, L., Kajarekar, S., Shriberg, E., Venkataraman, A.: MLLR transforms as features in speaker recognition. In: International Conference on Spoken Language Processing (ICSLP), Lisbon, Portugal, pp. 2425–2428 (September 2005)
34. Tan, X., Chen, S., Zhou, Z.H., Zhang, F.: Face recognition from a single image per person: A survey. Pattern Recognition 39(9), 1725–1745 (2006)
35. Tistarelli, M., Bicego, M., Grosso, E.: Dynamic face recognition: From human to machine vision. Image and Vision Computing 27(3), 222–232 (2009)

36. Villegas, M., Paredes, R.: Simultaneous learning of a discriminative projection and proto-types for nearest-neighbor classification. In: IEEE Conference on Computer Vision and Pattern Recognition, CVPR 2008, pp. 1–8 (2008)
37. Villegas, M., Paredes, R., Juan, A., Vidal, E.: Face verification on color images using local features. In: IEEE Computer Society Conference on Computer Vision and Pattern Recognition CVPR Workshops 2008, pp. 1–6 (June 2008)
38. Wolf, L., Hassner, T., Taigman, Y.: Descriptor based methods in the wild. In: Real-Life Images workshop at the European Conference on Computer Vision, (ECCV) (October 2008)
39. Zhang, W., Shan, S., Gao, W., Chen, X., Zhang, H.: Local Gabor Binary Pattern Histogram Sequence (LGBPHS): A Novel Non-Statistical Model for Face Representation and Recognition. In: Proc. ICCV, pp. 786–791 (2005)
40. Zhao, W., Chellappa, R., Phillips, P.J., Rosenfeld, A.: Face recognition: A literature survey. ACM Computing Surveys 35(4), 399–458 (2003)

Pattern Recognition in Histopathological Images: An ICPR 2010 Contest

Metin N. Gurcan[1,*], Anant Madabhushi[2], and Nasir Rajpoot[3]

[1] Department of Biomedical Informatics, The Ohio State University,
Columbus, OH 43210, USA
[2] Department of Biomedical Engineering, Rutgers The State University of New Jersey,
Piscataway, NJ 08854, USA
[3] Department of Computer Science, University of Warwick, Coventry CV4 7AL, UK
metin.gurcan@osumc.edu, anantm@rci.rutgers.edu,
nasir@dcs.warwick.ac.uk
http://bmi.osu.edu/cialab/ICPR_contest/

Abstract. The advent of digital whole-slide scanners in recent years has spurred a revolution in imaging technology for histopathology. In order to encourage further interest in histopathological image analysis, we have organized a contest called "Pattern Recognition in Histopathological Image Analysis." This contest aims to bring some of the pressing issues facing the advance of the rapidly emerging field of digital histology image analysis to the attention of the wider pattern recognition and medical image analysis communities. Two sample histopathological problems are explored: counting lymphocytes and centroblasts. The background to these problems and the evaluation methodology are discussed.

Keywords: histopathology, computerized image analysis, pattern recognition, follicular lymphoma.

1 Introduction

The advent of digital whole-slide scanners in recent years has spurred a revolution in imaging technology for histopathology. The large multi-Giga-pixel images produced by these scanners contain a wealth of information potentially useful for computer-assisted disease diagnosis, grading, and prognosis. Processing and analysis of such high-resolution images, however, remain non-trivial tasks, not just because of the sheer size of the images but also due to complexities of the underlying factors, including variable staining procedures and practices, illumination variations, diversity in imaging devices, and last but not the least the ultimate goal of the analysis. In order to encourage further interest in histopathological image analysis, we have organized a contest called "Pattern Recognition in Histopathological Image Analysis," as part of

---

[*] The authors are the organizers of this contest and have contributed equally to this article as well as to the design, preparation and evaluation of the contest.

D. Ünay, Z. Çataltepe, and S. Aksoy (Eds.): ICPR 2010, LNCS 6388, pp. 226–234, 2010.

the ICPR 2010. This contest aims to bring some of the pressing issues facing the advance of the rapidly emerging field of digital histology image analysis to the attention of the wider pattern recognition and medical image analysis communities.

We proposed two problems and provided the training dataset for each problem to the contestants. The problems are described in the following sections and Table 1 summarizes the data information:

Table 1. Summary of Datasets

Dataset Name	Number of training images	Number of test images
Problem 1	10	10
Problem 2	5	5

1.1 Problem 1: Counting Lymphocytes on Histopathology Images

Breast cancer (BC) is the second leading cause of cancer related deaths in women, with more than 182 000 new cases of invasive BC predicted in the United States for 2008 alone [1]. Although it is a common cancer diagnosis in women, the fact that BC exhibits an exceptionally heterogeneous phenotype in histopathology [2] leads to a variety of prognoses and therapies. One such phenotype is the presence of lymphocytic infiltration (LI) in invasive BC that exhibits amplification of the HER2 gene (HER2+ BC). Most HER2+ BC is currently treated with agents that specifically target the HER2 protein. Researchers have shown that the presence of LI in histopathology is a viable prognostic indicator for various cancers, including HER2+ BC [3]–[5]. The function of LI as a potential antitumor mechanism in BC was first shown by Aaltomaa et al. [4]. More recently, Alexe et al. [5] demonstrated a correlation between the presence of high levels of LI and tumor recurrence in early stage HER2+ BC.The ability to automatically detect and quantify extent of LI on histopathology imagery could potentially result in the development of an image based prognostic tool for Her2+ and ovarian cancer patients.

However, lymphocyte segmentation in Haemotoxylin (H) and Eosin (E)-stained histopathology images is complicated by the similarity in appearance between lymphocyte nuclei and other structures (e.g. cancer nuclei) in the image. Additional challenges include biological variability, histological artifacts, and high prevalence of overlapping objects. Although active contours are widely employed in image segmentation, they are limited in their ability to segment overlapping objects and are sensitive to initialization [6].

Hematoxylin and eosin (H&E) stained BC biopsy cores were scanned into a computer using a high resolution whole slide scanner (Aperio Systems) at 40x optical magnification at The Cancer Institute of New Jersey (CINJ). A total of 20 HER2+ BC images (from nine patients) exhibiting various levels of LI were used for this competition. The images were downsampled by a factor of 2 and saved as 200×200 pixels digital images. The ground truth for spatial presence of LI was obtained via manual detection and segmentation performed by a breast cancer oncologist from CINJ. The ground truth for LI detection evaluation was obtained in the form of highlighted pixels representing the approximate centers of each of the lymphocytes in all 100 images. Note that, since the 20 images comprised over 2000 individual lymphocytes, and on account of the effort involved in manual segmentation, only a few dozen lymphocytes randomly chosen from the set of 20 images were delineated by the expert to allow the evaluation of the segmentation performance of the model. The detection performance of the model, however, was evaluated on all lymphocytes across all 20 images. The H&E-stained histopathology images comprise of four main structures or entities, namely: 1) BC nuclei; 2) lymphocyte nuclei; 3) stroma; and 4) background, as illustrated in Figure 1. Note the extent of overlap between objects and the similarity between lymphocyte nuclei and BC nuclei. Lymphocyte nuclei tend to be stained deeper than BC nuclei and are often smaller in size.

Lymphocytic centers were indicated on all the images. A distinct set of testing images will be provided to the contestants on the day of the competition. These images may have been digitized on the same scanner and stained in a different lab compared to the training images.

1.2 Problem 2: Counting Centroblasts from Histology Images of Follicular Lymphoma

Follicular Lymphoma (FL), a common type of non-Hodgkins lymphoma, is a cancer of lymph system. According to World Health Organization's recommendations, FL has three histological grades indicating the degree of the malignancy of the tumor [7]. Histological grading of FL is based on the number of centroblasts, large malignant cells, in ten representative neoplastic follicle regions in a high power field (HPF) of 0.159 mm^2. Based on this method FL is stratified into three histological grades: FL grade I (0-5 centroblasts/HPF), FL grade II (6-15 centroblasts/HPF) and FL grade III (>15 centroblasts/HPF) ordered from the least to the most malignant subtypes, respectively. Further information about this problem and some previous work in this area can be found in the References [8-17].

There were a total of five images containing centroblasts which were H&E stained and digitized at 40 x resolution to serve as the training set. Centroblast centers were indicated on all the images, as marked by at least two expert pathologists. Figure 2 shows an example image. A distinct set of testing images will be provided to the contestants on the day of the competition. Characteristics of these images will be similar to those of the training images in terms of slide preparation and digitization.

2 Competition

Twenty three groups showed interest in the competition and were provided with the training dataset as well as the ground truth for Problems 1 and 2 as described in Sections 1.1 and 1.2. Five of these groups developed algorithms to solve these problems and submitted their results; three groups turned their efforts into papers, which are published in this volume.

3 Evaluation Methodology

All the submitted results were evaluated using a standard criteria and automatically. The following sections describe the evaluation methodologies.

3.1 Evaluation Methodology for Problem 1

3.1.1 Region-Based Measures

The region-based performance measures were defined as follow

- Dice Coefficient (DICE) $= \frac{2 \times |A(S) \cap A(G)|}{|A(S)| + |A(G)|}$;
- Overlap (OL) $= \frac{|A(S) \cap A(G)|}{|A(S) \cup A(G)|}$;
- Sensitivity (SN) $= \frac{|A(S) \cap A(G)|}{|A(G)|}$;
- Specificity (SP) $= \frac{|C - A(S) \cap A(G)|}{|C - A(G)|}$;
- Positive Predictive Value (PPV) $= \frac{|A(S) \cap A(G)|}{|A(S)|}$,

where C is the total number of pixels in the image and $|s|$ represents the cardinality of any set s. $|A(S)|$ and $|A(G)|$ are the areas of the closed boundary of segmentation results and manual delineation, respectively. The values shown in Table 1 are the values obtained by averaging across ten images. Note that higher values for each of the region-based measures indicates superior performance with a maximum value of 1.0 reflecting the best possible segmentation performance, while 0.0 reflecting the worst possible performance.

3.1.2 Boundary-Based Measures

The boundary-based performance measures are defined as follow

- Hausdorff distance (HD) $= \max_{w}[\min_{x} \|c_w - c_x\|], (c_w \in S, c_x \in G)$;
- Mean absolute distance (MAD) $= \frac{1}{M} \sum_{w=1}^{M} \|c_w - c_x\|$,

where S and G are closed boundaries of segmentation results and manual delineations, respectively. Each of S and G are represented as set of image pixels c_w and c_x respectively, where any pixel c is represented by its two dimensional Cartesian coordinates. M is the number of pixels on the closed boundaries of segmentation results. Note that lower values for each of the boundary-based measures indicates superior performance with a value of 0 reflecting perfect concordance between the boundary obtained via the segmentation algorithm and the expert delineated ground truth.

3.2 Evaluation Methodology for Problem 2

The ground-truth information regarding the centroblasts are the locations marked by a consensus of pathologists. Therefore the evaluation is based on counting the number of true/false detection by comparing the centroid locations of the cells detected by the proposed computerized systems. If the distance between the centroid of a detected cell and the ground-truth marking is less than a threshold (30 pixels, equivalent of ~7.5 microns), then it is considered as a true positive, otherwise it is counted as false positive. The threshold value is determined empirically by measuring the average size of a cell on the training set of images.

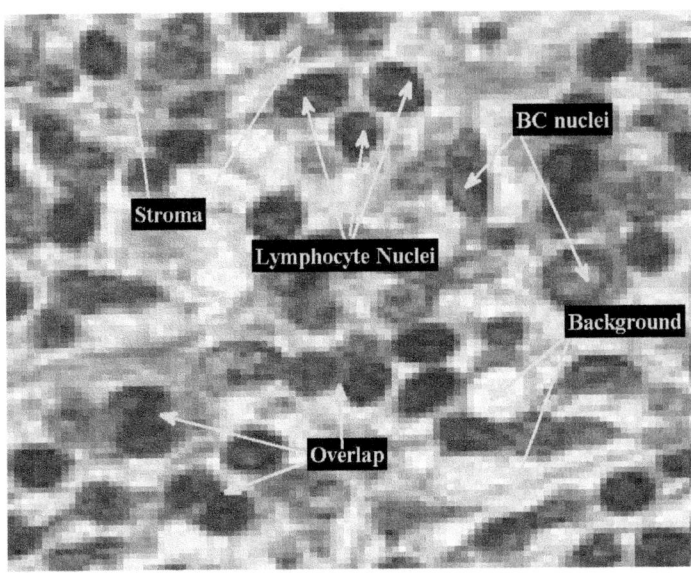

Fig. 1. Example of a HER2+ BC histopathology image showing lymphocyte nuclei, BC nuclei, stroma and the background. Note the overlap between adjacent nuclei and the similarity in appearance between cancer and lymphocyte nuclei [6].

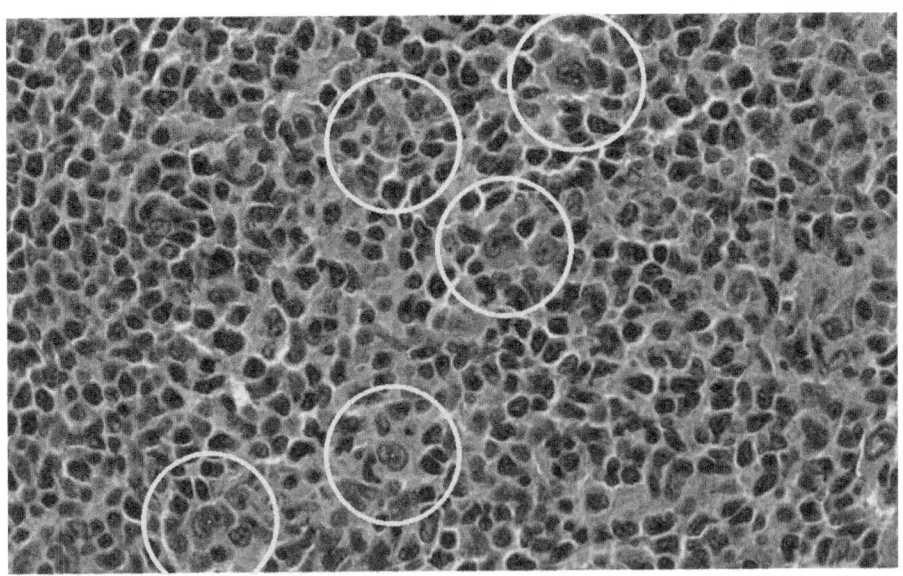

Fig. 2. An example of part of an H&E-stained follicular lymphoma image with centroblasts

4 Evaluation Results

Five groups participated in this competition and only four of these submitted papers. These groups are shown in Table 2. Below we summarize the results of evaluation for both off-line and on-line results for the training and testing data, respectively.

4.1 Evaluation of Off-Line Results for the Training Data

Submitted results of the five groups participating in this competition are summarized in Tables 3 and 4 below. As seen in these results, for Problem 1, Group 4's algorithm outperformed all the other methods in terms of both region-based and boundary-based measures of performance. For Problem 2, only two groups submitted their results, again Group 4's method producing impressive results. However, Group 4 chose not to submit details of their methods for publication in these proceedings. It is worth noting that these results were obtained using training data provided to these groups before the actual contest, where previously unseen test data was given to the contestants for on-site evaluation (please see Section 4.2).

Table 2. Groups participating in the competition

Group	People	Institute	Paper
1	Cheng, J, Veronika, M, Rajapakse, J	Singapore-MIT Alliance, Singapore	[18]
2	Gupta, S, Kuse, M, Sharma, T	The LNM Institute of Information Technology, Jaipur, India	[19]
3	Graf, F, Grzegorzek, M, Paulus, D	Institute for Computational Visualistics, University of Koblenz-Landau, Germany	[21]
4	Bruynooghe, M	Alkmaar, The Netherlands	-
5	Panagiotakis, C, Ramasso, E, Tziritas, G	Department of Computer Science, University of Crete, Greece	[20]

Table 3. Evaluation results for Problem 1 (Detecting Lymphocytes in Breast Histopathology Images); Best performance is shown in **bold**

Group#	Region-based Measures					Boundary-based Measures	
	DICE	OL	SN	SP	PPV	HD	MAD
Group 1 [18]	0.73	0.57	0.57	1	1	4.58	0.77
Group 2 [19]	0.74	0.58	0.58	1	1	3.63	0.65
Group 3	0.37	0.23	0.23	1	1	21.95	9.14
Group 4	**0.83**	**0.71**	**0.71**	1	1	3.73	**0.41**
Group 5 [20]	0.74	0.59	0.59	1	1	**3.51**	0.62

Table 4. Evaluation results for Problem 2 (Detecting Centroblasts in Follicular Lymphoma Histopathology Images); Best performance is shown in **bold**. Centroblast coordinates or contours were not submitted to evaluate more detailed region-based or boundary-based measures.

Group#	TPR	FPR
Group 1	0.38	0.83
Group 4	1	0

4.2 On-Site Evaluation

All the five groups participated at an on-site evaluation. The groups were given test images which were different from the training images and were asked to run their programs on these images and provide the organizers with results. The participating groups only attempted the first problem.

All groups were evaluated on (a) ability to identify lymphocytic centers and (b) the total number of lymphocytes identified. All groups were asked to provide segmentation results in the form of binary masks with just the centers of the lymphocytes identified. They were also asked to provide the contours of the individual cells, though these were not used for the evaluation (since ground truth evaluation for contours could not be obtained from a second independent expert).

For criterion (a) and (b) above, the mean and standard deviation errors were tabulated. For all 4 numbers reported, a smaller number represented a better result. In case of criterion (a) the Euclidean distance d between the ground truth and the result provided by the participants was calculated. In case of criterion (b) the absolute difference between the true number of cells and the number of cells N found by the participating group was identified. Table 5 shows the on-site evaluation results with the ranking of performance, where μ and σ denote the mean and standard deviation, respectively.

Table 5. Evaluation results for Problem 1 for on-site evaluation

Ranking	Group Number	μ_d	σ_d	μ_N	σ_N
1	2	3.04	**3.40**	**14.01**	**4.4**
2	5	**2.87**	3.80	14.23	6.3
3	3	7.60	6.30	24.50	16.2
4	1	8.10	6.98	26.67	12.5

No results were obtained for the Group 4 since they required feeding in the centers of the cells to their segmentation program and since contour evaluation was not performed during the on-site evaluation.

5 Conclusions

The main purpose of this contest was to encourage pattern recognition and computer vision researchers in getting involved in the rapidly emerging area of histopathology image analysis. Twenty three groups registered their interest in participating in this contest, while five of these groups actually submitted their results on training data released before the actual contest. Two of the groups submitted results for both the problems, detection of lymphocytes in breast histopathology images and detection of centroblasts in follicular lymphoma histophathology images. Of these, one group has produced quite promising results in terms of both types of performance measures, region-based and boundary-based. Given this was the first contest of its kind, we are encouraged by the level of enthusiasm and interest shown in this contest so far and look forward to the results of these groups' algorithms in the actual contest at the conference. Given that digital pathology is a nascent field and that application of pattern recognition and image analysis methods to digitized histopathology even more recent, there is not yet consensus on what level of performance would be acceptable in the clinic. While it is clear that most algorithms in this domain should produce an output which either directly (or via some transformation) correlates highly with clinical and patient outcome, it is not yet clear what level of algorithm performance would suffice towards this goal. Further versions of this competition will thus seek to explore, in a more quantitative fashion, the correlation between algorithmic performance and disease outcome.

Acknowledgments. This work was supported in part by Award Number R01CA 134451 from the National Cancer Institute (MG), the Wallace H. Coulter Foundation, the National Cancer Institute under Grants R01CA136535-01, R01 CA140772-01, Grant ARRA-NCI-3 R21 CA127186-02S1, R21CA127186-01, R03CA128081-01, and R03CA143991 - 01, and the Cancer Institute of New Jersey. The content is solely the responsibility of the authors and does not necessarily represent the official views of the National Cancer Institute, or the National Institutes of Health. The authors would like to thank Olcay Sertel, Jun Xu, Dr. Shridar Ganesan, and Ajay Basavanhally for their help in the preparation of the datasets and evaluation of the results.

References

1. Jemal, A., Siegel, R., Ward, E., Hao, Y., Xu, J., Murray, T., Thun, M.J.: Cancer statistics. CA Cancer J. Clin. 58(2), 71–96 (2008)
2. Bertucci, F., Birnbaum, D.: Reasons for breast cancer heterogeneity. J. Biol. 7(6) (2008)
3. van Nagell, J.R., Donaldson, E.S., Wood, E.G., Parker, J.C.: The significance of vascular invasion and lymphocytic infiltration in invasive cervical cancer. Cancer 41(1), 228–234 (1978)
4. Aaltomaa, S., Lipponen, P., Eskelinen, M., Kosma, V.M., Marin, S., Alhava, E., Syrjanen, K.: Lymphocyte infiltrates as a prognostic variable in female breast cancer. Eur. J. Cancer 28A(4/5), 859–864 (1992)
5. Alexe, G., Dalgin, G.S., Scanfeld, D., Tamayo, P., Mesirov, J.P., DeLisi, C., Harris, L., Barnard, N., Martel, M., Levine, A.J., Ganesan, S., Bhanot, G.: High expression of lymphocyte-associated genes in node negative her2+ breast cancers correlates with lower recurrence rates. Cancer Res. 67(22), 10669–10676 (2007)

6. Fatakdawala, H., Basavanhally, A., Xu, J., Bhanot, G., Ganesan, S., Feldman, M., Tomaszewski, J., Madabhushi, A.: Expectation Maximization Driven Geodesic Active Contour with Overlap Resolution: Lymphocyte Segmentation on Breast Cancer Histopathology. IEEE Transactions on Biomedical Engineering 57(7), 1676–1689 (2010) (PMID: 20172780)
7. Jaffe, E.S., Harris, N.L., Stein, H., Vardiman, J.W.: World Health Organization Classification of Tumours - Tumours of Haematopoietic and Lymphoid Tissues. IARC Press, Lyon (2001)
8. Gurcan, M.N., Boucheron, L., Can, A., Madabhushi, A., Rajpoot, N., Yener, B.: Histopathological Image Analysis: A review. IEEE Reviews in Biomedical Engineering 2, 147–171 (2009)
9. Sertel, O., Kong, J., Catalyurek, U.V., Lozanski, G., Saltz, J., Gurcan, M.N.: Histopathological image analysis using model-based intermediate representations and color texture: Follicular lymphoma grading. The Journal of Signal Processing Systems 55, 169–183 (2009)
10. Cooper, L., Sertel, O., Kong, J., Lozanski, G., Huang, K., Gurcan, M.N.: Feature-Based Registration of Histopathology Images with Different Stains: An Application for Computerized Follicular Lymphoma Prognosis. Computer Methods and Programs in Biomedicine 96(3), 182–192 (2009)
11. Sertel, O., Kong, J., Lozanski, G., Catalyurek, U., Saltz, J., Gurcan, M.N.: Computerized microscopic image analysis of follicular lymphoma. In: SPIE Medical Imaging 2008, San Diego, California, pp. 16–21 (February 2008)
12. Sertel, O., Kong, J., Catalyurek, U., Lozanski, G., Shanaah, A., Saltz, J., Gurcan, M.N.: Texture classification using nonlinear color quantization: Application to histopathological image analysis. In: IEEE ICASSP 2008, Las Vegas, NV, March 30-April 4 (2008)
13. Belkacem-Boussaid, K., Sertel, O., Lozanski, G., Shana'aah, A., Gurcan, M.N.: Extraction of color features in the spectral domain to recognize centroblasts in histopathology. In: IEEE EMBC 2009, Minneapolis, MN, September 2-6 (2009)
14. Samsi, S., Krishnamurthy, A.K., Groseclose, M., Caprioli, R.M., Lozanski, G., Gurcan, M.N.: Imaging Mass Spectrometry Analysis for Follicular Lymphoma Grading. In: IEEE EMBC 2009, Minneapolis, MN, September 2-6 (2009)
15. Teodoro, G., Sachetto, R., Sertel, O., Gurcan, M.N., Meira, W., Catalyurek, U., Ferreira, R.: Coordinating the use of GPU and CPU for improving performance of compute intensive applications. In: IEEE Cluster 2009, New Orleans, LA, August 31 – September 4 (2009)
16. Belkacem-Boussaid, K., Prescott, J., Lozanski, G., Gurcan, M.N.: Segmentation of follicular regions on H&E slides using matching filter and active contour models. In: SPIE Medical Imaging 2010, San Diego, California, February 13-18 (2010)
17. Belkacem-Boussaid, K., Pennell, M., Lozanski, G., Shana'ah, A., Gurcan, M.N.: Effect of pathologist agreement on evaluating a computer-assisted system: Recognizing centroblasts in follicular lymphoma cases. In: IEEE ISBI 2010, Rotterdam, The Netherlands, April 14-17 (2010)
18. Cheng, J., Veronika, M., Rajapakse, J.: Identifying Cells in Histopathological Images. In: Ünay, D., Çataltepe, Z., Aksoy, S. (eds.) ICPR 2010. LNCS, vol. 6388, pp. 247–255. Springer, Heidelberg (2010)
19. Gupta, S., Kuse, M., Sharma, T.: A Classification Scheme for Lymphocyte Segmentation in H&E Stained Histology Images. In: Ünay, D., Çataltepe, Z., Aksoy, S. (eds.) ICPR 2010. LNCS, vol. 6388, pp. 237–245. Springer, Heidelberg (2010)
20. Panagiotakis, C., Ramasso, E., Tziritas, G.: Lymphocyte Segmentation using the Transferable Belief Model. In: Ünay, D., Çataltepe, Z., Aksoy, S. (eds.) ICPR 2010. LNCS, vol. 6388, pp. 256–265. Springer, Heidelberg (2010)
21. Graf, F., Grzegorzek, M., Paulus, D.: Counting Lymphocytes in Histopathology Images Using Connected Components. In: Ünay, D., Çataltepe, Z., Aksoy, S. (eds.) ICPR 2010. LNCS, vol. 6388, pp. 267–273. Springer, Heidelberg (2010)

A Classification Scheme for Lymphocyte Segmentation in H&E Stained Histology Images

Manohar Kuse, Tanuj Sharma, and Sudhir Gupta

The LNM Institute of Information Technology, Jaipur, India
kusemanohar.08@lnmiit.ac.in, tanuj.08@lnmiit.ac.in, sudhir@lnmiit.ac.in

Abstract. A technique for automating the detection of lymphocytes in histopathological images is presented. The proposed system takes Hematoxylin and Eosin (H&E) stained digital color images as input to identify lymphocytes. The process involves segmentation of cells from extracellular matrix, feature extraction, classification and overlap resolution. Extracellular matrix segmentation is a two step process carried out on the HSV-equivalent of the image, using mean shift based clustering for color approximation followed by thresholding in the HSV space. Texture features extracted from the cells are used to train a SVM classifier that is used to classify lymphocytes and non-lymphocytes. A contour based overlap resolution technique is used to resolve overlapping lymphocytes.

Keywords: Lymphocytes, Classification, Contour Overlap Resolution.

1 Introduction

A lymphocyte is a type of blood cell in the immune system. Lymphocyte count is carried out to help diagnose many ailments. The infiltration of lymphocyte has been correlated with the disease outcome in cases of breast and ovarian cancer, leukemia, acquired immuno deficiency syndrome, viral infection, etc [10]. The ability to automatically detect and quantify extent of lymphocyte infiltration on histopathology imagery could potentially result in the development of a computer assisted diagnosis tool for Her2+ and ovarian cancer [9].

A study showed that the Lymphocytic Infiltration is relevant prognostic indicators and might be used as markers for an appropriate treatment strategy in patients with stage I carcinomas [13]. Another study claims to find strong correlations between the infiltration of lymphocytes and occurence of cancer[5].

Early detection of breast cancer is the key for its prognosis. Mammography has been one of the most reliable method of detection of breast cancers. However, enormous sizes of mammogram data had made it is difficult to manually detect breast cancer [2]. Qualitative pathological examination of the images leads to inexact classification of the cells and is subject to observer variation and variability based on the spatial focus of observation rendering the derived high level information subjective. Computer assisted diagnosis can provide objective description of the cells and assist pathologists for finding disorders associated with lymphocyte count.

D. Ünay, Z. Çataltepe, and S. Aksoy (Eds.): ICPR 2010, LNCS 6388, pp. 235–243, 2010.

Visual inspection of the histology slides does not allow one to distinguish between lymphocyte nuclei and cancer nuclei (see figure 1). Other challenges associated with automation of lymphocyte detection are the ability of the method to accommodate variability in staining procedures, differing scales of image digitization, varying illumination conditions and high occurrence of overlapping objects.

This paper describes a clinically relevant classification scheme of Hematoxylin and Eosin (H&E) stained histology slides to detect lymphocytes. The scheme is based on automated image processing, supervised learning of texture features and contour based overlap resolution. This work was done as part of a contest titled "Pattern Recognition in Histopathological images" held during International Conference on Pattern Recognition, 2010. There were a total of 10 images comprising lymphocytic infiltration that were H&E stained and digitized at 20 X resolution. 6 images were used as the training set and the other 4 were used as testing images for the results obtained in this paper. The images also came with expert annotations of representative lymphocytes. The expert annotations provided the approximate locations of centers of the lymphocytes (see figure 2(b)), and a few boundary anotations were also provided to get an idea of the shapes of lymphocytes.

Figure 2(a) shows one such histology image. Figure 2(b) and 2(c) shows the annotated centers and boundaries respectively, provided by the organizers. While the annotation of lymphocyte centers was complete, only five lymphocyte boundary annotations were provided per image.

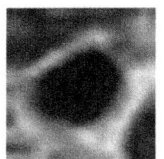
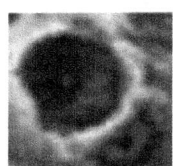

(a) Lymphocyte by
Ground Truth

(b) Non-Lymphocyte
by Ground Truth

Fig. 1. Visual inspection of the histology slides does not allow one to distinguish between lymphocyte nuclei and cancer nuclei

2   Related Work

Various techniques have been proposed to detect lymphocytes based on color, texture and shape features. Hybrid segmentation methods have been used to detect nuclei from images of histology slides stained under different conditions [12,16].

The watershed transformation is one of the most powerful tools for segmenting images [6] but the problem with watershed segmentation is that noisy and

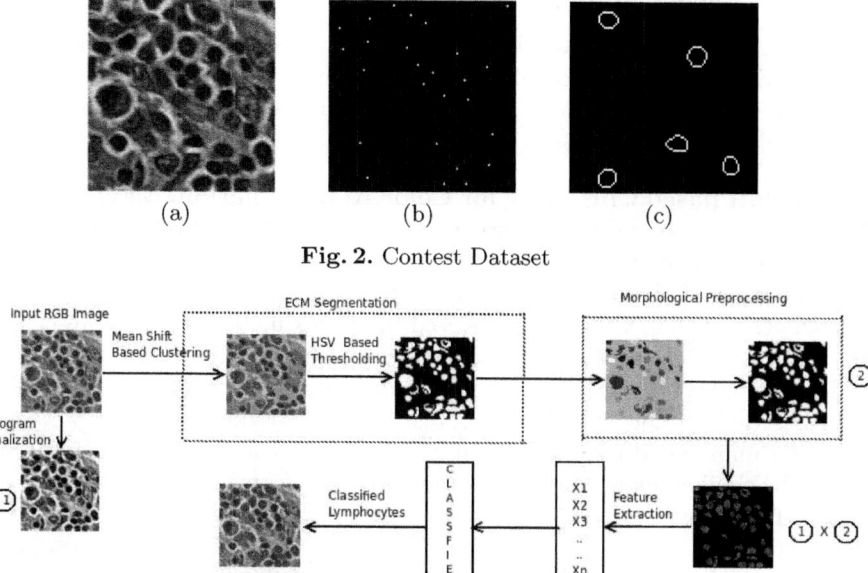

Fig. 2. Contest Dataset

Fig. 3. Overview Diagram

textured images have many minima, most of them being irrelevant for segmentation. Using the watershed on a gray tone image without any preparation leads to a strong over segmentation. The best solution to this problem consists in initially determining markers for each region of interest, including the background of the image. This makes it semi automated with subjectivity creeping in because of the choice of markers.

Active contour based models for lymphocyte segmentation have also been proposed [7], but the choice of seed points affects its segmentation performance. Bikhet et al [1] have used hierarchical thresholding to localize white blood cells, followed by extraction of gray level and morphological features to train a supervised classifier. Thresholding works well on a given set of images but fails with variability in the image set. Ongun et al [14] have used morphological pre-processing to segment the cells followed by fuzzy patch labeling.

3   Proposed Classification Scheme

The main stages of the proposed classification scheme are: 1) Extracellular matrix (ECM) segmentation, 2) Morphological pre processing, 3) Contour based overlap resolution, 4) Feature extraction, 5) Classification using a trained SVM classifier. MATLAB was used for prototyping of the scheme designed for this contest. Figure 3 shows the overview diagram of the proposed classification scheme.

3.1   ECM Segmentation

The H&E stain dyes DNA-rich cell nuclei blue and collagen-rich extracellular matrix (ECM) pink, allowing differentiation of cell from the surrounding ECM based on color [19]. Two steps are involved in the ECM segmentation. 1) Mean Shift Clustering, 2) HSV Based Thresholding.

Mean Shift based Clustering for Color Approximation. Mean Shift Clustering is a non parametric clustering technique based on density estimation for analysis of complex feature space. Dense regions in feature space correspond to local maxima of the probability density function, i.e to the modes of the unknown density [4,8]. Clustering was used to approximate the colors present (see figure 4) in the image to reduce computational efforts.

For example there are 4061 distinct colors present in figure 4(a). After mean shift based clustering, the number of distinct colors reduced to 172.

As an unintended consequence, it also lead to some structures being represented by similar colors which could then be easily segmented using a thresholding in HSV space.

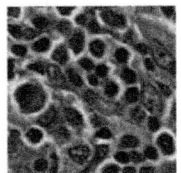 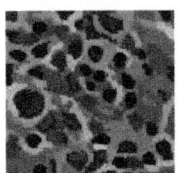

(a) Before Colour Approximation   (b) After Colour Approximation

Fig. 4. Colour Approximation using Mean Shift Clustering

HSV Based Thresholding. The HSV color space corresponds closely to the human perception of color and it has been proven more accurate and effective in distinguishing colored objects. The values of the thresholding to separate pink hue from blue hue, were obtained by 3D visualization of the distribution of these colors (as shown in figure 5) using an open source software ImageJ [15]. Extracellular matrix was segmented from the cells using equation 1. Where M represents the binary mask which is being formed after thresholding.

$$M(i,j) = 1 \text{ , if } 0.6667 \leq \text{ hue}(\text{ i , j }) \leq 0.7292$$
$$0 \text{ , otherwise} \tag{1}$$

3.2   Morphological Pre-processing

Connected components analysis (CCA) labels the the blobs in a binary image, as per its connectivity. The labels thus formed were used to iterate through each of the blobs thus formed, to extract the blob features. Overlap resolution is applied to blobs which satisfy some threshold on area and perimeter as discussed in section 3.3.

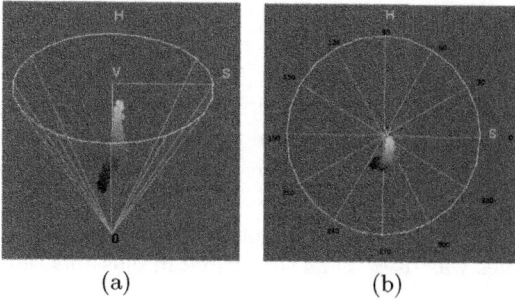

(a) (b)

Fig. 5. 3D Visualization of HSV Colour Space

3.3 Contour Based Overlap Resolution

A novel contribution of this paper is in resolving cell overlaps. The importance of resolving overlaps in lymphocyte detection and grading is discussed in [7]. Overlaping of lymphocytes, sometimes makes it difficult to segment them.

Here we have used a contour based heuristic for revolving the overlap among the lymphocytes. Contours are defined by those pixels that are at an equal distance from the detected cell boundary. Further, those closed contours which cover an area that approximates to the area of an average lymphocyte are retained while ignoring other contours. Figure 6 shows an example to illustrate the overlap resolution.

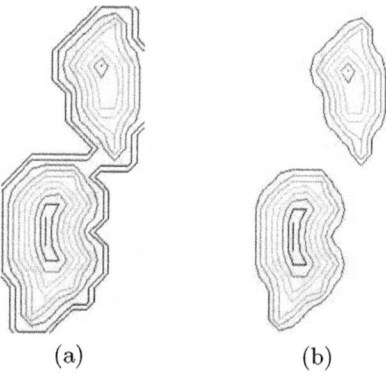

(a) (b)

Fig. 6. Overlap Resolution

3.4 Feature Extraction

The mask obtained from the previous steps represented lymphocytes as digital number 1 and other areas as digital number 0. This mask was multiplied with the histogram equalized grayscale image of the RGB image shown in figure 2(a).

Histogram equalization was performed to normalize varying illumination conditions. Eighteen texture features were extracted for every detected cell region [17,18,3]. These are – Autocorrelation, Contrast, Correlation, Cluster Prominence, Cluster Shade, Dissimilarity, Energy, Entropy, Homogeneity, Maximum probability, Variance, Sum average, Sum variance, Sum entropy, Difference variance, Difference entropy, Information measure of correlation, Normalized inverse difference moment [17,18,3]. These features were derived from the gray level co-occurence matrix for four values of offset and four values of direction. Average of these eight values was used as feature value in classification.

3.5   Supervised Classification

Supervised classification was performed to classify the cells into two classes – lymphocytes and non-lymphocytes. For training the classifier, the labels for every feature pattern were obtained from the annotated dataset and a training dataset was constructed that consisted of 80 patterns for lymphocytes and 98 patterns for non lymphocytes. A support vector machine classifier was trained using this training dataset [11].

4   Results

The classification scheme described in section 3 was applied on 4 testing images that had 94 lymphocytes and 74 non lymphocytes as per expert annotation. A correct detection of lymphocyte in the confusion matrix tabulated in table 1 meant that a lymphocyte centre marked by the expert existed in the region classified as lymphocyte by the proposed scheme.

Table 1. Confusion Matrix

		Ground Truth	
		Lymphocytes	Non-Lymphocytes
Classifier	Lymphocytes	161	25
	Non-Lymphocytes	55	133

It can be observed from the confusion matrix that the proposed classification scheme is able to achieve a classification accuracy of 78% at a false positive rate of 14.7%. Figure 7(a) shows the lymphocytes detected by the proposed classification scheme as cells that are delineated with a red boundary. Figure 7(b) shows the lymphocytes annotated by the expert that were delineated with the help of given lymphocyte centers. Visual inspection of the results shows that there is good agreement between the derived results and the ground truth. The results sent by us were also evaluated by the organizers using two region based measures and two boundary based measures.

The region based measures are defined as follow

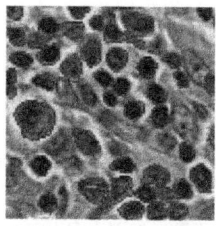   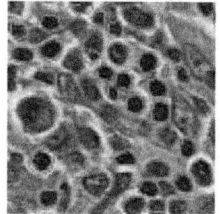

(a) Classifier Result   (b) Expert Anno-
                                tated

Fig. 7. Comparison between Classifier Results and Ground Truth

1) Dice coefficient $DICE = \frac{2 \times |A(S) \cap A(G)|}{|A(S)| + |A(G)|}$

2) Sensitivity $SN = \frac{|A(S) \cap A(G)|}{|A(G)|}$

The boundary based performance measures are defined as follow

1) Hausdorff distance

$$HD = max_w[\ min_x ||c_w - c_x||\ ](c_w \in S, c_x \in G)$$

2) Mean absolute distance

$$MAD = \frac{\sum_{w=1}^{M} ||c_w - c_x||}{M}$$

Where C is the total number of pixels in the image and $|s|$ represents cardinality of any set s. $A(s)$ and $A(G)$ is the area of the close boundary of segmentation results and manual delineation. For boundary based measures S and G are closed boundaries of segmentation results and manual delineations. M is the number of pixels on the closed boundaries of segmentation results.

Table 2 shows the results summarized by the organizers using the above mentioned metrics. The results of group 2 correspond to the results obtained from the work mentioned in this paper. DICE coefficient is a measure of similarity of images. Our method gives 74% means that, the actual result is 74% similar to the output provided by our method. Our sensitivity is 58% means that 58% of the positives are correctly identified. It can be observed that DICE coefficient is only 0.9 less than the best reported result. There is a scope for improvement in sensitivity by the introduction of newer features related to shape and color.

Table 2. Performance Comparison

Group	DICE	SN	HD	MAD
1	0.73	0.57	4.58	0.77
2	0.74	0.58	3.63	0.65
3	0.37	0.23	21.95	9.14
4	0.83	0.71	3.73	0.41
5	0.74	0.59	3.51	0.62

5   Conclusions

We have developed a classification scheme for automatically detecting lymphocytes from H&E stained histopathology slides. However, the proposed scheme
needs extensive testing on different images that are truly representative of the
various scenarios in the real world. Such a dataset will also help to build a good
knowledge base for supervised classification of images. Without such an extensive evaluation, a prognosis tool for lymphocyte count related disorders cannot
be developed especially when the risk associated with misclassification is high.

6   Future Work

As of now, the system does not require user interaction or parameter tuning and
produces classification results that are better than most methods used in the
contest. The size of the lymphocytes can be determined automatically for use
with overlap resolution given the scale at which the image was acquired.

Classification results are largely based on the training of the classifier and
thus there is a scope of using incremental learning to keep the knowledge base
updated. A dimensionality reduction exercise can help find those features that
aid in classification. Further, use of relatively higher resolution images than those
used in this contest can lead to a better quantification of the texture features
and it is our belief that this will further increase the ability of the classifier to
distinguish between lymphocytes and non lymphocytes.

References

1. Bikhet, S.F., Darwish, A.M., Tolba, H.A., Shaheen, S.I.: Segmentation and classification of white blood cells. In: IEEE International Conference on Acostics, Speech
   and Signal Processing, ICASSP, pp. 550–553 (1992)
2. Cheng, H.D., Shi, X.J., Min, R., Hu, L.M., Cai, X.P., Du., H.N.: Approaches
   for automated detection and classification of masses in mammograms. Pattern
   Recogn. 39(4), 646–668 (2006)
3. Clausi, D.A.: An analysis of co-occurrence texture statistics as a function of grey
   level quantization. Can. J. Remote Sensing 28, 45–62 (2002)
4. Comaniciu, D.: Mean shift: A robust approach toward feature space analysis. IEEE
   Transactions on Pattern Analysis and Machine Intelligence 58, 71–96 (2002)
5. Deschoolmeester, V., Baay, M., Van Marck, E., Weyler, J., Vermeulen, P., Lardon,
   F., Vermorken, J.: Tumor infiltrating lymphocytes: an intriguing player in the
   survival of colorectal cancer patients. BMC Immunology 11(1), 19 (2010)
6. Doughri, R., M'hiri, S., Romdhane, K.B., Ghorbel, F., Essafi, S.: Segmentation
   and classification of breast cancer cells in hostological images. Information and
   Communication Technology (2006)
7. Fatakdawala, H., Basavanhally, A., Xu, J., Bhanot, G., Ganesan, S., Feldman,
   M., Tomaszewski, J., Madabhushi, A.: Expectation maximization driven geodesic
   active contour with overlap resolution: Application to lymphocyte segmentation on
   breast cancer histopathology. In: International Conference on Bioinformatics and
   Bioengineering (2009)

8. Fukunaga, Keinosuke, Hostetler, L.D.: The estimation of the gradient of a density function, with applications in pattern recognition. IEEE Transactions on Information Theory (IEEE) 21(1), 32–40 (1975)
9. PR in HIMA, http://bmi.osu.edu/cialab/ICPR_contest
10. Jemal, A., Siegel, R., Ward, E., Hao, Y., Xu, J., Murray, T., Thun, M.J.: Cancer statistics 2008. CA Cancer J. Clin. 58, 71–96 (2008)
11. Joachims, T.: Making large-scale svm learning practical. advances in kernel methods - support vector learning. MIT-Press, Cambridge (1999)
12. Latson, L., Sebek, B., Powell, K.A.: Automated cell nuclear segmentation in colour images of he stained breast biopsy. Anal. Quant. Cytol. Histol. 25(321-331) (2003)
13. Losi, L., Ponti, G., Di Gregorio, C., Marino, M., Rossi, G., Pedroni, M., Benatti, P., Roncucci, L., de Leon, M.P.: Prognostic significance of histological features and biological parameters in stage i (pt1 and pt2) colorectal adenocarcinoma. Pathology - Research and Practice 202(9), 663–670 (2006)
14. Ongun, G., Halici, U., Leblebiciogl, K., Atalay, V.: Feature extraction and classification of blood cells for an automated differential blood count system. IEEE IJCNN, 2461–2466 (2001)
15. Rasband, W.S.: Imagej. U.S. National Laboratory of Health, Bethesda (1997-2005)
16. Schnorrenberg, F., Pattichis, C., Kyriacou, K., Schizas, C.: Computer-aided detection of breast cancer nuclei. IEEE Transaction on Information Technology in Bio-medicine, 128–140 (1997)
17. Shanmugam, K., Haralick, R.M., Dinstein, I.: Textural features of image classification. IEEE Transactions on Systems, Man and Cybernetics SMC-3, 6 (1973)
18. Soh, L., Tsatsoulis, C.: Texture analysis of sar sea ice imagery using gray level co-occurrence matrices. IEEE Transactions on Geoscience and Remote Sensing 37, 2 (1999)
19. Scarff, R.W., Torloni, H.: Histological typing of breast tumors. international classification of tumors. World Health Organization, 13–20 (1968)

Identifying Cells in Histopathological Images

Jierong Cheng[1,2,*], Merlin Veronika[1,2], and Jagath C. Rajapakse[1,2,3]

[1] Computation and Systems Biology, Singapore-MIT Alliance, Singapore 637460
[2] BioInformatics Research Centre,Nanyang Technological University,
Singapore 637553
[3] Department of Biological Engineering, Massachusetts Institute of Technology,
Cambridge, MA 02139, USA
{jrcheng,merl0001,asjagath}@ntu.edu.sg

Abstract. We present an image analysis pipeline for identifying cells in histopathology images of cancer. The analysis starts with segmentation using multi-phase level sets, which is insensitive to initialization and enables automatic detection of arbitrary objects. Morphological operations are used to remove small spots in the segmented images. The target cells are then identified based on their features. The detected cells were compared with the manual detection performed by pathologists. The quantitative evaluation shows promise and utility of our technique.

Keywords: Histopathological images, lymphocytes, centroblasts, level sets, feature selection.

1 Introduction

Diagnostic medicine has taken a huge leap with the availability of novel imaging modalities and automated analysis. Traditionally, cancer samples are analyzed by manual measurement of clinical markers like progesterone receptor, estrogen receptor, HER2+ for breast cancer, and prostate-specific antigen for prostate cancer. Digital whole slide scanners have set a revolution in histological evaluation of tissues, and processing of large images from these modalities and data generated thereof demands efficient downstream analysis. This remains non-trivial due to the size of information, the variability of staining and illumination procedures, and the diversity in acquisition process.

Histopathological images provide an important tool for accurate diagnosis of various types of cancers and other diseases. However, identification of different types of cells and the cells at different pathological stages has been difficult: visual examination of histopathology images is complicated by target cells' similarity with other structures, which has an impact on the discretion of clinicians to predict survival and disease outcome. Manual identification suffers from lack of reproducibility among pathologists, primarily due to the subjective nature in identifying large transformed cells which are morphologically heterogeneous.

---

* Corresponding author.

D. Ünay, Z. Çataltepe, and S. Aksoy (Eds.): ICPR 2010, LNCS 6388, pp. 244–252, 2010.

Moreover, overlap among lymphocytic nuclei and other structures demands algorithms that have the ability to accurately estimate true extent of lymphocytic infiltration [1].

Our work focuses on the analysis of histopathological images of breast cancer and follicular lymphoma. Breast cancer (BC) is one of the common cancers in women of United States with an estimated in situ cases of 62,280, invasive cases of 192,370, and mortality of 40,170 in 2009 [2]. Because of the large number of cases involved, examining biopsy tissue specimens has become an integral part of BC diagnosis and prognosis where a pathologist looks for indicative features and patterns. For example, a positive correlation of human epidermal growth factor receptor 2 (HER2+) amplification and lymphocytic infiltration has been identified, which may aid in future therapy [3]. Most HER2+ BC are currently treated with agents that specifically target HER2+ protein.

B-cell lymphomas make up most (about 85%) of non-Hodgkin lymphomas (NHL) in the United States [4]. Follicular lymphoma (FL) is a common subtype of B-cell lymphoma comprising approximately 22% of all NHLs and 70% of indolent lymphomas [5]. Its incidence is increasing, with over 24,000 new cases diagnosed each year [6]. FL grading is based on the average number of centroblasts or large transformed cells in ten representative neoplastic follicles at 40x high-power field (HPF): for grade 1, 0-5 centroblasts per HPF; grade 2, 6-15 centroblasts per HPF; and grade 3, >15 centroblasts per HPF.

In this paper, we propose a framework for identifying lymphocytic cells and centroblasts in Hematoxylin and Eosin (H&E) stained images. Our method was tested on two different datasets: breast cancer and follicular lymphoma tissues, provided as part of Pattern Recognition in Histopathological Image Analysis contest by ICPR 2010. In section 2, we describe the analysis framework. Results are presented in section 3. Section 4 concludes the paper with future directions.

2   Method

2.1   Multi-phase Level Sets

The initial segmentation of cells was performed by using multi-phase level set framework [7]. The multi-phase model is a generalization of an active contour model without edges based on a two-phase segmentation [8]. The active contours are implicitly represented by level set functions and the changes in objects appear automatically as level set functions evolve. This model enables automatic detection of an arbitrary number of objects from an arbitrary initial front [8]. We chose to initialize level set functions to be multiple small circles spread over the whole image. This type of initial condition has been proven to have tendency to converge to a global minimizer and at much faster rate [7]. It needs only $\log_2 n$ level set functions for n phases or segments in the piecewise constant intensity profile. In the original H&E stained images, there are roughly four regions of different intensity level (whitish, pink, purple, and deep purple). To separate the target region more accurately, we chose $n = 4$ phase segmentation over two-phase segmentation.

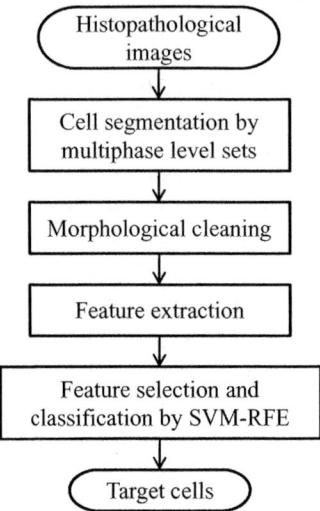

Fig. 1. Image processing pipeline for cell identification

The four phase energy function is given by

$$F_4(\Phi, c) = \int_\Omega (f - c_{11})^2 H(\phi_1) H(\phi_2) \mathrm{d}x\mathrm{d}y + \int_\Omega (f - c_{10})^2 H(\phi_1)(1 - H(\phi_2)) \mathrm{d}x\mathrm{d}y$$
$$+ \int_\Omega (f - c_{01})^2 (1 - H(\phi_1)) H(\phi_2) \mathrm{d}x\mathrm{d}y + \int_\Omega (f - c_{00})^2 (1 - H(\phi_1))(1 - H(\phi_2)) \mathrm{d}x\mathrm{d}y$$
$$+ \nu \int_\Omega |\nabla H(\phi_1)| \mathrm{d}x\mathrm{d}y + \nu \int_\Omega |\nabla H(\phi_2)| \mathrm{d}x\mathrm{d}y$$

$$(1)$$

where $(x, y) \in \Omega \subset \mathbf{R}^2$ is the 2-D domain of image f. Zero-level set of pair of level set functions, $\Phi = (\phi_1, \phi_2)$, defines the segmentation. $c = (c_{11}, c_{10}, c_{01}, c_{00})$ is the constant vector of averages of image f in each phases. H is the Heaviside Dirac function and ν is a fixed positive parameter.

After the evolution of two level set functions, the average intensities within every phases were calculated. The phase with the lowest average intensity was considered the region of target cells which appear relatively dark in original images.

2.2   Morphological Cleaning

The holes of the segmented binary images were filled by a hole-filling algorithm based on morphological reconstruction. Morphological opening with a disk-shaped structuring element of a radius of r pixel was applied in order to remove the small spots on the images. This operation removes pixels of objects thinner than r pixels. Morphological opening of a binary image b is defined as the erosion of that image followed by the dilation of the eroded image, by a structuring element S:

$$b \circ S = (b \ominus S) \oplus S \tag{1}$$

where $\ominus$ and $\oplus$ denote erosion and dilation, respectively.

Binary regions composed of multiple objects were separated by using the watershed algorithm [9], [10].

2.3   Feature Selection and Classification

For each candidate object, we extracted 173 features as described in [11]. These features include 49 Zernike features, 30 Daubechies 4 wavelet features, 60 Gabor features, 5 skeleton features, 13 Haralick features and 16 morphological features. These features have been shown to be effective in recognition of protein subcellular localization images [11].

In order to select relevant features that give optimal classification of cells into different types, we used a wrapper-type feature selection method: Support Vector Machine Recursive Feature Elimination (SVM-RFE) criteria [12]. This uses Support Vector Machine (SVM) as the classifier and recursively eliminates irrelevant features. The rank of a particular feature was determined by the corresponding weight of the SVM. For more details about the SVM-RFE method, readers are referred to [12].

Given the centroids of target cells, provided as ground truth by the contest organizers, the cells were identified into two classes: the targeted cells and the rest. These class labels along with the input features were used as training data for identifying relevant features by SVM-RFE and validated by five fold cross-validation. SVM-RFE essentially ranks the features and top-ranked features were then used for testing.

The entire proposed framework is depicted in Fig. 1.

3   Results

3.1   Datasets

The first dataset we used in the experiments is H&E stained breast biopsy images of size 100×100 pixels, representing HER2+ breast cancer exhibiting lymphocytic infiltration. Ten images were used for training and the ground truths of these images were obtained via manual detection performed by a pathologist (provided by the organizers as centroids and boundary segmentations of respective lymphocytes). We used a structuring element of size $r = 1$ for all the images in this dataset. Detected lymphocytes and lymphocytic centers on representative images are displayed in Fig. 2.

The second dataset is H&E stained follicular lymphoma images of size 2068 $\times$ 1253 pixels digitized at 40 $\times$ resolution. Five images containing centroblasts were used for training and the ground truths for these images were obtained via manual detection performed by a pathologist (provided by the organizers

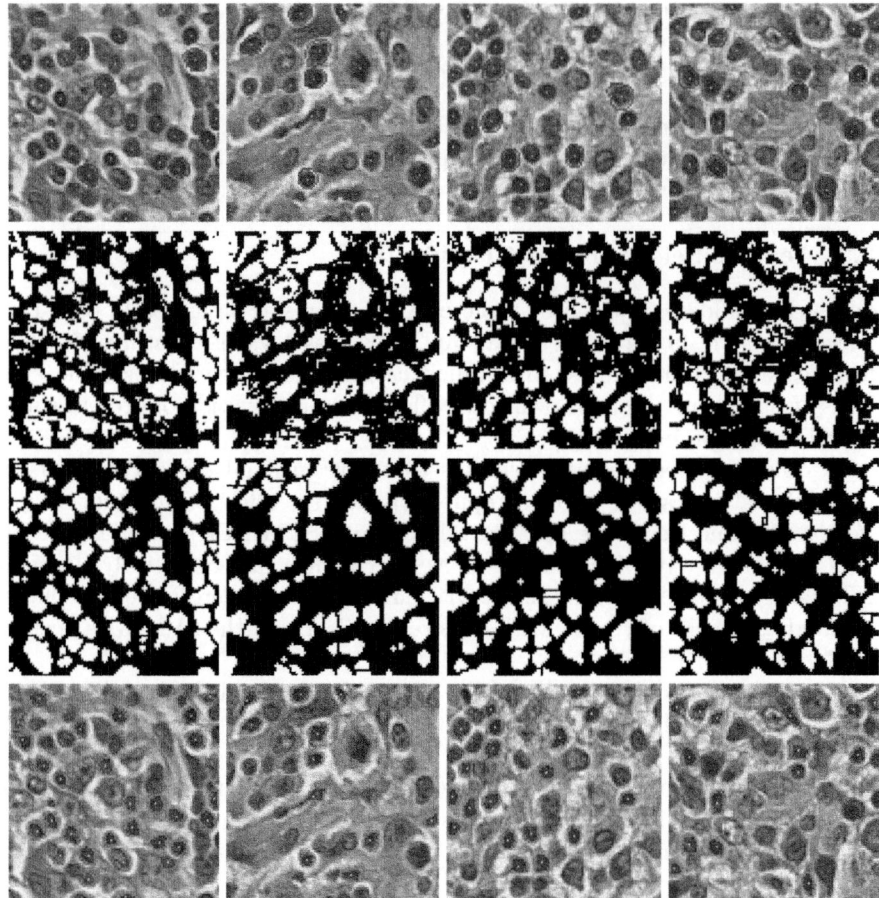

Fig. 2. Lymphocyte detection from breast cancer tissues. First row: expert annotations of representative lymphocytes and all lymphocytic centers. Second row: binary images after multi-phase level set segmentation. Third row: binary images after morphological cleaning. Forth row: automatically detected lymphocytes and lymphocytic centers.

as centroids and boundary segmentations of respective centroblasts). We used a structuring element of size $r = 12$ for all the images in this dataset. Detected centroblasts on representative images are displayed in Fig. 3.

3.2   Feature Selection

All 173 features were extracted from each cell and then relevant features were identified by SVM-RFE method. We performed ten trials of five-fold cross-validation. For the first dataset, the five features that appear most frequently in top 20 ranking are:

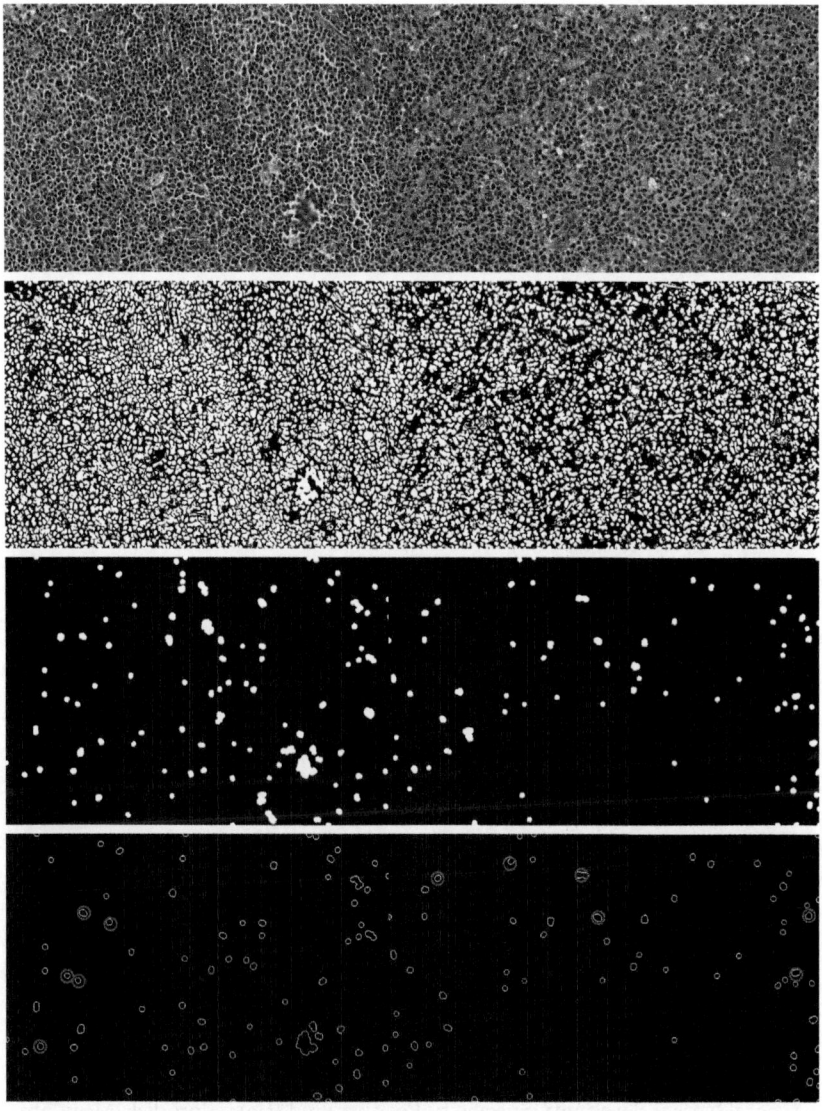

Fig. 3. Centroblast detection from follicular lymphoma tissues. First row: centroblast centers marked by expert pathologists. Second row: binary images after multi-phase level set segmentation. Third row: binary images after morphological cleaning. Forth row: automatically detected centroblasts (true positives are marked by red circles).

1. object skeleton to object fluorescence ratio;
2. Zernike moment feature 12, 0;
3. Gabor texture feature 43;
4. edges to area fraction;
5. number of branch points per length of object skeleton.

The average accuracy achieved from 10 trials on bootstrap samples of five-fold cross-validation is $73.6 \pm 0.77\%$.

Because of large size of the second dataset, we used only the standard deviation of intensity which is the feature found to be the best for classification.

3.3   Evaluation

In terms of evaluation of lymphocyte detection, the following region-based performance measures were employed:

- Dice coefficient $= \frac{2 \times |A \cap M|}{|A| + |M|}$
- Overlap $= \frac{|A \cap M|}{|A \cup M|}$
- Sensitivity $= \frac{|A \cap M|}{|M|}$
- Specificity $= \frac{N - |A \cup M|}{N - |M|}$
- Positive predictive value $= \frac{|A \cap M|}{|A|}$

where A and M are the areas of the closed boundary of segmentation results and manual delineation, respectively. N is the total number of pixels in the image.

The boundary-based performance measures are

- Haussdorf distance $= \max_w [\min_x \|n_w - n_x\|], (n_w \in A, n_x \in M)$
- Mean absolute distance $= \frac{1}{N_A} \sum_{w=1}^{N_A} \|n_w - n_x\|$

where N_A is the total number of pixels on the closed boundaries of segmentation results.

For evaluation of centroblast detection, true positives and false positives were calculated based on the closeness of the detected centroids of centroblasts to the expert markings. If the distance is less than 30 pixels (equivalent of about

Table 1. Quantitative evaluation results for lymphocyte dataset

Dice coef.	Overlap	Sensitivity	Specificity	Positive pred.	Haussdorf dist.	Mean abs. dist.
0.73	0.57	0.57	1	1	4.58	0.77

Table 2. Quantitative evaluation results for centroblast dataset

Sensitivity	False positive rate
0.38	82.85

7.5 μm), then it is considered as a true positive, otherwise it is counted as false positive.

The evaluation results are shown in Tables 1 and 2.

Our algorithms were implemented in MATLAB and deployed on an Intel Quad Core 2.0 GHz processor and 3.25 GB RAM. The running time to process one image comprising lymphocytic infiltration is around 5 seconds. The running time to process one image containing centroblasts is around 35 seconds.

4   Conclusion

We presented an effective image analysis framework for identifying cells from H&E stained images. The method was evaluated on cell identification of two types of histopathological tissue images: breast cancer and follicular lymphoma. From the evaluation results on our experiments, we conclude that the image analysis pipeline described in this paper is effective in detecting true locations of lymphocytes and centroblasts on histopathology images with a high true positive rate. The proposed method has the advantages over manual analysis in terms of speed and reproducibility. Our future direction will be to develop a comprehensive automated grading system for histopathological diagnosis and prognosis.

Acknowledgments. The authors would like to thank Mundra Piyushkumar Arjunlal and Liu Song of BioInformatics Research Centre, Nanyang Technological University, for help on applying feature selection.

References

1. Fatakdawala, H., Xu, J., Basavanhally, A., Bhanot, G., Ganesan, S., Feldman, M., Tomaszewski, J., Madabhushi, A.: Expectation maximization driven geodesic active contour with overlap resolution (emagacor): Application to lymphocyte segmentation on breast cancer histopathology. IEEE Trans. Biomedical Engineering 99, 1–8 (2010)
2. A. C. Society: Breast cancer facts and figures. American Cancer Society, Inc., Atlanta (2009-2010)
3. Alexe, G., Dalgin, G.S., Scanfeld, D., Tamayo, P., Mesirov, J.P., DeLisi, C., Harris, L., Barnard, N., Martel, M., Levine, A.J., Ganesan, S., Bhanot, G.: High expression of lymphocyte-associated genes in node-negative her2+ breast cancers correlates with lower recurrence rates. Cancer Research 67, 10669–10676 (2007)
4. Griffin, N.R., Howard, M.R., Quirke, P., O'Brien, C.J., Child, J.A., Bird, C.C.: Prognostic indicators in centroblastic-centrocytic lymphoma. Journal of Clinical Pathology 41, 866–870 (1988)
5. The non-hodgkin's lymphoma classification project, a clinical evaluation of the international lymphoma study group classification of non-hodgkin's lymphoma. Blood, 3909–3918 (1997)
6. Friedberg, J.: Treatment of follicular non-hodgkin's lymphoma: the old and the new. Semin Hematology 2, s2–s6 (2008)

7. Vese, L.A., Chan, T.F.: A multiphase level set framework for image segmentation using the mumford and shah model. International Journal of Computer Vision 50(3), 271–293 (2002)
8. Chan, T.F., Vese, L.A.: Active contours without edges. IEEE Trans. Image Processing 10(2), 266–277 (2001)
9. Meyer, F.: Topographic distance and watershed lines. Signal Processing 38(1), 113–125 (1994)
10. Cheng, J., Rajapakse, J.C.: Segmentation of clustered nuclei with shape markers and marking function. IEEE Trans. Biomedical Engineering 56(3), 741–748 (2009)
11. Huang, K., Murphy, R.F.: Boosting accuracy of automated classification of fluorescence microscope images for location proteomics. BMC Bioinformatics 5(78) (2004)
12. Mundra, P.A., Rajapakse, J.C.: SVM-RFE with MRMR filter for gene selection. IEEE Transactions on NanoBioscience 9(1), 31–37 (2010)

Lymphocyte Segmentation Using the Transferable Belief Model

Costas Panagiotakis[1], Emmanuel Ramasso[2], and Georgios Tziritas[1]

[1] Department of Computer Science, University of Crete, P.O. Box 2208,
Heraklion, Greece
{cpanag,tziritas}@csd.uoc.gr
[2] FEMTO-ST Institute, UMR CNRS 6174 - UFC / ENSMM / UTBM,
Automatic Control and Micro-Mechatronic Systems Department, France
emmanuel.ramasso@femto-st.fr

Abstract. In the context of several pathologies, the presence of lymphocytes has been correlated with disease outcome. The ability to automatically detect *lymphocyte nuclei* on histopathology imagery could potentially result in the development of an image based prognostic tool. In this paper we present a method based on the estimation of a mixture of Gaussians for determining the probability distribution of the principal image component. Then, a post-processing stage eliminates regions, whose shape is not similar to the *nuclei* searched. Finally, a Transferable Belief Model is used to detect the *lymphocyte nuclei*, and a shape based algorithm possibly splits them under an equal area and an eccentricity constraint principle.

1 Introduction

Recently, there is an increasing activity on analysing histopathological images, as a potential prognostic tool for cancer patients. One important step for the diagnosis is the cell segmentation. Demir and Yener [1] review the different approaches classified in two categories: region-based and boundary-based methods. Lymphocyte segmentation in histopathology images is complicated by the similarity in appearance between *lymphocyte* and *cancer nuclei* in the image [2]. In [2], a computer-aided diagnosis (CADx) scheme is proposed to automatically detect and grade the extent of lymphocytic infiltration in digitized HER2+ BC histopathology. Lymphocytes are automatically detected by a combination of region growing and Markov random field algorithms using the luminance channel in Lab color space. Finally, a support vector machine classifier is used to discriminate samples with high and low lymphocytic infiltration. In [3], lymphocytes are automatically detected via a segmentation scheme comprising a Bayesian classifier and template matching, using the Saturation color channel in HSV color space.

In [4], a segmentation scheme, Expectation Maximization driven Geodesic Active Contour with Overlap Resolution (EMaGACOR), is proposed for automatically detecting and segmenting lymphocytes on HER2+ Breast Cancer

D. Ünay, Z. Çataltepe, and S. Aksoy (Eds.): ICPR 2010, LNCS 6388, pp. 253–262, 2010.

histopathology images. EMaGACOR utilizes the Expectation-Maximization (EM) algorithm for automatically initializing a geodesic active contour and includes a scheme for resolving overlapping structures. EMaGACOR was evaluated on a total of 100 HER2+ breast biopsy histology images and was found to have a detection sensitivity of over 86% and a positive predictive value (PPV) of over 64%.

Our method addresses the problem of lymphocyte detection and should be considered as a region-based approach. The first step of our method consists of a likelihood classification based on the estimation of the parameters of a mixture of Gaussians. A post-processing step eliminates regions with size or shape that differ greatly from a typical shape of *lymphocyte nuclei*. For the remaining regions the following features are extracted: mean value, variance, eccentricity and size. A Transferable Belief Model is then trained and used in order to detect the *lymphocyte nuclei*. Finally, a shape based algorithm possibly splits the detected regions under an equal area and an eccentricity constraint principle.

The organisation of the paper is as follows: Section 2 describes the segmentation stage with the estimation of a mixture of Gaussians and the shape-based detection of candidate *leymphocyte nuclei*; Section 3 presents the Transferable Belief Model used and the results of training based on the ground-truth; in Section 4 is presented our technique for solving possible overlaps. Then, the results on the ICPR contest data set are given in Section 5.

2   Segmentation

There are three possible classes corresponding to *stroma, cancer nuclei* and *lymphocyte nuclei*. We admit Gaussian distributions for the three classes and use the EM algorithm for estimating the parameters of the model. We observe that the three colour channels are strongly correlated. Therefore we start by applying principal component analysis (PCA) in order to select only one image component. Let us note $x(s)$ this component at a site s of the image grid. Let $p(x)$ be the probability density function for the principal image component. According to the mixture of Gaussians model we have:

$$p(x) = \sum_{k=1}^{3} \frac{P_k}{\sigma_k \sqrt{2\pi}} e^{-\frac{(x-\mu_k)^2}{2\sigma_k^2}} = \sum_{k=1}^{3} P_k p_k(x|\mu_k, \sigma_k^2). \tag{1}$$

The unknown parameters are the *a priori* probabilities (P_k), the mean (μ_k) and the variance (σ_k^2) values.

At first, the Max-Lloyd algorithm is used for obtaining initial parameter values. The empirical probability density function is used for the estimation. Let N denotes the number of image pixels. At i-th iteration of the EM algorithm we have:

– E-step: calculate the posterior probabilities

$$P^{(i+1)}(k|x, \theta^{(i)}) = \frac{P_k^{(i)} e^{-\frac{(x-\mu_k^{(i)})^2}{2\sigma_k^{2(i)}}}}{\sqrt{2\pi}\sigma_k^{(i)} p^{(i)}(x)}, \tag{2}$$

where θ is the set of all the unknown parameters.

- M-step: estimate the prior probabilities, the mean and the variance values as follows

$$P_k^{(i+1)} = \frac{1}{N} \sum_{s \in G} P^{(i+1)}(k|x(s), \theta^{(i)}) \tag{3}$$

$$\mu_k^{(i+1)} = \frac{1}{N P_k^{(i+1)}} \sum_{s \in G} P^{(i+1)}(k|x(s), \theta^{(i)}) x(s) \tag{4}$$

$$\sigma_k^{2(i+1)} = \frac{1}{N P_k^{(i+1)}} \sum_{s \in G} P^{(i+1)}(k|x(s), \theta^{(i)})(x(s) - \mu_k^{(i)})^2 \tag{5}$$

The above steps are implemented using the empirical probability density for limiting the computational time. A stopping threshold of 10^{-6} is given on the relative gain per iteration for the log likelihood value.

Having the estimation of the probability density functions for the three classes the image sites are classified according to the maximum likelihood principle. Therefore, for classifying the site s to class k, the likelihood $p_k(x(s)|\mu_k, \sigma_k^2)$ is maximized.

Then, a post-processing stage follows on the regions detected as candidate *lymphocyte nuclei*, which being darker are identified by the mean value. Three region parameters are measured: the area, the eccentricity and the solidity. The area of region r (denoted A_r) is given by the number of pixels that belong to region r. Very small regions are eliminated.

The eccentricity of region r (denoted E_r) is defined by the ratio between the two principal axes of the best fitting ellipse, measuring how thin and long a region is. It holds that $E_r \geq 1$. The eccentricity can be defined by the three second order moments $m_r(1,1), m_r(2,0)$ and $m_r(0,2)$. Let (c_{rx}, c_{ry}) denote the centroid of region r (given by the set of sites O_r, (s_x, s_y) being the coordinates of a point).

$$c_{rx} = \frac{1}{A_r} \sum_{s \in O_r} s_x \tag{6}$$

$$c_{ry} = \frac{1}{A_r} \sum_{s \in O_r} s_y \tag{7}$$

$$m_r(p, q) = \sum_{s \in O_r} (s_x - c_{rx})^p (s_y - c_{ry})^q \tag{8}$$

$$E_r = \sqrt{\frac{m_r(2,0) + m_r(0,2) + \sqrt{(m_r(2,0) - m_r(0,2))^2 + 4m_r^2(1,1)}}{m_r(2,0) + m_r(0,2) - \sqrt{(m_r(2,0) - m_r(0,2))^2 + 4m_r^2(1,1)}}} \tag{9}$$

The eccentricity criterion is intended to filter line segments.

The solidity criterion measures the proportion of the pixels in the convex hull of the region that are also in the region. Therefore, it is relevant to the region shape. In our implementation a value of $2/3$ is required for accepting a region as *lymphocyte nucleus* candidate.

3   Transferable Belief Model

Image and shape features are computed for each candidate region. The mean value M_i and the variance V_i of the image of a candidate region i are extracted. In order to be independent from scaling and variability in appearance, the mean and the variance of each region are normalized by division with the corresponding median values obtained on the set of all regions.

The two extracted features are combined within the *Transferable Belief Model* (TBM) framework [5] [6] in order to perform *lymphocyte nuclei* detection. The TBM is an alternative to probability measure for knowledge modelling and the main advantage and power of the TBM is the capacity to explicitly model doubt and conflict. TBM has been successfully applied on object detection and tracking problems [7] combined with shape and motion based features.

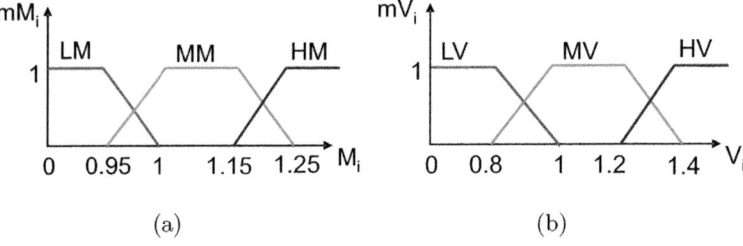

(a)   (b)

Fig. 1. From numerical features to belief. **(a)** Mean value. LM, MM and HM correspond to low, medium and high values of the normalized mean of image intensity, respectively. **(b)** Variance. LV, MV and HV correspond to low, medium and high values of the normalized variance of image intensity, respectively.

The mean value and the variance can be adequately converted into beliefs (symbolic representation). This is the first step of the TBM framework. We have proposed the numeric-to-symbolic conversion presented in Fig. 1, where L is used for low value, M for medium values and H for high values. Let us note $f_k(m)$ and $g_k(v)$ the two belief functions, where $k = 1, 2, 3$ corresponds respectively to low, medium and high values. Using symbolic representation, the *lymphocyte nuclei* detection can be performed based on appropriate table rules (see Table 1). The values of Table 1 (values of $T(k, l)$) have been estimated using the ground truth images of the ICPR 2010 contest, by estimating the probability of *lymphocyte nuclei* detection for each belief pair.

Having estimated the table of rules, we compute the the plausibility B_i of each candidate *lymphocyte nucleus* region i as follows:

$$B_i = \sum_{k=1}^{3} \sum_{l=1}^{3} f_k(M_i) g_l(V_i) T(k, l) \tag{10}$$

A region i will be detected as *lymphocyte nucleus*, if $B_i > 0.55$. We have selected the threshold of 0.55, since it gives the highest accuracy results on the ground truth data set.

Table 1. Table rules providing $T(k,l)$ used in Equation (10)

	LV	MV	HV
LM	0.999	0.988	0.958
MM	0.565	0.669	0.701
HM	0.052	0.1	0.032

4   Region Splitting

Having detected the *lymphocyte nuclei* based on appearance features, we have to resolve possible overlaps using shape features. Finally, the area (A_i) and the eccentricity (E_i) [7] are used in the decision of splitting a detected region to more than one regions plausibly corresponding to *lymphocyte nuclei.*

The area and the eccentricity are normalized with report to their respective median values. According to the feature A_i, the region i can be splitted into N_i regions, where $N_i \in \{1, \ldots, \lceil A_i \rceil\}$. We split a region i into N_i possible sub-regions selecting the more appropriate splitting as described hereafter.

The proposed algorithm splits the region i into N_i equal area regions minimizing the maximum eccentricity of the resulting sub-regions j, $j \in \{1, \ldots, N_i\}$, since the *lymphocyte nuclei* are circular-like regions. A circular-like region has minimum eccentricity, close to one. Similar to the minimization of maximum error on polygonal approximation problem using equal errors criterion [8], the problem of minimizing the maximum eccentricity can be sub-optimally solved under the equal area criterion and the above eccentricity constraint. We have implemented this criterion using the following algorithm. The pseudo-code of the Region Splitting to N_i sub-regions is given in Algorithm 1.

– Initially, we sequentially select N_i seed-points $p_j, j \in \{1, \ldots, N_i\}$ of region i from which N_i parallel region growing algorithms start. The seeds should follow the next constraint so that the growing algorithms start from the farthest sub-regions: the minimum distance between all pairs of these points should be maximized.
– The optimal algorithm that solves this problem has $O(\binom{R_i}{N_i})$ computation cost, where R_i denotes the number of pixels of region i. We have used the next approximate algorithm that sub-optimally solves this problem in $O(R_i^2)$ based on the optimal solution for two regions. p_1 and p_2 are given as the two farthest points of region i (optimal solution for two regions) (lines 1-11 of Algorithm 1). The next points $p_j, j \in \{3, \ldots, N_i\}$ are sequentially computed by getting the point p of region i that maximizes the minimum of distances from p to $p_{j-1}, p_{j-2}, \ldots, p_1$ (lines 12-27 of Algorithm 1).
– Then N_i parallel growing algorithms start from seeds p_j, $j \in \{1, \ldots, N_i\}$ (lines 28-30, 31-36 of Algorithm 1). In each step, the growing algorithm j, $j \in \{1, \ldots, N_i\}$ adds the most close point to p_j from the set of non-visiting boundary points of sub-region j that minimizes eccentricity of sub-region j, yielding equal area regions that uniformly grow with a circular-like shape having minimal eccentricity (line 33 of Algorithm 1).

Finally, we select splitting to N_i regions, where N_i maximizes the following criterion:

$$C(N_i) = \begin{cases} \dfrac{B_i}{\sqrt{\max(A_i, \dfrac{1}{A_i}) \cdot E_i}}, & N_i = 1 \\[4mm] \dfrac{(1 - b(N_i)) \max_{j \in \{1,...,N_i\}} B_{i,j}}{\sqrt{\max(\bar{A}, \dfrac{1}{\bar{A}}) \cdot \bar{E}}}, & N_i > 1 \end{cases} \quad (11)$$

where $\bar{A}$ and $\bar{E}$ denote the mean area and the mean eccentricity of the N_i split regions. $b(N_i)$ denotes the percentage of boundary pixels between the resulting sub-regions (intrinsic boundary pixels) of splitting. $B_{i,j}$ denotes the plausibility of *lymphocyte nuclei* sub-region for the sub-region j of region i estimated by TBM framework. This criterion is maximized when the mean area and mean eccentricity is close to one (that corresponds to most appropriate shape for *lymphocyte nucleus* region) and the maximum probability of *lymphocyte nucleus* sub-region is high.

Fig. 2 illustrates an example of region splitting algorithm execution for $N_i = 2$ and $N_i = 3$. According to ground truth, the algorithm successfully gives three partitions, since for $N_i = 3$ the proposed criterion was maximized, $C(1) = 0.47, C(2) = 0.24, C(3) = 0.53, C(4) = 0.35$.

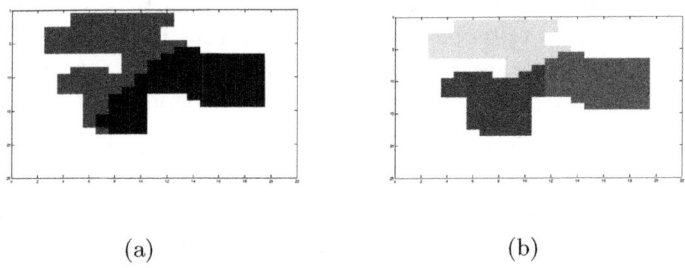

(a)   (b)

Fig. 2. An example of Region Splitting into **(a)** $N_i = 2$. **(b)** and $N_i = 3$ sub-regions.

5   Experimental Results

We have tested our method on the data of the Pattern Recognition in Histopathological Images contest (ICPR 2010). Fig. 3 illustrates results of the proposed scheme for image *im8.tif* of the data set. Figs. 3(a) and 3(b) illustrate the original image and the principal image component, respectively. Fig. 3(c) illustrates final results of the method with ground truth. Red boundaries correspond to candidate regions that are detected as *lymphocyte nuclei* regions (see Section 3). Blue boundaries correspond to candidate regions that are not detected as *lymphocyte nuclei* regions (see Section 3). Green and white squares are the centroids of real *lymphocyte nuclei* and detected regions, respectively. Fig. 3(d)

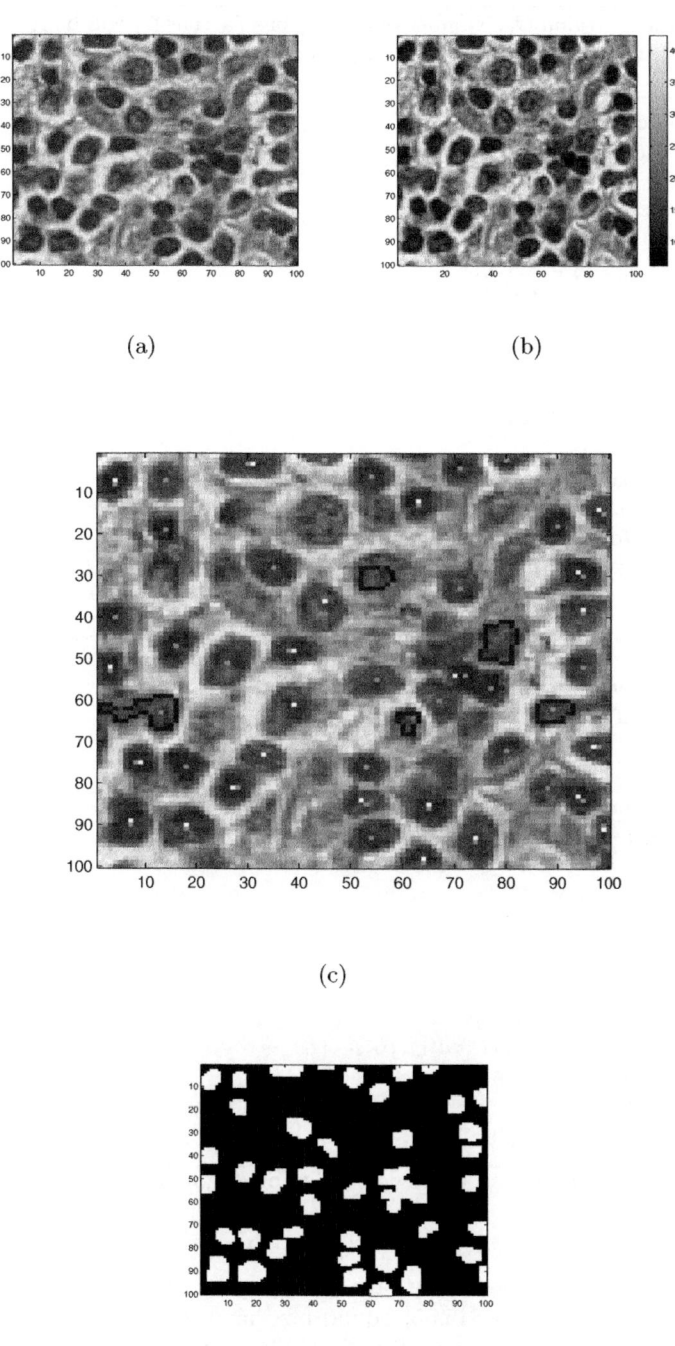

(a)

(b)

(c)

(d)

Fig. 3. (a) The original image. (b) The one channel image after PCA. (c) The final detection with ground truth. (d) The final detected regions.

input : Region O_i. Number of sub-regions N_i that O_i will be split.
output: The N_i sub-regions R_i^j, $j \in \{1, ..., N_i\}$.

```
1  d_max = 0
2  foreach (x₁,y₁) ∈ O_i do
3      foreach (x₂,y₂) ∈ O_i do
4          d = (x₁ - x₂)² + (y₁ - y₂)²
5          if d > d_max then
6              d_max = d
7              p₁ = (x₁,y₁)
8              p₂ = (x₂,y₂)
9          end
10     end
11 end
12 for j = 3 to N_i do
13     d_max = 0
14     foreach (x₁,y₁) ∈ O_i do
15         d_min = ∞
16         for n = 1 to j - 1 do
17             d = (p_n.x - x₁)² + (p_n.y - y₁)²
18             if d < d_min then
19                 d_min = d
20             end
21         end
22         if d_min > d_max then
23             d_max = d_min
24             p_j = (x₁,y₁)
25         end
26     end
27 end
28 for j = 1 to N_i do
29     R_i^j = {p_j}
30 end
31 repeat
32     for j = 1 to N_i do
33         p̂_j = getNextPoint(j, R_i, p_j, O_i)
34         R_i^j = R_i^j ∪ {p̂_j}
35     end
36 until ∀j ∈ {1, ..., N_i} ⇒ p̂_j = ∅
```

Algorithm 1. Region Splitting Algorithm

illustrates final detection of the proposed method (white regions). The region that belongs in $[75,85] \times [45,55]$ bound box has been successfully splitted into two sub-regions. Similarly with Fig. 3(c), Fig. 4 depicts the final results of the method with ground truth for the rest images of dataset.

Table 2 depicts the Sensitivity and the PPV for each image of the tested data set. According to this table, Sensitivity and PPV take values in range

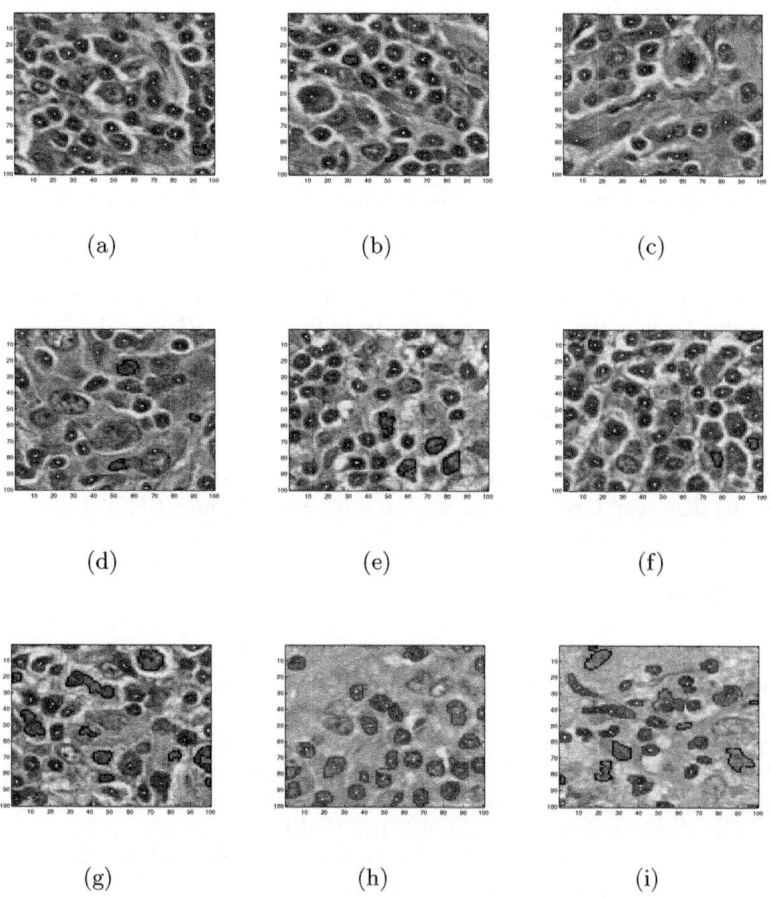

(a)   (b)   (c)

(d)   (e)   (f)

(g)   (h)   (i)

Fig. 4. The final detection with ground truth

Table 2. Sensitivity and PPV

Image	Sensitivity	PPV
im1	0.968	0.815
im2	0.961	0.714
im3	0.900	0.720
im4	0.950	0.791
im5	0.965	0.933
im6	0.944	0.756
im7	0.928	0.928
im8	0.883	0.926
im9	0.941	0.592
im14	0.952	0.869

[0.928, 0.968] and [0.714, 0.926], respectively. The mean values of Sensitivity and PPV are 0.938 and 0.807, respectively.

6 Conclusion

We have proposed an appearance and shape based method for automatic detection of *lymphocyte nuclei* on histopathology images. We have used a mixture of Gaussians for determining the probability distribution of the principal image component and the TBM framework with a region splitting method to detect and split the *lymphocyte nuclei* regions. The proposed algorithm gives high accuracy results on the whole data set: Sensitivity of 0.938 and PPV of 0.807.

Acknowledgements

The work of Costas Panagiotakis has been supported by postdoctoral scholarship (2009-10) from the Greek State Scholarships Foundation (I.K.Y.).

References

1. Demir, C., Yener, B.: Automated cancer diagnosis based on histopathological images: a systematic survey. Technical Report 05-09, Rensselaer Polytechnique Institute (2005)
2. Basavanhally, A., Ganesan, S., Agner, S., Monaco, J., Feldman, M., Tomaszewski, J., Bhanot, G., Madabhushi, A.: Computerized image-based detection and grading of lymphocytic infiltration in her2+ breast cancer histopathology. IEEE Transactions on Biomedical Engineering 57(3), 642–653 (2010)
3. Basavanhally, A., Agner, S., Alexe, G., Ganesan, G.B.S., Madabhushi, A.: Manifold learning with graph-based features for identifying extent of lymphocytic infiltration from high grade breast cancer histology. In: Workshop on Microscopic Image Analysis with Applications in Biology (in conjunction with MICCAI) (2008)
4. Fatakdawala, H., Basavanhally, A., Xu, J., Bhanot, G., Ganesan, S., Feldman, M., Tomaszewski, J., Madabhushi, A.: Expectation maximization driven geodesic active contour with overlap resolution (EMaGACOR): Application to lymphocyte segmentation on breast cancer histopathology. IEEE Transactions on Biomedical Engineering (to appear, 2010)
5. Smets, P., Kennes, R.: The Transferable Belief Model. Artificial Intelligence 66(2), 191–234 (1994)
6. Smets, P.: Decision making in the TBM: the necessity of the pignistic transformation. Int. Jour. of Approximate Reasoning 38, 133–147 (2005)
7. Panagiotakis, C., Ramasso, E., Tziritas, G., Rombaut, M., Pellerin, D.: Shape-based individual/group detection for sport videos categorization. Intern. J. Pattern Recognition Artificial Intelligence 22(6), 1187–1213 (2008)
8. Panagiotakis, C., Tziritas, G.: Any dimension polygonal approximation based on equal errors principle. Pattern Recogn. Lett. 28(5), 582–591 (2007)

Counting Lymphocytes in Histopathology Images Using Connected Components

Felix Graf, Marcin Grzegorzek, and Dietrich Paulus

Institute of Computational Visualistics, University of Koblenz-Landau
Universitätsstrasse 1, 56070 Koblenz, Germany
{felixgraf,marcin,paulus}@uni-koblenz.de

Abstract. In this paper, a method for automatic counting of lymphocytes in histopathology images using connected components is presented. Our multi-step approach can be divided into two main parts: processing of histopathology images, and recognition of interesting regions. In the processing part, we use thresholding and morphology methods as well as connected components to improve the quality of the images for recognition. The recognition part is based on a modified template matching method. The experimental results achieved for our algorithm prove its high robustness for this kind of applications.

1   Introduction

In the context of several pathologies including breast and ovarian cancer, the presence of lymphocytes has been correlated with disease outcome. For instance for Her2+ breast cancer, the presence of lymphocytic infiltration (LI) has been correlated with nodal metastasis and tumor recurrence.

The ability to automatically detect and quantify extent of LI on histopathology imagery could potentially result in the development of an image based prognostic tool for Her2+ and ovarian cancer patients.

However, lymphocyte segmentation in H & E-stained histopathology images is complicated by the similarity in appearance between lymphocyte nuclei and other structures (e.g. cancer nuclei) in the image. Additional challenges include biological variability, histological artifacts, and high prevalence of overlapping objects. Although active contours are widely employed in image segmentation, they are limited in their ability to segment overlapping objects and are sensitive to initialization.

In this paper, we introduce a new method to tackle the problem of counting lymphocytes in histopathology images. Our approach can be divided into two main parts: processing of histopathology images (Section 3), and recognition of interesting regions (Section 4). In the processing part, we use thresholding and morphology methods as well as connected components to improve the quality of the images for recognition. The recognition part is based on a modified template matching method. The experimental results achieved for our algorithm prove its high robustness for this kind of applications (Section 5). The paper is closed by some conclusions in Section 6.

D. Ünay, Z. Çataltepe, and S. Aksoy (Eds.): ICPR 2010, LNCS 6388, pp. 263–269, 2010.

2   Related Work

Counting lymphocytes in histopathology images is a difficult and important task and there is some related work in this area.

A good overview with a systematic survey of the computational steps in automated cancer diagnosis based on histopathology gives [3]. In this paper, the computational steps are detailed, their challenges addressed and the remedies to overcome the challenges are discussed. The first computational step is the image preprocessing to determine the focal areas. Usually it is preceded by noise reduction to improve its success and in the case of cellular-level diagnosis it also comprises nucleus/cell segmentation. Step two is the feature extraction to quantify the properties of these focal areas. It defines appropriate representations of the focal areas that provide distinctive objective measures. Classifying the focal areas as malignant or not or identifying their malignancy levels is step three. Automated diagnostic systems that operate on quantitative measures are designed. This step also estimates the accuracy of the system.

The most related work has been published by Sertel et al. in [6] and [5]. In [6] a novel color texture classification approach is introduced and applied to computer-assisted grading of follicular lymphoma from whole-slide tissue samples. In [5] a model-based intermediate representation (MBIR) of cytological components that enables higher level semantic description of tissue characteristics and a novel color-texture analysis approach that combines the MBIR with low level texture features, which capture tissue characteristics at pixel level is introduced.

Further very related works have been published by scientists from the Laboratory for Computational Imaging & Informatics from Rutgers. In [1] an approach for computerized image-based detection and grading of lymphocytic infiltration in HER2+ breast cancer histopathology is presented. Fatakdawala et al. present in [4] an interesting method for an expectation maximization driven geodesic active contour with overlap resolution and apply their algorithm to lymphocyte segmentation on breast cancer histopathology.

3   Processing Histopathology Images

This section describes the processing steps of the images before the recognition of patterns (Figure 1). The images are processed to get the best possible results and necessary parameters are calculated.

3.1   Groundtruth

Each image of the ground truth contains the contour of one lymphocyte annotated by experts (Figure 2). The three parameters arclength L, area A and compactness c are calculated which describes the elliptic shaped lymphocytes. The compactness is defined as

$$c = \frac{L^2}{4\pi A}$$

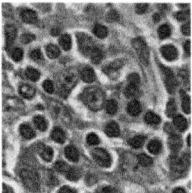

Fig. 1. H&E stained training data image comprising lymphocytic infiltration

Fig. 2. Groundtruth image containing the contour of one lymphocyte

and describes how similar the object is to a circle. The arclength is the length of the contour and the area is the amount of pixel inside the contour.

For these parameters the mean value and the variance for all contours in all images of the ground truth is calculated.

3.2 Thresholding and Morphology

The images of the training data are RGB color images. They are divided into three single channel images. The red channel contains the most information because of the staining and therefore is weighted more. Using a threshold according to [5] the possible lymphocytes are separated from the background and are shown in the resulting binary images (Figure 3).

To separate touching lymphocyte candidates the morphological operation opening with an ellipse as structuring element is used. By applying an erosion followed by a dilatation the separation is done without changing the size of the lymphocytes.

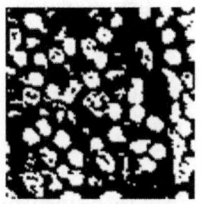

Fig. 3. Image after thresholding but before morphology

3.3 Connected Components

Connected components, called blobs, are labeled in the binary images using contour tracing technique described in [2]. Blobs have the advantage that many parameters can be calculated easily. Some of them will be used in Section 4.1 for the recognition.

4   Recognition

After processing the images to prepare them for the recognition, candidates for lymphocytes have to be tested if they are really lymphocytes or not (Figure 4(a)). A classification into two classes has to be done. It will be distinguished between candidates that are lymphocytes and candidates that were wrongly detected as lymphocytes and therefore have to be sorted out.

4.1 Parameter and Selection Criteria

For each blob the same parameters as for the ground truth in Section 3.1 are calculated, which were arclength, area and compactness. The parameters for the blobs are compared to those of the ground truth. Candidates for the selection criteria are minimum value, maximum value, mean value, variance or combinations of the once mentioned before. An optimal criterion has still to be found by comparing the results of different possible combinations. Until know, if a blob does not conform to the criterion it is not a lymphocyte (Figure 4(b)). Further differentiations will be mentioned in Section 4.2.

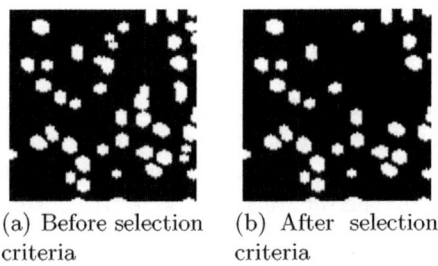

(a) Before selection criteria   (b) After selection criteria

Fig. 4. Binary image showing blobs before and after applying selection criteria (here: mean value)

4.2 Template Matching

Blobs give us many possibilities to achieve different parameters or properties, e. g., contour, bounding volume and ellipse fitting. One attempt to find lymphocytes is an easy template matching. A static ellipse with fixed parameters is moved pixel by pixel over the image. The ellipse is described by its two axes. If there is a match better than 80 percent a lymphocyte is detected. Using blobs similar results can be achieved in less time by fitting the largest ellipse possible into each blob (Figure 5). Additionally the parameters of the ellipses are easy to compute, e. g., center, area, contour, arclength, axes and angle.

Lymphocytes that overlap in a bigger region and not only in a few pixels cannot be separated using morphological operators. Here the calculated parameters of the blobs can be used. If the selection criteria is not fulfilled the blob is classified as a non-lymphocyte. To make the classification more accurate in another iteration multiple ellipses are fitted into the blob that does not conform to the criterion. The blob is divided into as much parts as the criteria finds overlapping lymphocytes in it.

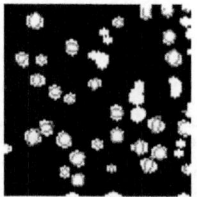

Fig. 5. Fitting largest ellipse possible into blobs

5   Experiments and Results

The performed experiments to count lymphocytes in histopathology images and the achieved results will be summarized in this section.

5.1   Experiments

The experiments were performed on H&E stained images digitized at 20x resolution. Color space of the training data is RGB, the image format is TIF and the image size is 100x100 pixels. Depending on the desired output the resulting images contain contour, center or blobs of the lymphocytes (Figure 6). The optimal output depends on the used evaluation methods.

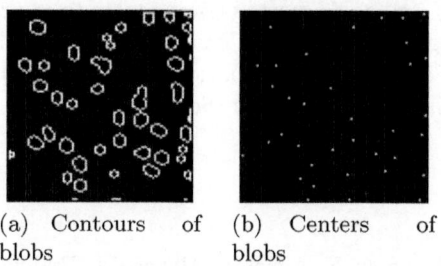

(a) Contours of blobs   (b) Centers of blobs

Fig. 6. Contours and centers of blobs possible as resulting images

5.2   Results

Table 1 shows a quantitative result for ten images. The first column contains the image number and the second column contains the ground truth. Groundtruth means the number of lymphocytes in the image annotated by experts. The third column shows the number of possible candidates found using blobs and the fourth column contains the number of possible lymphocytes found by an easy template matching algorithm using an elliptic shaped template.

For further experiments quantitative evaluations of the results have to be used as well. Therefore the region-based measures dice coefficient, overlap, sensitivity, specificity and positive predictive value and the boundary-based measures hausdorff distance and mean absolute distance are calculated.

Table 1. Quantitative results for ten images of training data

image	training data	blobs	template matching
1	32	36	32
2	26	34	29
3	20	22	21
4	19	21	18
5	29	29	29
6	36	32	37
7	28	30	26
8	43	38	41
9	17	7	5
10	20	11	6

6   Conclusions

In this paper, a new approach for counting lymphocytes in histopathology images has been introduced. The algorithm has been divided into two main parts: processing of histopathology images (Section 3), and recognition of interesting image regions (Section 4). For processing, we have used thresholding and morphology methods as well as connected components to improve the quality of the images. In the recognition phase, we have applied a modified template matching algorithm. The quantitative evaluation has shown a high robustness of our approach for this kind of applications (Section 5).

References

1. Basavanhally, A., Ganesan, S., Agner, S., Monaco, J., Feldman, M., Tomaszewski, J., Bhanot, G., Madabhushi, A.: Computerized image-based detection and grading of lymphocytic infiltration in her2+ breast cancer histopathology. IEEE Transactions on Biomedical Engineering 57(3), 642–653 (2010)

2. Chang, F., Chen, C.-J., Lu, C.-J.: A linear-time component-labeling algorithm using contour tracing technique. Computer Vision and Image Understanding 93, 206–220 (2004)
3. C. Demir and B. Yener. Automated cancer diagnosis based on histopathological images: a systematic survey. Technical report, Rensselaer Polytechnic Institute, Department of Computer Science (2005)
4. Fatakdawala, H., Xu, J., Basavanhally, A., Bhanot, G., Ganesan, S., Feldman, M., Tomaszewski, J., Madabhushi, A.: Expectation maximization driven geodesic active contour with overlap resolution (emagacor): Application to lymphocyte segmentation on breast cancer histopathology. IEEE Transactions on Biomedical Engineering 57(7), 1676–1689 (2010)
5. Sertel, O., Kong, J., Catalyurek, U.V., Lozanski, G., Saltz, J.H., Gurcan, M.N.: Histopathological image analysis using model-based intermediate representations and color texture: follicular lymphoma grading. Journal of Signal Processing 55(1-3), 169–183 (2009)
6. Sertel, O., Kong, J., Lozanski, G., Shana'ah, A., Catalyurek, U., Saltz, J., Gurcan, M.: Texture classification using nonlinear color quantization: application to histopathological image analysis. Acoustics, Speech and Signal Processing, 597–600 (March 2008)

An Overview of Contest on Semantic Description of Human Activities (SDHA) 2010

M.S. Ryoo[1,2], Chia-Chih Chen[1], J.K. Aggarwal[1], and Amit Roy-Chowdhury[3]

[1] Computer and Vision Research Center, the University of Texas at Austin, USA
[2] Robot/Cognition Research Department, ETRI, Korea
[3] Video Computing Group, Dept. of EE, University of California, Riverside, USA
mryoo@etri.re.kr, {ccchen,aggarwaljk}@mail.utexas.edu, amitrc@ee.ucr.edu
http://cvrc.ece.utexas.edu/SDHA2010/

Abstract. This paper summarizes results of the 1st Contest on Semantic Description of Human Activities (SDHA), in conjunction with ICPR 2010. SDHA 2010 consists of three types of challenges, High-level Human Interaction Recognition Challenge, Aerial View Activity Classification Challenge, and Wide-Area Activity Search and Recognition Challenge. The challenges are designed to encourage participants to test existing methodologies and develop new approaches for complex human activity recognition scenarios in realistic environments. We introduce three new public datasets through these challenges, and discuss results of the state-of-the-art activity recognition systems designed and implemented by the contestants. A methodology using a spatio-temporal voting [19] successfully classified segmented videos in the UT-Interaction datasets, but had a difficulty correctly localizing activities from continuous videos. Both the method using local features [10] and the HMM based method [18] recognized actions from low-resolution videos (i.e. UT-Tower dataset) successfully. We compare their results in this paper.

Keywords: Activity recognition contest, human activity recognition, video analysis.

1   Introduction

Human activity recognition is an area with an increasing amount of interest, having a variety of potential applications. An automated recognition of human activities from videos is essential for the construction of smart surveillance systems, intelligent robots, human-computer interfaces, quality of life devices (e.g. elderly monitoring), and military systems. Developments of spatio-temporal feature extraction, tracking, and high-level activity analysis are leading today's computer vision researchers to explore human activity recognition methodologies practically applicable for real world applications.

In this contest, we propose three types of activity recognition challenges which focus on different aspects of human activity recognitions: High-level Human Interaction Recognition Challenge, Aerial View Activity Classification Challenge,

D. Ünay, Z. Çataltepe, and S. Aksoy (Eds.): ICPR 2010, LNCS 6388, pp. 270–285, 2010.
© Springer-Verlag Berlin Heidelberg 2010

Table 1. A table summarizing the results of SDHA 2010 contest. We made the authors of the teams who decided not to submit their results anonymous. The 'success' indicates that the team successfully submitted their results meeting the requirements. We invited three teams who showed the best results to submit their papers [9,17,21].

Challenge	TeamName	Authors	Institution	Success	Paper
Interaction	Team BIWI	Yao et al.	ETH	△	Variations of a Hough-Voting Action Recognition System
	TU Graz	-	TU Graz	X	-
	SUVARI	-	Sabanci Univ.[1]	X	-
	Panopticon	-	Sabanci Univ.[1]	X	-
Aerial-view	Imagelab	Vezzani et al.	Univ. of Modena and Reggio Emilia	O	HMM based Action Recognition with Projection Histogram Features
	ECSI_ISI	Biswas et al.	Indian Statistical Institute	O	-
	BU_Action	Guo et al.	Boston University	O	Aerial View Activity Classification by Covariance Matching of Silhouette Tunnels
	Team BIWI	Yao et al.	ETH	O	Variations of a Hough-Voting Action Recognition System
Wide-area	Vistek	-	Sabanci Univ.[2], Univ. of Amsterdam	X	-

and Wide-Area Activity Search and Recognition Challenge. The three types of datasets named UT-Interaction, UT-Tower, and UCR-Videoweb are introduced for each challenge respectively. The objective of our challenges is to provide videos of human activities which are of practical interests, and make researchers evaluate their existing/new activity recognition methodologies.

In the interaction challenge, contestants are asked to correctly localize ongoing activities from continuous video streams containing multiple human-human interactions (i.e. a high-level surveillance setting). The aerial view challenge requires the participants to develop recognition methodologies that handles low-resolution videos where each person's height is of approximately 20 pixels. This challenge is particularly motivated by military applications such as unmanned aerial vehicles taking videos from an aerial view. The wide-area challenge asks contestants to retrieve videos similar to query events using a multi-camera dataset. This dataset consists of videos obtained from multiple camera covering different regions of a wide area, which is a very common situation in many surveillance scenarios (e.g. airport).

The challenges are designed to encourage researchers to test their new state-of-the-art recognition systems on the three datasets with different characteristics (Table 2). Even though there exist other public datasets composed of human action videos [16,8,20,13] (Fig. 3 (a-e)), most of them focus on recognition of simple actions (e.g. walking, jogging, ...) in controlled environments (e.g. only one actor appears in the videos, taken from a single camera). Several baseline methods have been implemented by the contest organizers as well, comparing contestants' results with well-known previous methodologies. The contest and its datasets will provide impetus for future research in many related areas.

Table 2. A table summarizing the characteristics of the contest datasets. '# Executions' describes the total number of activity executions in the entire dataset. '# Actors' is the number of actors appearing in the scene simultaneously, and 'Multi-person' describes whether the dataset involves multi-person activities or not. 'Continuous' indicates whether the dataset consists of continuous video sequences involving multiple occurrences of human activities.

Dataset Name	# Activities	# Executions	# Cameras	# Actors	Resolution	Multi-person	Continuous
UT-Interaction	6	120+	1	2~4	720*480	O	O
UT-Tower	9	108	1	1	360*240	X	X
UCR-Videoweb	52	Multiple	4~8	2~10	640*480	O	O

2 Previous Datasets

Several public datasets have been introduced in the past 10 years, encouraging researchers to explore various action recognition directions. The KTH dataset [16] and the Weizmann dataset [8] are the typical examples of these datasets. These two single-camera datasets have been designed for research purposes, providing a standard for researchers to compare their action classification performances. The datasets are composed of videos of relatively simple periodic actions, such as walking, jogging, and running. The videos are segmented temporally so that each clip contains no more than one action of a single person. They were taken in a controlled environment; their backgrounds and lighting conditions are mostly uniform. In general, they have a good image resolution and little camera jitters. The I-XMAS dataset [20] was similar, except that they provided videos from multiple cameras for a 3-D reconstruction.

Recently, more challenging datasets were constructed by collecting realistic videos from movies [13,12,14]. These movie scenes are taken from varying view points with complex backgrounds, in contrast of the previous public datasets [16,8]. These dataset encourages the development of recognition systems that are reliable under noise and view point changes. However, even though these videos were taken in more realistic environments, the complexity of the actions themselves were similar to [16,8]: the datasets contain simple instantaneous actions such as kissing and hitting. They were not designed to test recognition of high-level human activities from continuous sequences.

There also are datasets motivated by surveillance applications. PETS datasets [1] and i-LIDS datasets [6] belong to this category. The videos in these datasets were taken in uncontrolled environments (e.g. subway stations), and they contain few application specific activities (e.g. leaving a baggage). Videos from multiple cameras watching the same site with different view points are provided.

Each of the datasets introduced in SDHA 2010 has its unique characteristics that distinguish it from other previous datasets. The UT-Interaction dataset is designed to encourage detection of interaction-level human activities (e.g. pushing and hugging). Instead of asking to classify simple periodic actions, it encourages localization of the multiple activities from continuous video streams spatially and temporally. The UT-Tower dataset contains very low-resolution videos, which makes their recognition challenging. The UCR-Videoweb dataset introduces continuous videos taken from multiple cameras observing different areas of a place (e.g. CCTV cameras for a university campus building).

Up to our knowledge, SDHA 2010 is the first computer vision contest designed to compare performances of activity recognition methodologies. There have been previous competitions for recognizing objects (e.g. PASCAL-VOC [7]) or recognizing a specific scene (e.g. abandoned baggage detection [6]), but no previous contest attempted to measure general accuracies of systems on recognizing various types of human activities. Our objective is to evaluate the state-of-the-arts in activity recognition and establish standard datasets for future exploration.

3   High-Level Human Interaction Recognition Challenge

In the "High-level Human Interaction Recognition Challenge", contestants are asked to recognize ongoing human activities from continuous videos. The objective of the challenge is to encourage researchers to explore the recognition of complex human activities from continuous videos, taken in realistic settings. Each video contains several human-human interactions (e.g. hand shaking and pushing) occurring sequentially and/or concurrently. The contestants must correctly annotate which activity is occurring when and where for all videos. Irrelevant pedestrians are also present in some videos. Accurate detection and localization of human activities are required, instead of a brute force classification of videos.

The motivation is that many of real-world applications require high-level activities performed by multiple individuals to be recognized. Surveillance systems for airports and subway stations are typical examples. In these environments, continuous sequences provided from CCTV cameras must be analyzed toward correct detection of multi-human interactions such as two persons fighting. In contrast to previous single-person action classification datasets discussed in Section 2, the challenge aims to establish a new public dataset composed of continuous executions of multiple real-world human interactions.

3.1   Dataset Description

The UT-Interaction dataset[1] contains videos of continuous executions of 6 classes of human-human interactions: hand-shake, point, hug, push, kick and punch. Fig. 1 shows example snapshots of these multi-person activities. Ground truth labels for all interactions in the dataset videos are provided, including time intervals and bounding boxes. There is a total of 20 video sequences whose lengths are around 1 minute (e.g. Fig. 2). Each video contains at least one execution per interaction, providing us about 8 executions of human activities per video on average. Several actors with more than 15 different clothing conditions appear in the videos. The videos are taken with the resolution of 720*480, 30 fps, and the height of a person in the video is about 200 pixels.

We divide videos into two sets. The set #1 is composed of 10 video sequences taken on a parking lot. The videos of the set #1 are taken with slightly different

---

[1] http://cvrc.ece.utexas.edu/SDHA2010/Human_Interaction.html

Fig. 1. Example snapshots of the six human-human interactions

zoom rate, and their backgrounds are mostly static with little camera jitter. The set #2 (i.e. the other 10 sequences) are taken at a lawn on a windy day. Background is moving slightly (e.g. tree moves), and they contain more camera jitters. From sequences 1 to 4 and from 11 to 13, only two interacting persons appear in the scene. From sequences 5 to 8 and from 14 to 17, both interacting persons and pedestrians are present in the scene. In sets 9, 10, 18, 19, and 20, several pairs of interacting persons execute the activities simultaneously. Each set has a different background, scale, and illumination. The UT-Interaction set #1 was first introduced in [15], and we are extending it with this challenge.

For each set, we selected 60 activity executions that will be used for the training and testing in our challenge. The contestant performances are measured using the selected 60 activity executions. The other executions, marked as 'others' in our dataset, are not used for the evaluation.

3.2   Results

The interaction challenge consists of two types of tasks: the classification task and the continuous detection (i.e. localization) task. The contestants are requested to evaluate their systems with these two different experimental settings:

For the 'classification', 120 video segments (from 20 sequences) cropped based on their ground truth bounding boxes and ground truth time intervals are provided. The video sequences were segmented spatially and temporally to contain only one interaction performed by two participants, and the classification accuracies are measured with these video segments in a way similar to the previous settings [16,8]. That is, the performance of classifying a testing video segment into its correct category is measured.

In the 'detection' setting, the entire continuous sequences are used for the continuous recognition. The activity recognition is measured to be correct if and only if the system correctly annotates an occurring activity's time interval (i.e. a pair of starting time and ending time) and its spatial bounding box. If the annotation overlaps with the ground truth more than 50% spatially and temporally, the detection is treated as a true positive. Otherwise, it is treated as a false positive. Contestants are requested to submit a Precision-Recall curve for each set, summarizing the detection results.

Fig. 2. Example video sequences of the UT-Interaction dataset

In both tasks, the contestants were asked to measure the performances of their systems using 10-fold leave-one-out cross validation per set as follows: For each round, contestants leave one among 10 sequences for the testing and use the other 9 for the training. Contestants are required to count the number of true positives, false positives, and false negatives obtained through the entire 10 rounds, which will provide a particular precision and recall rate of the system (i.e. a point on a PR curve). Various PR rates will be obtained by changing the system parameters, and the PR curve is drawn by plotting them.

A total of four teams showed their intent to participate the challenge. However, only one among them succeeded to submit results for the classification task, which we report with Tables 3 and 4. The team BIWI [19] used a Hough transform-based method to classify interaction videos. Their method is based on [21], which uses a spatio-temporal voting with extracted local XYT features. A pedestrian detection algorithm was also adopted for the better classification. Particularly for the interaction challenge, they have modeled each actor's action using their voting method, forming a hierarchical system consisting of 2-levels.

In addition, in order to compare the participating team's result with previous methodologies, we have implemented several existing well-known action classification methods. Two different types of features (i.e. spatio-temporal features from [16] and 'cuboids' from [4]) are adopted, and three types of elementary classifiers, {k-nearest neighbor classifiers (k-NNs), Bayesian classifiers, and support vector machines (SVMs)}, are implemented. Their combinations generate six baseline methods as specified in Tables 3 and 4.

The baseline classifiers rely on a feature codebook generated by clustering the feature vectors into several categories. Codebooks were generated 10 times using k-means algorithm, and the systems' performances have been averaged for the 10 codebooks. SVM classification accuracies with the best codebook is also provided for the comparison. In the baseline methods, video segments have been normalized based on the ground truth so that the main actor of the activity (e.g. the person punching the other) always stands on the left-hand side.

The classification results shows that the pointing interaction composed of least number of feature and the hugging interaction composed of the largest

Table 3. Activity classification accuracies of the systems tested on the UT-Interaction dataset #1. The 1st, 2nd, and 3rd best system accuracies are described per activity: the blue color is for the 1st, the orange color suggests the 2nd, and the green color is for the 3rd.

	Shake	Hug	Kick	Point	Punch	Push	Total
Laptev + kNN	0.18	0.49	0.57	0.88	0.73	0.57	0.57
Laptev + Bayes.	0.38	0.72	0.47	0.9	0.5	0.52	0.582
Laptev + SVM	0.49	0.79	0.58	0.8	0.6	0.59	0.642
Latpev + SVM (best)	0.5	0.8	0.7	0.8	0.6	0.7	0.683
Cuboid + kNN	0.56	0.85	0.33	0.93	0.39	0.72	0.63
Cuboid + Bayes.	0.49	0.86	0.72	0.96	0.44	0.53	0.667
Cuboid + SVM	0.72	0.88	0.72	0.92	0.56	0.73	0.755
Cuboid + SVM (best)	0.8	0.9	0.9	1	0.7	0.8	0.85
Team BIWI	0.7	1	1	1	0.7	0.9	0.88

Table 4. Activity classification accuracies of the systems tested on the UT-Interaction dataset #2

	Shake	Hug	Kick	Point	Punch	Push	Total
Laptev + kNN	0.3	0.38	0.76	0.98	0.34	0.22	0.497
Laptev + Bayes.	0.36	0.67	0.62	0.9	0.32	0.4	0.545
Laptev + SVM	0.49	0.64	0.68	0.9	0.47	0.4	0.597
Latpev + SVM (best)	0.5	0.7	0.8	0.9	0.5	0.5	0.65
Cuboid + kNN	0.65	0.75	0.57	0.9	0.58	0.25	0.617
Cuboid + Bayes.	0.26	0.68	0.72	0.94	0.28	0.33	0.535
Cuboid + SVM	0.61	0.75	0.55	0.9	0.59	0.36	0.627
Cuboid + SVM (best)	0.8	0.8	0.6	0.9	0.7	0.4	0.7
Team BIWI	0.5	0.9	1	1	0.8	0.4	0.77

number of distinctive features was recognized with a high accuracy in general. Punching was confused with pushing in many systems because of their similarity. The participating team, BIWI, showed the highest recognition accuracy. The performances of the "Cuboid + SVM" with the best codebook were comparable.

3.3 Discussions

No team was able to submit a valid result for the detection task with continuous videos. There were 4 teams intended to participate challenge, but only one team succeeded to classify human interactions successfully and none succeeded to performed the continuous recognition. This implies that the recognition of high-level human activities from continuous videos still is an unsolved problem. Despite the demands from various applications including surveillance, the problem remains largely unexplored by researchers.

Applying the 'sliding windows' technique together with the classifier used above will be a straight forward solution. However, given the reported classification accuracies, such method is expected to generate many false positives. Using a voting-based methodology (e.g. [15,21]) is a promising direction for the detection task, and they must be explored further. In addition, we were able to observe that the hierarchical approach obtained better performances than the other baseline methods. Developing hierarchical approaches for continuous detection and localization of complex human activities will be required.

4   Aerial View Activity Classification Challenge

The ability to accurately recognize human activities at a distance is essential for several applications. Such applications include automated surveillance, aerial or satellite video analysis, and sports video annotation and search, etc. However, due to perspective distortion and air turbulence, the input imagery is presented in low-resolution and the available action patterns tend to be missing and blurry. In addition, shadows, time-varying lighting conditions, and unstabilized videos can all add up to the difficulty of this task. Therefore, without explicitly addressing these issues, most existing work in activity recognition may not be appropriate under the scenario.

In this "Aerial View Activity Classification Challenge", we aim to motivate researchers to explore techniques that achieve accurate recognition of human activities in videos filmed from a distant view. To simulate the video settings, we took image sequences of a single person performing various activities from the top of the University of Texas at Austin's main tower. We name it UT-Tower dataset[2]. The average height of a human figure in this dataset is about 20 pixels. The contest participants are expected to classify 108 video clips from a total of 9 categories of human activities. The performance of each participating team is evaluated by their leave-one-out accuracy on the dataset.

As described in Section 2, there exist several public datasets that are widely referred and tested in the literature of human activity recognition [8,16,13,5,20]. However, all these datasets (except the Soccer dataset) are taken from an approximate side view and they have human figures presented in high-resolution imagery (Fig. 3). The Soccer dataset contains low-resolution videos similar to ours, but the action categories of the Soccer dataset are defined by the proceeding directions of the players, and nearly half of the video sequences are the mirrors of the other half. These issues limit their applicability to the evaluation of activity recognition algorithms that focus on low-resolution video settings. Therefore, with this challenge, we distribute a new dataset for the assessment of general and surveillance oriented applications.

4.1   Dataset Description

Filmed top-down from a 307-foot high tower building, the UT-Tower dataset is composed of low-resolution videos similar to the imagery taken from an aerial vehicle. There are 9 classes of human actions: 'pointing', 'standing', 'digging', 'walking', 'carrying', 'running', 'wave1', 'wave2', 'jumping'. Algorithm performance on both *still* and *moving* types of human activities are to be examined. A total of 6 individuals acted in this dataset. We let each performer repeat every activity twice so that there are 108 sequences in the dataset. To add to the

---

[2] http://cvrc.ece.utexas.edu/SDHA2010/Aerial_View_Activity.html

Fig. 3. The widely used public datasets are mostly in medium- to high-resolution, for example, (a) Weizmann dataset, (b) KTH dataset, (c) HOHA dataset, and (d) I-XMAS dataset. Low-resolution datasets include (e) Soccer dataset and the proposed (f) UT-Tower dataset. The sizes of the images are proportional to their actual resolutions.

variety of the dataset, we recorded the activities under two types scene settings: concrete square and lawn. The videos were taken in 360×240 pixels resolution at 10fps. In addition to the low-resolution setup, the UT-Tower dataset also poses other challenges. For example, the direct sunlight causes salient human cast shadows and the rooftop gust brings continuous jitters to the camera. Fig. 4 shows the example video sequences of the dataset.

We manually segmented the original video into short clips so that each clip contains one complete track of human activity. In order to alleviate segmentation and tracking issues and make participants focus on the classification problem, we provide ground truth bounding boxes as well as foreground masks for each video. Contestants are free to take advantages of them or apply their own preprocessing techniques.

4.2   Results

In the aerial challenge, the contestants were asked to classify video clips in the UT-Tower dataset into the above-mentioned 9 action categories. Similar to the classification task of the interaction challenge, a leave-one-out cross validation setting is used. Here, one among 108 videos are used for the testing, and the

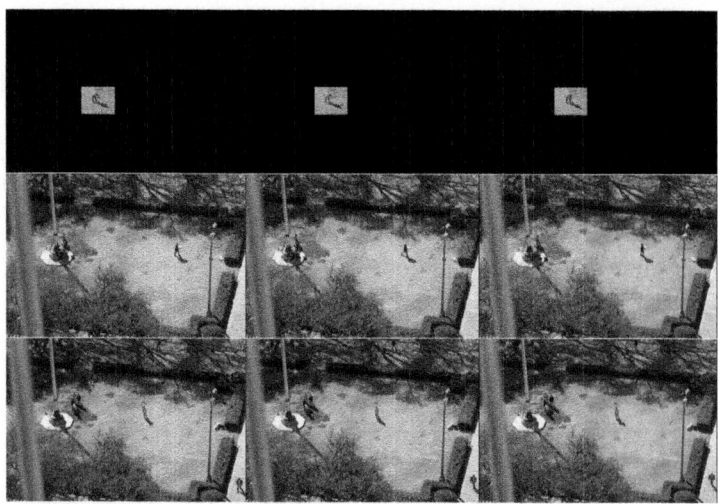

Fig. 4. The examples of 'digging', 'carrying', and 'wave1' in the UT-Tower dataset

others are used for the training. That is, a 108-fold cross validation is performed to evaluate the performances of the systems.

There are totally 4 university teams participated in this contest. Each team has tested their proposed algorithm on the UT-Tower dataset and reported the results. We briefly summarize the submitted methodologies and our baseline technique as follows.

Team BIWI. BIWI team from ETH Zurich proposes to use a Hough transform-based voting framework for action recognition [19]. They separate the voting into two stages to bypass the inherent high dimensionality problem in Hough transform representation. Random trees are trained to learn a mapping between densely-sampled feature patches and their corresponding votes in a spatio-temporal-action Hough space. They perform recognition by voting with a collection of learned random trees.

BU Action Covariance Manifolds. Boston University team represents a track of human action as a temporal sequence of local shape-deformations of centroid-centered object silhouettes [10], i.e., the shape of the silhouette tunnel. The empirical covariance matrix of a set of 13-dimensional feature is extracted as feature from the silhouette tunnel. The silhouette tunnel of a test video is broken into short overlapping segments and each segment is classified using a dictionary of labeled action covariance matrices with the nearest neighbor rule.

ECSU_ISI. The team from Indian Statistical Institute adopts a bag-of-word-based approach, which represents actions by the chosen key poses. The key poses are extracted from an over-complete codebook of poses using the theory of graph

Table 5. System accuracies (%) of the aerial-view challenge

	Point	Stand	Dig	Walk	Carry	Run	Wave1	Wave2	Jump	Total
Team BIWI	100	91.7	100	100	100	100	83.3	83.3	100	95.4
BU	91.7	83.3	100	100	100	100	100	100	100	97.2
ECSU_ISI	100	83.3	91.7	100	100	100	100	91.7	91.7	95.4
Imagelab	83.3	83.3	100	100	100	100	100	100	100	96.3
Baseline	100	83.3	100	100	100	100	83.3	100	100	96.3

connectivity. They train a Support Vector Machines (SVM) classifier to perform action classification.

Imagelab. The University of Modena and Reggio Emilia team applies a hidden Markov model (HMM) based technique [18] on the dataset. Their action descriptor is a K-dimensional feature set extracted from the projection histograms of the foreground masks. They train a HMM per action and is able to perform recognition on-line.

Baseline. We consider a baseline approach as a simple combination of a commonly used feature and a linear classifier. For this purpose, we use time serious of Histogram of Oriented Gradients (HOG) [3] to characterize successive human poses and a linear kernel SVM classifier for classification. A track of human action is divided into overlapped spatio-temporal volumes, from which we extract and concatenate sequences of HOG vectors as the baseline action descriptors.

We tabulate the average accuracy per activity as well as the overall accuracy of each team and the baseline method in Table 5. Note that prior to this competition, Chen and Aggarwal [2] have tested their method on part of this dataset (60 sequences of the lawn scene). They were able to achieve 100% accuracy on the partial dataset. For the sake of fairness, we did not include their latest results in this paper.

4.3 Discussions

As shown in Table 5, all the contestants achieve very similar accuracies on this low-resolution dataset. The BU team using a silhouette-based method performed the best among four participating teams. In addition, we are surprised to find out that the baseline method was comparable; it obtained the 2nd best performance. 'pointing', 'standing', 'wave1', and 'wave2' are the most common activities that caused misclassifications. The action pairs of <pointing, standing>, <pointing, wave1>, and <wave1, wave2> can be confusing to a recognition algorithm in the sense that one action can only be distinguished from the other by a very short period of hand motion. In low-resolution imagery, vague and sparse action features, salient human cast shadow, and unstabilized videos can all make the discerning task even more challenging. Therefore, we believe a more elaborate preprocessing procedure and the employment of multiple features in classification may further the performance on this dataset.

5   Wide-Area Activity Search and Recognition Challenge

The objective of the "Wide-Area Activity Search and Recognition Challenge" is to search a video given a short query clip in a wide-area surveillance scenario. Our intention is to encourage the development of activity recognition strategies that are able to incorporate information from a network of cameras covering a wide-area. The UCR-Videoweb dataset[3] introduced in this paper has activities that are viewed from 4-8 cameras and allows us to test performance in a camera network. For each query, a clip video containing a specific activity was provided, and the contestants are asked to search for similar videos.

In contrast to the other two challenges, the wide-area challenge was an open challenge: The contestants were free to choose particular types of human activities from the dataset for the recognition, and they were allowed to explore a subset of entire videos.

5.1   Dataset Description

The Videoweb dataset consists of about 2.5 hours of video observed from 4-8 cameras. The data is divided into a number of scenes that were collected over several days. Each scene is observed by a camera network where the actual number of cameras changes depending on the scene due to its nature. For each scene, the videos from the cameras are available. Annotation is available for each scene and the annotation convention is described in the dataset. It identifies the frame numbers and camera ID for each activity that is annotated. The videos from the cameras are approximately synchronized.

The videos contain several types of activities including throwing a ball, shaking hands, standing in a line, handing out forms, running, limping, getting into/out of a car, and cars making turns. The number for each activity varies widely. The data was collected in 4 days and the number of scenes are: {day1: 7 scenes}, {day2: 8 scenes}, {day3: 18 scenes}, and {day4: 6 scenes}. Each scene are on average 4 minutes long and there are 4-7 cameras in each scene. Each scene contains multiple activities. Figure 6 shows example sequences of the dataset.

5.2   Results

In the wide-area challenge, the contestants were asked to formulate their own activity search problem with the dataset, and report their results. That is, each contestant must choose query clips from some scenes in the dataset and use them to retrieve similar scenes in another parts of the dataset. The 'correctly identified clip' is defined as the clip in which the overlap in the range of frame numbers obtained by the search engine for an activity is at least 50% of the range in the annotation and not more than 150% of that range.

There was a single team who showed an intention to participate the wide-area challenge. However, unfortunately, no team succeeded to submit valid results for the wide-area challenge. Here, we report results of systems implemented by the

---

[3] http://vwdata.ee.ucr.edu/

Fig. 5. Example images from the UCR-Videoweb dataset. Each image shows a snapshot obtained from one of 8 different cameras at a particular time frame.

contest organizer [11], so that they can be served as a baseline for the future research. We formulate three types of problems, where each of them focuses on the search of different types of human activities, and report the system performances on these tasks.

Query-Based Complex Activity Search. In this task, we searched for interactions in videos using a single video clip as a query. We worked with 15 minutes of video where up to 10 different actors take part in any given complex activity which involves interaction of humans with other humans, objects, or vehicles. We have used 6 scenes from day 3 data as the test set. The problem was very similar to the human-human interaction detection problem in Section 3, recognizing three types of interactions: shaking hands, hugging, and pointing. Table 6 shows the detection accuracies together with false positive rates.

Table 6. Recognition accuracy on three complex human-human interactions

Interaction	Our recognition accuracy	False positive rate
Shake hands	0.68	0.57
Hug	0.74	0.55
Point	0.63	0.25

Modeling and Recognition of Complex Multi-person Interactions in Video. This task is to examine the formation and dispersal of groups and crowds from multiple interacting objects, which is a fast-growing area in video search. We search for activities involving multiple objects and analyze group formations and interactions. For this task, four scenes have been used for the testing (more details can be found at [11]). We apply a modeling-based methodology to test the implemented system within a query-based retrieval framework. Table 7 shows the types of interactions searched and the precision/recall values of the system.

Table 7. Precision/Recall Values for DB query and retrieval of two-object and complex, multi-object motions

Activity	Precision	Recall	Total Fetched	True Pos.	Ground Truth
Person Entering Building	1	1	4	4	4
Person Exiting Building	1	1	2	2	2
Person Entering Vehicle	0.75	0.75	4	3	3
Person Exiting Vehicle	1	1	3	3	3
People Walking Together	1	0.6	3	3	5
People Coming Together	0.7	0.7	7	5	5
People Going Apart	0.8	1	5	4	5
People Milling Together	0.78	0.92	14	11	13
People Meandering Together	0.85	0.92	27	23	25
Group Formation	1	0.78	7	7	9
Group Dispersal	0.8	0.8	5	4	4
Person Joining Group	1	0.95	18	18	19
Person Leaving Group	1	1	11	11	11

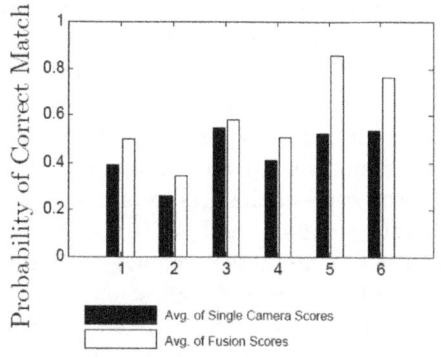

Different Action Classes

Fig. 6. This figure shows the comparison of the recognition scores of our overall approach with single camera action recognition scores. For action class 3, the single view action recognition was almost flat over all action classes, so the fusion could not improve the result much. On the other hand, in action class 5, at least one of the cameras got a good shot of the action and the fused scores went up. In this experiment, each of the targets was viewed by 1-3 cameras simultaneously.

Activity Recognition Based on Multi-camera Data-Fusion. In the past few years, multi-camera installations have rapidly positioned themselves in many applications, e.g., video surveillance, national and homeland security, assisted living facilities, environmental monitoring, disaster response etc. The automated analysis of human actions from these video streams has gained a lot of importance recently. The goal of this task is to search for human actions in such an environment, integrating information from multiple cameras.

We used 8 scenes from day3: the segments of videos having at least one of our defined action classes were selected from these 8 scenes. 10 minutes of the UCR-Videoweb data-set was used for training and another 10 minutes was used for testing. We trained our system for six different action classes, i.e. 1 - Sit, 2 - Walk, 3 - Picking up object, 4 - Shake hand, 5 - Hug and 6 - Wave one hand. Approximately 15 video clips of 2-3 seconds each were used to train our classifier per activity. For each action class, 10-20 instances of each action were used for testing and about 30 different scenarios of multiple actions occurring in multiple cameras were used for testing.

We show the statistics of the performance gain of our method over single-view action recognition scores in Fig. 6. That is, we show that data association and information fusion among multiple cameras improves recognition performance.

6   Conclusion

In this overview paper, we have summarized the results of the first Contest on Semantic Description of Human Activities (SDHA) 2010. SDHA 2010 is one of the very first activity recognition contest, consists of three types of challenges. The challenges introduced three new public datasets (UT-Interaction, UT-Tower, and UCR-Videoweb), which motivated contestants to develop new approaches for complex human activity recognition scenarios in realistic environments. Researchers from various universities participated in SDHA 2010, proposing new activity recognition systems and discussing their results. In addition, several baseline methods were implemented and compared with contestants' results. SDHA 2010 evaluated the state-of-the-arts in human activity recognition.

Table 1 summarizes the results of SDHA 2010. A total of four teams showed their intent to participate the interaction challenge. However, only a single team succeeded to submit results for the classification task, and no team submitted correct detection results. There were four teams participated in the aerial-view challenge, and all teams submitted results with high recognition accuracies (>0.95). One team intended to participated the wide-area challenge, but the team decided not to submit the results. This is due to the fact that the activities used in the aerial-view challenge were relatively simple compared to the others. Simple one-person actions were classified in the challenge, while the activities in the other two challenges include high-level multi-person interactions. We are able to observe that localization of ongoing activities from continuous video streams is a challenging problem, which remains open for future investigations.

References

1. PETS 2006 benchmark data, http://www.cvg.rdg.ac.uk/PETS2006/data.html
2. Chen, C.C., Aggarwal, J.K.: Recognizing human action from a far field of view. In: IEEE Workshop on Motion and Video Computing, WMVC (2009)
3. Dalal, N., Triggs., B.: Histograms of oriented gradients for human detection. In: CVPR (2005)

4. Dollar, P., Rabaud, V., Cottrell, G., Belongie, S.: Behavior recognition via sparse spatio-temporal features. In: IEEE VS-PETS Workshop, pp. 65–72 (2005)
5. Efros, A.A., Berg, A.C., Mori, G., Malik., J.: Recognizing action at a distance. In: ICCV (2003)
6. Everingham, M., Van Gool, L., Williams, C.K.I., Winn, J., Zisserman, A.: i-LIDS abandoned baggage detection challenge dataset, http://www.elec.qmul.ac.uk/staffinfo/andrea/avss2007_ss_challenge.html
7. Everingham, M., Van Gool, L., Williams, C.K.I., Winn, J., Zisserman, A.: The PASCAL Visual Object Classes Challenge 2009 (VOC2009) Results (2009), http://www.pascal-network.org/challenges/VOC/voc2009/workshop/index.html
8. Gorelick, L., Blank, M., Shechtman, E., Irani, M., Basri, R.: Actions as space-time shapes. PAMI 29(12), 2247–2253 (2007)
9. Guo, K., Ishwar, P., Konrad, J.: Action change detection in video by covariance matching of silhouette tunnels. In: ICASSP (2010)
10. Guo, K., Ishwar, P., Konrad, J.: Action recognition in video by sparse representation on covariance manifolds of silhouette tunnels. In: Ünay, D., Çataltepe, Z., Aksoy, S. (eds.) ICPR 2010. LNCS, vol. 6388, pp. 298–309. Springer, Heidelberg (2010)
11. Kamal, A., Sethi, R., Song, B., Fong, A., Roy-Chowdhury, A.: Activity recognition results on UCR Videoweb dataset. In: Technical Report, Video Computing Group, University of California, Riverside (2010)
12. Ke, Y., Sukthankar, R., Hebert, M.: Spatio-temporal shape and flow correlation for action recognition. In: CVPR (2007)
13. Laptev, I., Perez, P.: Retrieving actions in movies. In: ICCV (2007)
14. Rodriguez, M.D., Ahmed, J., Shah, M.: Action MACH: A spatio-temporal maximum average correlation height filter for action recognition. In: CVPR (2008)
15. Ryoo, M.S., Aggarwal, J.K.: Spatio-temporal relationship match: Video structure comparison for recognition of complex human activities. In: ICCV (2009)
16. Schuldt, C., Laptev, I., Caputo, B.: Recognizing human actions: A local SVM approach. In: ICPR (2004)
17. Vezzani, R., Piccardi, M., Cucchiara, R.: An efficient bayesian framework for online action recognition. In: ICIP (2009)
18. Vezzani, R., Baltieri, D., Cucchiara, R.: HMM based action recognition with projection histogram features. In: Ünay, D., Çataltepe, Z., Aksoy, S. (eds.) ICPR 2010. LNCS, vol. 6388, pp. 290–297. Springer, Heidelberg (2010)
19. Waltisberg, D., Yao, A., Gall, J., Gool, L.V.: Variations of a Hough-voting action recognition system. In: Ünay, D., Çataltepe, Z., Aksoy, S. (eds.) ICPR 2010. LNCS, vol. 6388, pp. 309–315. Springer, Heidelberg (2010)
20. Weinland, D., Ronfard, R., Boyer, E.: Free viewpoint action recognition using motion history volumes. CVIU 104(2), 249–257 (2006)
21. Yao, A., Gall, J., Gool, L.V.: A hough transform-based voting framework for action recognition. In: CVPR (2010)

HMM Based Action Recognition with Projection Histogram Features

Roberto Vezzani, Davide Baltieri, and Rita Cucchiara

Dipartimento di Ingegneria dell'Informazione - University of Modena and Reggio Emilia, Via Vignolese, 905 - 41100 Modena - Italy
{roberto.vezzani,davide.baltieri,rita.cucchiara}@unimore.it

Abstract. Hidden Markov Models (HMM) have been widely used for action recognition, since they allow to easily model the temporal evolution of a single or a set of numeric features extracted from the data. The selection of the feature set and the related emission probability function are the key issues to be defined. In particular, if the training set is not sufficiently large, a manual or automatic feature selection and reduction is mandatory. In this paper we propose to model the emission probability function as a Mixture of Gaussian and the feature set is obtained from the projection histograms of the foreground mask. The projection histograms contain the number of moving pixel for each row and for each column of the frame and they provide sufficient information to infer the instantaneous posture of the person. Then, the HMM framework recovers the temporal evolution of the postures recognizing in such a manner the global action. The proposed method have been successfully tested on the UT-Tower and on the Weizmann Datasets.

Keywords: HMM, Projection Histograms, Action Classification.

1 Introduction

Action classification is a very important task for a lot of automatic video surveillance applications. The main challenge relies on developing a method that is able to cope with different types of action, even if they are very similar to each other and also in the case of cluttered and complex scenarios. Occlusions, shadows and noise are the main problems to be faced.

In video surveillance applications the actions should usually be recognized by means of an image stream coming from a single camera. Common 2D approaches analyze the action in the image plane relaxing all the environmental constraints of 3D approaches but lowering the discriminative power of the action-classification task. The action classification can be performed in the image plane by explicitly identifying feature points [1], or considering the whole silhouette [2, 3]. Other approaches directly map low-level image features to actions, preserving spatial and temporal relations. To this aim, feature choice is a crucial aspect to obtain a discriminative representation. An interesting approach that detects human action in videos without performing motion segmentation

D. Ünay, Z. Çataltepe, and S. Aksoy (Eds.): ICPR 2010, LNCS 6388, pp. 286–293, 2010.

was proposed by Irani et al. in [4]. They analyzed spatio- temporal video patches to detect discontinuities in the motion-field directions. Despite the general applicability of this method, the high computational cost makes it unusable for real-time surveillance applications.

After their first application in speech recognition [5], HMMs have been widely used for action recognition tasks. In a recent and comprehensive survey on action recognition [6] several HMM based methods are presented. Yamato *et al* in [7] used HMMs in their most simpler shape: a set of HMM, one for each action, is trained. The observation probability function is modeled as a discrete distribution adopting a mesh feature computed frame by frame on the data [8], and finally, the learning was based on the well known Baum-Welch approach. Similarly, Li [9] proposed a simple and effective motion descriptor based on oriented histograms of optical flow field sequence. After dimensional reduction by principal component analysis, it was applied to human action recognition using the hidden Markov model schema. Recently, Martinez *et al* [10] proposed a framework for action recognition based on HMM and a silhouette based feature set. Differently from the other proposals, their solution lies on an 2D modeling of human actions based on motion templates, by means of motion history images (MHI). These templates are projected into a new subspace using the Kohonen self organizing feature map (SOM), which groups viewpoint (spatial) and movement (temporal) in a principal manifold, and models the high dimensional space of static templates. The higher level is based on a Baum-Welch learned HMM.

In this work we adopt the common HMM framework with a feature set particularly suitable for low quality images. We firstly segment and track the foreground images by means of the Ad-Hoc system [11]. Thus, the projection histograms of the foreground blobs are computed and adopted as feature set [2]. To avoid the course of dimensionality we sub-sampled the histograms, in order to obtain a feature set with a reasonably limited number of values. Ad-Hoc includes a shadow removal algorithm [12]; nevertheless shadows can contain information about the current posture and can be adopted as additional data to recover missing one.

In Section 2 the traditional HMM action classification framework is reported. Section 3 describes the Projection Histogram feature set as well as a shape based feature set used as reference. Finally, comparative tests and the results of the proposed schema over the UT-Tower dataset are reported in Section 4.

2   HMM Action Classification

Given a set of C action classes $\Lambda = \lambda^1 \ldots \lambda^C$, our aim is to find the class λ^* which maximise the probability $P(\lambda|O)$, where $O = \{o_1 \ldots o_T\}$ is the entire sequence of frame-wise observations (features). In his famous tutorial [5], Rabiner proposed to use hidden Markov models to solve this kind of classification problems. An HMM should be learned for each action; the classification of an observation sequence O is then carried out selecting the model whose likelihood is highest, $\lambda^* = \arg\max_{1 \le c \le C} [P(O|\lambda^c)]$. If the classes are equally likely, this solution is optimal also in a Bayesian sense.

$$\lambda^* = \arg\max_{1 \leq c \leq C} [P(O|\lambda^c)] \tag{1}$$

Since the decoding of internal state sequence is not required, the recursive forward algorithm with the three well known initialization, induction and termination equations have been applied.

$$\begin{aligned}
\alpha_1(j) &= \pi_i b_j(o_1), 1 \leq i \leq N \\
\alpha_{t+1}(j) &= \left[\sum_{i=1}^{N} \alpha_t(i)a_{ij}\right] b_j(o_{t+1}) \\
P(O|\lambda) &= \sum_{j=1}^{N} \alpha_T(j)
\end{aligned} \tag{2}$$

The term $b_j(o_t)$ depends on the type of the observations. We adopted the K-dimensional feature set described in the following, which requires to model the observation probabilities by means of density functions. As usual, we adopt a Gaussian Mixture Model, which simplifies the learning phase allowing a simultaneous estimation of both the HMM and the Mixtures parameters using the Baum-Welch algorithm, given the numbers N and M of hidden states and Gaussians per state respectively. In this case, the term $b_j(o_t)$ of Eq. 2 can be approximated as:

$$b_j(o_t) = \sum_{m=1}^{M} c_{jm} \mathcal{N}^K(o_t|\mu_{jm}, \Sigma_{jm}) \tag{3}$$

where $\mathcal{N}^K(\mu, \Sigma)$ is a K-dimensional Gaussian distribution having mean vector μ and covariance matrix Σ; μ_{jm}, Σ_{jm} and c_{jm} are the mean, the covariance matrix and the mixture weight of the m-th component for the action j.

3   Feature Sets

The selection of the feature set to use is very important for the final classification rate. In particular, the adopted features should capture and follow the action peculiarities, but, at the same time, they should allow the action generalization.

In this paper we propose and compare two different feature sets. The first is based on the so called Projection Histograms and it is based on the shape of the foreground mask only; position and global motion of the person are not considered. The projection histograms have been used in the past for frame by frame posture classification [2]. The second feature set, instead, is composed by a mix of different measures, some of them based on the appearance and some on the person position and speed [13]. Independently from the semantics and the computation schema, the input for the HMM framework is a K-dimensional vector $o_t^1 \ldots o_t^K \in \mathbb{R}^K$.

3.1   Projection Histograms Feature Set

Since the videos were acquired by a fixed camera, each frame $I_t(x, y)$ is processed to extract the foreground mask (F) by means of a background subtraction step [12]. For this contest, we directly used the foreground images furnished within

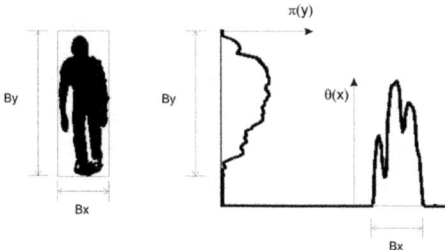

Fig. 1. Vertical and Horizontal Projection histograms of a sample blob

the dataset [14]. The feature vectors o_t are then obtained from the projection histograms of the foreground mask [2], i.e. projections of the person's silhouette onto the principal axes x and y.

Examples of projection histograms are depicted in Fig. 1.

Given the boolean foreground mask $F(x, y)$, the projection histograms θ and π can be mathematically defined as:

$$\theta(x) = \sum_{y=0}^{F_y} \phi(F(x,y)) \; ; \; \pi(y) = \sum_{x=0}^{F_x} \phi(F(x,y)) \tag{4}$$

where the function ϕ is equal to 1 if $F(x, y)$ is true, 0 otherwise, while F_x and F_y are the width and the height of the foreground mask F respectively.

In practice, θ and π can be considered as two feature vectors and the final feature vector $O_t \in \mathbb{R}^K$ used to describe the current frame is obtained from θ and π normalizing each value such as they sum up to 1, resampling the two projection histograms to a fixed number $S = K/2$ of bins, and concatenating them into a unique vector.

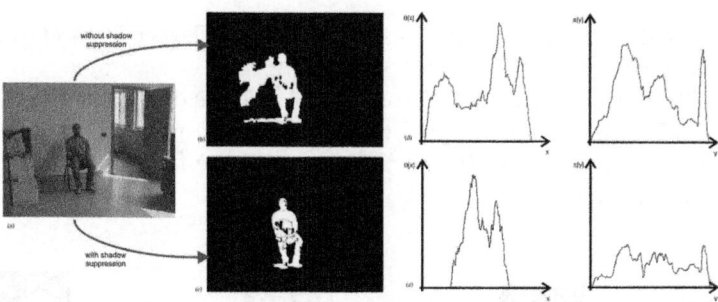

Fig. 2. Comparison of the projection histograms achieved by preserving (top) or removing (bottom) shadows

3.2   Model Based Feature Set

Projection histograms do not depend on any assumption on the people shape and they can be used to describe a generic object. We propose another simple feature set, which is based on a simplified body model, discriminative enough to obtain reasonable classification rates, but not too complex to permit fast processing. The foreground silhouettes are divided into five slices $S^1 \ldots S^5$ using a radial partitioning centered in the gravity center $\{x_c(t), y_c(t)\}$. These slices should ideally correspond to the head, the arms and the legs. Calling A_t and $\{A_t^i\}_{i=1\ldots5}$ the areas of the whole silhouette and of each slice $\{S^i\}$ respectively, the 17-dimensional feature set is obtained as reported in Fig. 3. The features contain both motion (o^1 and o^2) and shape information ($o^3 \ldots o^{17}$).

$$o_t = \{o_t^1 \ldots o_t^{17}\}, = \begin{cases} o_t^1 = x_c(t) - x_c(t-1); \\ o_t^2 = y_c(t) - y_c(t-1); \\ o_t^{3\ldots7} = \frac{A_t^i}{A_t}, i = 1 \ldots 5; \\ o_t^{8\ldots12} = \frac{\max_{(x,y)\in S_i} x}{\sqrt{A_t^i}}, i = 1 \ldots 5; \\ o_t^{13\ldots17} = \frac{\max_{(x,y)\in S_i} y}{\sqrt{A_t^i}}, i = 1 \ldots 5; \end{cases} \tag{5}$$

Fig. 3. Model-based 17-dimensional Feature set

4   Experimental Results

The proposed method have been tested on the UT-Tower Dataset [14] and on the Weizmann dataset [15].

The **UT-Tower Dataset** [14] contains 112 videos of 9 actions performed 12 times each, Some actions are performed in different ways, thus in the on-line recognition we used all the 16 specific classes (Some frames of the dataset are reported in Fig. 4).

Fig. 4. Sample input frame of the UT-Tower dataset

We tested the system using the projection histogram feature set. The number of bins have been sub-sampled to 10 for each direction, obtaining a 20-dimensional feature set. The classification precision achieved using a leave-one-out test schema is around 96%. The confusion matrix is reported in table 5.

The low quality of the segmentation masks and the too limited size of the blobs make the alternative feature set ineffective. Moreover, shadows play an important role in the classification results. In Fig. 2(d) and 2(e) the projection histograms obtained by including shadows or removing them are shown: shadows strongly affect projection histograms based on blob's silhouette, and thus they usually must be removed. Anyway, if the shadow characteristics (i.e., size, position, direction) are not changing among sequences, they can be leaved; on the contrary, information about the performed action are also embedded in the shadow. Thus, we can avoid any shadow removal step if the shadows are always in the same direction and if the adopted feature set is not model based (such as the projection histograms). The model-based feature set described in section 3.2, instead, starts with the estimation of the body center. Shadows strongly compromise this estimation and the overall action classification rate, achieving performance around the 60% on the same dataset.

		Ground thruth - Action ID								
		1	2	3	4	5	6	7	8	9
Recognized Action ID	1	10	0	0	0	0	0	2	0	0
	2	0	10	0	1	0	0	1	0	0
	3	0	0	12	0	0	0	0	0	0
	4	0	0	0	12	0	0	0	0	0
	5	0	0	0	0	12	0	0	0	0
	6	0	0	0	0	0	12	0	0	0
	7	0	0	0	0	0	0	12	0	0
	8	0	0	0	0	0	0	0	12	0
	9	0	0	0	0	0	0	0	0	12

Fig. 5. Confusion matrix of the Projection Histograms Feature set on the UT-Tower dataset

The **Weizmann dataset** [15] contains 90 videos of 10 main actions performed by 9 different people. Some actions are performed in different ways, thus in the on-line recognition we used all the 16 specific classes. Example frames of this well known dataset are shown in Figure 7.

With this dataset the model based feature set performs better than the projection histograms one. The confusion matrix obtained using the model based feature set is shown in Figure 6.

We empirically tuned the HMM parameters. In particular the number N of hidden states and the number M of Gaussians of the mixture model of Eq. 3 have been set to 5 and 3 respectively to maximize the recognition rates based on some experiments we carried out on the Weizmann dataset.

	Ground thruth - Action ID									
	1	2	3	4	5	6	7	8	9	10
1	100	0	0	0	0	0	0	0	0	0
2	0	99	0	0	0	0	0	0	1	0
3	0	0	68	0	4	0	27	1	0	0
4	0	12	0	87	0	0	0	0	1	0
5	0	0	0	0	81	0	19	0	0	0
6	0	0	0	0	5	95	0	0	0	0
7	0	0	12	0	31	0	57	0	0	0
8	0	0	0	0	0	0	0	100	0	0
9	0	0	0	0	0	0	0	0	86	14
10	0	0	0	0	0	0	0	0	6	94

(Recognized Action ID)

Fig. 6. Confusion matrix of the Model Based Feature set on the Weizmann dataset

Fig. 7. Sample input frame of the Weizmann dataset

The complete system, including the background subtraction and updating step, the object tracking, feature extraction and action classification is working in real time, processing about 15 frames per second.

5 Conclusions

In this paper, a traditional HMM framework for action recognition is presented. We proposed and compared two different feature sets, based on projection histograms and shape descriptors respectively. The framework was initially developed for the participation to the ICPR 2010 Contest on Semantic Description of Human Activities - "Aerial View Activity Classification Challenge" [14]. Using the projection histogram feature set the classification precision is around 96%. The system was also tested on the **Weizmann dataset** [15], on which the shape descriptors performs better than projection histograms. Given the temporal segmentation of the actions and a well representative training set, the Hidden Markov Model approach still guarantees good performances both in terms of precision and computational load.

Acknowledgments

This work has been done within the THIS project with the support of the Prevention, Preparedness and Consequence Management of Terrorism and other Security-related Risks Programme European Commission - Directorate-General Justice, Freedom and Security. The authors also thank Massimo Piccardi for several discussions on the work and for his valuable advice.

References

1. Laptev, I., Lindeberg, T.: Space-time interest points. In: IEEE Int. Conf. on Computer Vision (ICCV 2003), Nice, France (2003)
2. Cucchiara, R., Grana, C., Prati, A., Vezzani, R.: Probabilistic posture classification for human behaviour analysis. IEEE Transactions on Systems, Man, and Cybernetics, Part A: Systems and Humans 35, 42–54 (2005)
3. Ke, Y., Sukthankar, R., Hebert, M.: Spatiotemporal shape and flow correlation for action recognition. In: Proc. of IEEE Intl Conference on Computer Vision and Pattern Recognition, pp. 1–8 (2007)
4. Shechtman, E., Irani, M.: Space-time behavior-based correlation -or- how to tell if two underlying motion fields are similar without computing them? IEEE Transactions on Pattern Analysis and Machine Intelligence 29, 2045–2056 (2007)
5. Rabiner, L.R.: A tutorial on hidden markov models and selected applications in speech recognition. Proc. of the IEEE 77, 257–286 (1989)
6. Turaga, P., Chellappa, R., Subrahmanian, V., Udrea, O.: Machine recognition of human activities: A survey. IEEE Trans. on Circuits and Systems for Video Technology 18, 1473–1488 (2008)
7. Yamato, J., Ohya, J., Ishii, K.: Recognizing human action in time-sequential images using hidden markov model. In: Proceedings of IEEE Computer Society Conference on Computer Vision and Pattern Recognition, pp. 379–385 (1992)
8. Umeda, M.: Recognition of multi-font printed chinese. characters. In: Proc. of 6th International Conference on Pattern Recognition, pp. 793–796 (1982)
9. Li, X.: Hmm based action recognition using oriented histograms of optical flow field. Electronics Letters 43, 560–561 (2007)
10. Martinez-Contreras, F., Orrite-Urunuela, C., Herrero-Jaraba, E., Ragheb, H., Velastin, S.: Recognizing human actions using silhouette-based hmm. In: Sixth IEEE International Conference on Advanced Video and Signal Based Surveillance, AVSS 2009, pp. 43–48 (2009)
11. Vezzani, R., Cucchiara, R.: Ad-hoc: Appearance driven human tracking with occlusion handling. In: First International Workshop on Tracking Humans for the Evaluation of their Motion in Image Sequences (THEMIS 2008), Leeds, UK (2008)
12. Cucchiara, R., Grana, C., Piccardi, M., Prati, A.: Detecting moving objects, ghosts and shadows in video streams. IEEE Transactions on Pattern Analysis and Machine Intelligence 25, 1337–1342 (2003)
13. Vezzani, R., Piccardi, M., Cucchiara, R.: An efficient bayesian framework for online action recognition. In: Proceedings of the IEEE International Conference on Image Processing, Cairo, Egypt (2009)
14. Chen, C.C., Ryoo, M.S., Aggarwal, J.K.: UT-Tower Dataset: Aerial View Activity Classification Challenge (2010), http://cvrc.ece.utexas.edu/SDHA2010
15. Gorelick, L., Blank, M., Shechtman, E., Irani, M., Basri, R.: Actions as space-time shapes. IEEE Trans. on Pattern Analysis and Machine Intelligence 29, 2247–2253 (2007)

Action Recognition in Video by Sparse Representation on Covariance Manifolds of Silhouette Tunnels

Kai Guo, Prakash Ishwar, and Janusz Konrad*

Department of Electrical and Computer Engineering, Boston University
8 Saint Mary's St., Boston, MA USA 02215

Abstract. A novel framework for action recognition in video using empirical co-variance matrices of bags of low-dimensional feature vectors is developed. The feature vectors are extracted from segments of silhouette tunnels of moving objects and coarsely capture their shapes. The matrix logarithm is used to map the segment covariance matrices, which live in a nonlinear Riemannian manifold, to the vector space of symmetric matrices. A recently developed sparse linear representation framework for dictionary-based classification is then applied to the log-covariance matrices. The log-covariance matrix of a query segment is approximated by a sparse linear combination of the log-covariance matrices of training segments and the sparse coefficients are used to determine the action label of the query segment. This approach is tested on the Weizmann and the UT-Tower human action datasets. The new approach attains a segment-level classification rate of 96.74% for the Weizmann dataset and 96.15% for the UT-Tower dataset. Additionally, the proposed method is computationally and memory efficient and easy to implement.

Keywords: video analysis; action recognition; silhouette tunnel; covariance manifold; sparse linear representation.

1   Introduction

Algorithms for recognizing human actions in a video sequence are needed in applications such as video surveillance, where the goal is to look for typical and anomalous patterns of behavior, and video search and retrieval in large, potentially distributed, video databases such as YouTube. Developing algorithms for action recognition in video that are not only accurate but also efficient in terms of computation and memory-utilization is challenging due to the complexity of the task and the sheer size of video.

The action recognition problem, in its full generality, is challenging due to the complexity of the scene (multiple interacting moving objects, clutter, occlusions, illumination variability, etc.), the camera (imperfections, motion and shake, and viewpoint), and the complexity of actions (non-rigid objects and intra- and inter- class action variability). Even when there is only a single uncluttered and unoccluded object[1] and

---

* This material is based upon work supported by the US National Science Foundation (NSF) under awards CNS–0721884 and (CAREER) CCF–0546598, and National Geospatial-Intelligence Agency (NGA) under award HM1582-09-1-0037.

[1] Such footage may be obtained by detecting, tracking, and isolating object trajectories.

D. Ünay, Z. Çataltepe, and S. Aksoy (Eds.): ICPR 2010, LNCS 6388, pp. 294–305, 2010.

the camera and illumination conditions are perfect (a typical assumption in the literature), the complexity and variability of actions makes action recognition a difficult problem.

The accuracy and efficiency of an action recognition algorithm critically depends on 1) *how actions are modeled and represented* and 2) *how distances between action representations are measured for classification*. To date, various action models and representations have been proposed, from those based on Hidden Markov Models [14,12], through interest-point models [10,4,9,8] which are sparse (relative to the number of pixels) yet highly discriminative, e.g., corners and SIFT features, and local motion models, e.g., kinematic characteristics from optical flow [1] and 3D local steering kernels [11], to silhouette tunnel shape models [6,7]. Similarly, various metrics have been proposed to measure distances between action representations, from the Hausdorff distance between sets of action feature vectors in Euclidean space extracted from multiple action instances (e.g., see [6]) to the matrix cosine similarity measure (Frobenius inner product) between matrices of action feature vectors [11]. The methods developed to-date are either computationally and/or memory intensive and/or their accuracy varies significantly across different data sets.

In [7] we developed a nearest-neighbor (NN) supervised classification algorithm for human action recognition using a labeled dictionary of empirical feature-covariance matrices. These were obtained from bags of low-dimensional feature vectors extracted from the object silhouette tunnels and coarsely captured their shape. A Riemannian metric on the manifold of covariance matrices was used for determining nearest neighbors. In this paper, we apply the recently developed sparse linear representation framework for dictionary-based classification [13] to the matrix logarithm of the feature-covariance matrices as an alternative to NN-classification. We report the performance of this new approach on the Weizmann human action dataset [6] and the UT-tower dataset [3] provided by the ICPR 2010 "Aerial View Activity Classification Challenge". We also compare its performance with the method we previously developed in [7] that uses the same action representation (covariance matrix of silhouette shape features) but a different classification rule (NN-classifier).

2   Framework

We view action recognition as a supervised classification problem where the goal is to classify a query video segment using a dictionary of previously labeled training video segments. Video segments are typically high dimensional, e.g., a 20-frame video segment with a 128×128 frame resolution is, roughly, a 3×10^5-dimensional vector, whereas the number of training video segments is meager in comparison. It is therefore impractical to learn the global structure of training video segments by building classifiers directly in high-dimensional space. Graphical models, which attempt to capture global dependencies through local structure, are powerful; but training classifiers based on these models is challenging.

2.1   Action Representation Using Low-Dimensional Feature-Covariance Matrices

We adopt a "bag of dense local feature vectors" modeling approach wherein a video segment is represented by a *dense* set of low-dimensional local feature vectors which describe the action. The local features, described in detail in Sec. 3, coarsely capture the shape of an object's silhouette tunnel (see Fig. 3). The advantage of this approach is that even a single video segment provides a very large number of local feature vectors (one per pixel) from which their statistical properties can be reliably estimated. However, the dimensionality of a bag of dense local feature vectors is still very high as there are as many feature vectors as pixels. This motivates the need for dimensionality reduction.

Estimating the distribution of the local feature vectors, though ideal, is computation-intensive and may not lead to a lower-dimensional representation. On the other hand, the mean feature-vector, which is low-dimensional, can be learned reliably and rapidly but may not be sufficiently discriminative. In the recent work [7] we discovered that if the features are well-chosen, then the feature-covariance matrix, which captures the second-order statistical properties of a bag of feature vectors, provides a remarkably discriminative representation for action recognition. In addition to their simplicity and effectiveness, covariance matrices have low storage and processing requirements. The action representation based on the covariance matrix of a bag of low-dimensional local feature vectors that coarsely capture the shape of an object's silhouette tunnel is depicted in Fig. 1. The operator which transforms an input video segment into an output feature-covariance matrix representation is denoted by Ψ.

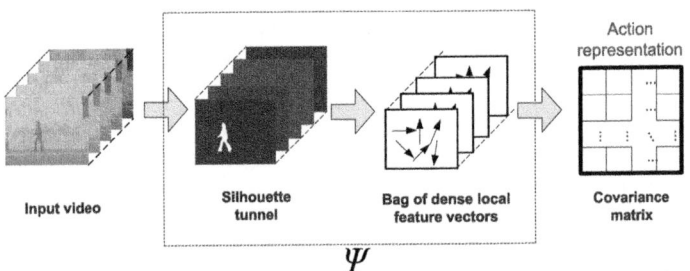

Fig. 1. Transformation of a video segment into a feature covariance matrix representation

2.2   Classification on a Covariance Manifold

The set of all covariance matrices of a specified size do not form a vector space (they are not closed under multiplication by negative scalars); they form a Riemannian manifold. Classification problems on covariance manifolds can be converted into vector-space classification problems via the matrix logarithm: if $C = UDU^T$ is the eigendecomposition of the covariance matrix C, where D is the diagonal matrix of eigenvalues, then $\log(C) := U \log(D) U^T$, where $\log(D)$ is the diagonal matrix whose diagonal entries are the natural logarithms of the corresponding entries of D. The matrix logarithm maps the Riemannian manifold of symmetric non-negative definite matrices to the vector space of symmetric matrices [2].

Recently, in [13] Wright et al. developed a powerful framework (closely related to compressive sensing) for supervised classification in vector spaces based on finding a *sparse* linear approximation of a query vector in an overcomplete dictionary of training vectors. The key idea underlying this approach is that if the training vectors of *all* the classes are pooled together and a query vector is expressed as a linear combination of the *fewest possible* training vectors, then the majority of the training vectors in the linear combination are likely to be of the same class as the query vector. The pooling together of the training vectors of all the classes is important for classification because the training vectors of each individual class may well span the space of all query vectors. The pooling together induces a "competition" among the training vectors of different classes to approximate the query using the fewest possible number of training vectors. The sparse representation approach has been successfully applied to many vision tasks such as face recognition, image super-resolution, and image denoising. We extend this approach to action recognition by applying it to (column) vectorized log-covariance matrices that we refer to as samples. Specifically, we approximate the log-covariance matrix of a query segment by a sparse linear combination of log-covariance matrices of all training segments.

The overall framework for action recognition is depicted in Fig. 2.

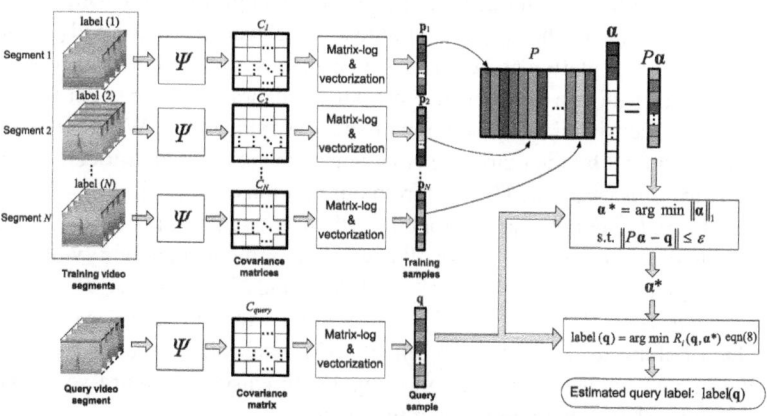

Fig. 2. Overview of the proposed action recognition framework (see Secs. 2.2 and 4)

3   Silhouette Tunnel Shape Features

In this section, we describe the low-dimensional local features that we use to describe actions. The sequence of 2-D silhouettes of a moving and deforming object (see Fig. 3) is particularly attractive for action recognition because (i) it accurately captures object dynamics, (ii) it can be reliably, robustly, and efficiently computed in real-time using state-of-the-art background subtraction techniques, and (iii) it is largely invariant to chromatic, photometric, and textural properties of objects which are independent of their actions. Under ideal conditions, each frame in the silhouette sequence would contain a white mask (white = 1) which exactly coincides with the 2-D silhouette of the moving and deforming object against a "static" black background (black = 0). A

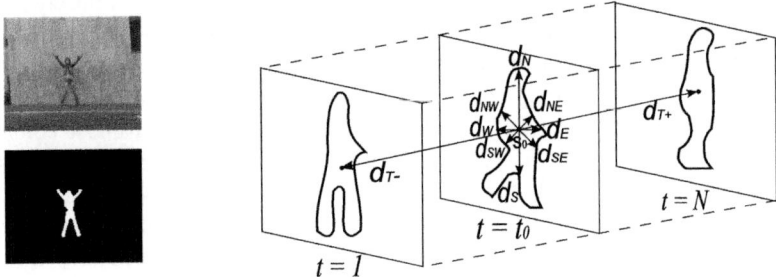

Fig. 3. *Left:* One frame of the "jumping-jack" human action sequence (top row) and the corresponding silhouette (bottom row) computed using background subtraction from the Weizmann human action dataset. *Right:* Each point $\mathbf{s}_0 = (x_0, y_0, t_0)^T$ of a silhouette tunnel within an N-frame action segment has a 13-dimensional feature vector associated with it: 3 position features x_0, y_0, t_0, and 10 shape features given by distance measurements from (x_0, y_0, t_0) to the tunnel boundary along 10 different spatio-temporal directions shown in the figure.

sequence of such object silhouettes in time forms a spatio-temporal volume in x-y-t space that we refer to as a silhouette tunnel. Action recognition may then be viewed as recognizing the *shape* of the silhouette tunnel. There is an extensive body of literature devoted to the representation and comparison of shapes of volumetric objects. Our goal is to reliably discriminate between shapes; not to accurately reconstruct them. Hence a coarse, low-dimensional representation of shape would suffice. We capture the shape of a silhouette tunnel by the empirical covariance matrix of a bag thirteen-dimensional local shape features (described below) from our previous work [7].

Let $\mathbf{s} = (x, y, t)^T$ denote the horizontal, vertical, and temporal coordinates of a pixel. Let $\mathcal{A}$ denote the set of coordinates of all pixels belonging to an action (video) segment which is W pixels wide, H pixels tall, and N frames long, i.e., $\mathcal{A} := \{(x, y, t)^T : x \in [1, W], y \in [1, H], t \in [1, N]\}$. Let $\mathcal{S}$ denote the subset of pixel-coordinates in $\mathcal{A}$ which belong to the silhouette tunnel. With each pixel located at $\mathbf{s}$ within the silhouette tunnel, we associate the following 13-dimensional feature vector $\mathbf{f}(\mathbf{s})$ that captures certain shape characteristics of the tunnel:

$$\mathbf{f}(x, y, t) := [x, y, t, d_E, d_W, d_N, d_S, d_{NE}, d_{SW}, d_{SE}, d_{NW}, d_{T+}, d_{T-}]^T, \qquad (1)$$

where $(x, y, t)^T \in \mathcal{S}$ and d_E, d_W, d_N, and d_S are Euclidean distances from (x, y, t) to the nearest silhouette boundary point to the right, to the left, above and below the pixel, respectively. Similarly, d_{NE}, d_{SW}, d_{SE}, and d_{NW} are Euclidean distances from (x, y, t) to the nearest silhouette boundary point in the four diagonal directions, while d_{T+} and d_{T-} are similar measurements in the temporal direction. Fig. 3 depicts these features graphically. The 13×13 "shape" covariance matrix representation $C_{\mathcal{S}}$ of silhouette tunnel $\mathcal{S}$ in the action segment $\mathcal{A}$ is given by

$$C_{\mathcal{S}} := \frac{1}{|\mathcal{S}|} \sum_{\mathbf{s} \in \mathcal{S}} (\mathbf{f}(\mathbf{s}) - \boldsymbol{\mu}_F)(\mathbf{f}(\mathbf{s}) - \boldsymbol{\mu}_F)^T, \qquad (2)$$

where $\mu_F = \sum_{s \in \mathcal{S}} \frac{1}{|\mathcal{S}|} \mathbf{f}(\mathbf{s})$ is the mean feature vector. Note that the size of an action segment $|\mathcal{A}|$ is typically on the order of 10^5 whereas a 13×13 covariance matrix, being symmetric, has only 91 independent entries. This provides a low-dimensional representation of the feature vectors no matter how numerous they may be.

4 Classification via Sparse Linear Representation

In this section, we first explain how the log-covariance matrix of a query action segment can be approximated by a sparse linear combination of log-covariance matrices of all training action segments by solving an l^1-minimization problem. We then discuss how the locations of large non-zero coefficients in the sparse linear approximation can be used to determine the label of the query.

The logarithm of a 13×13 covariance matrix C is a 13×13 symmetric matrix $\log(C)$ which has only 91 independent entries (elements on and above the main diagonal). We use $\mathbf{p} \in \mathbb{R}^{91}$ to denote the (column) vectorized matrix of the entries in $\log(C)$ that are on or above the main diagonal. For convenience of exposition, we will refer to such column vectorized log-covariance matrices as simply 'samples'. Let $\mathbf{p}_{i,j} \in \mathbb{R}^{91}$ denote the j-th training sample in the i-th class where $i = 1, \ldots, K$, and $j = 1, \ldots, n_i$. Thus there are K action classes, n_i training samples in action class i, and the total number of training samples is given by $M = \sum_{i=1}^{K} n_i$. We can stack up all the training samples from class i, column by column, to form the $91 \times n_i$ matrix $P_i := [\mathbf{p}_{i,1}\ \mathbf{p}_{i,2}\ \cdots\ \mathbf{p}_{i,n_i}]$. The $91 \times M$ matrix of all training samples is then given by $P := [P_1\ P_2\ \cdots\ P_K]$.

A given query sample $\mathbf{q}$ can be expressed as a linear combination of training samples by solving the matrix-vector equation given by

$$\mathbf{q} = P\alpha, \tag{3}$$

where $\alpha \in \mathbb{R}^M$ is the vector of coefficients. Since $M \gg 91$, the system (3) is underdetermined and has a solution except in the highly unlikely circumstance in which there are less than 91 linearly independent samples across all classes and $\mathbf{q}$ is outside of their span. If a solution to (3) exists, it is necessarily nonunique unless additional prior information, e.g., sparsity, restricts the set of feasible α.

We seek a sparse solution to (3) where, under ideal conditions, the only non-zero coefficients in α are those which correspond to the class of the query sample. If (3) has a solution α^* with $r < 91/2$ non-zero coefficients and every set of 91 columns of P is linearly independent, then α^* is the unique sparsest solution to (3) (see [5]) which can be found, in principle, by solving the following NP-hard optimization problem:

$$\alpha^* = \arg \min \|\alpha\|_0, \quad s.t. \quad \mathbf{q} = P\alpha, \tag{4}$$

where $\|\alpha\|_0$ is the so-called l^0-norm: the number of non-zero entries in α. A key result in the theory of compressive sensing (see [5]) is that if the optimal solution α^* is sufficiently sparse, then solving the l^0-minimization problem (4) is equivalent to solving the following l^1-minimization problem

$$\alpha^* = \arg \min \|\alpha\|_1, \quad s.t. \quad \mathbf{q} = P\alpha. \tag{5}$$

Unlike (4), this problem is a convex optimization problem that can be solved in polynomial time. In practice, estimates of $\mathbf{p}_{i,j}$ may be noisy and (3) may not hold exactly. In practice one therefore solves the following ϵ-robust l^1-minimization problem

$$\boldsymbol{\alpha}^* = \arg\min \ \|\boldsymbol{\alpha}\|_1, \quad s.t. \quad \|P\boldsymbol{\alpha} - \mathbf{q}\|_2 \leq \varepsilon. \tag{6}$$

It turns out that even when not all sets of 91 columns of P are linearly independent, the solution $\boldsymbol{\alpha}^*$ to (6) is still very sparse in the sense that its components, arranged in decreasing order of magnitude, decay very rapidly.

We now discuss how the locations of large non-zero components of $\boldsymbol{\alpha}^*$ can be used to determine the label of the query. Each component of $\boldsymbol{\alpha}^*$ weights the contribution of its corresponding training sample to the representation of the query sample. Ideally, the sparse non-zero coefficients should only be associated with training samples that come from the same class as the query sample. In practice, however, non-zero coefficients will be spread across more than one action class. To decide the label of the query sample, we follow Wright et al. [13] and use a reconstruction residual error (RRE) measure to decide the query class. Let $\boldsymbol{\alpha}_i^* := [\alpha_{i,1}^* \ \alpha_{i,2}^* \ \cdots \ \alpha_{i,n_i}^*]^T$ denote the coefficients associated with training samples from class i, i.e., columns of P_i. The RRE measure of class i is then defined as:

$$R_i(\mathbf{q}, \boldsymbol{\alpha}^*) := \|\mathbf{q} - P_i\boldsymbol{\alpha}_i^*\|_2. \tag{7}$$

To the query sample $\mathbf{q}$ we assign the class label that leads to the minimum RRE, i.e.,

$$\text{label}(\mathbf{q}) := \arg\min_i R_i(\mathbf{q}, \boldsymbol{\alpha}^*). \tag{8}$$

5 Some Practical Considerations and the Overall Algorithm

One important aspect of human action recognition is the repetitive nature of actions. Many actions, such as walking, running and jumping, consist of multiple, roughly periodic, "repetitions" of shorter action segments which describe the essential action characteristics. Long video sequences of the same action may exhibit large differences due to action variability. In addition, the frame-boundaries where one action ends and another begins may not be available in some practical scenarios. This motivates the need to break a long query video sequence into a sequence of overlapping action segments and classify each segment. Short overlapping action segments can also increase the number and diversity of the training set so that the action can be classified more reliably. Ideally, the duration of an action segment should be long enough to contain at least one "period" of an action. The typical period of many moderately-paced human actions is on the order of 0.4-0.8 seconds. For a camera operating at 25 frames per second (fps), this corresponds to an action segment which contains 10–20 frames.

The motion of the centroid of an object's silhouette across frames is of secondary importance for action recognition. It is the sequence of deformations of the silhouettes about their centroids that is crucial. We can remove the motion of the centroids by aligning them to the same spatial coordinates. It is also possible to make the silhouette tunnel shape covariance matrix C_S invariant to spatial scaling (e.g., due to zoom) and temporal scaling (e.g., due to temporal subsampling) by normalizing the feature vectors before computing C_S via (2). We refer to [7] for the details.

The overall framework for action recognition can be summarized as follows (see Figs. 1 and 2). We start with a raw query video sequence which has only one moving object. We compute the silhouette sequence by background subtraction and then parse it into a sequence of overlapping N-frame-long segments (we used 8-frame segments with a 4-frame overlap in our experiments). We map the silhouette tunnel of each N-frame-long action segment to its shape covariance matrix, take its logarithm and column-vectorize the upper-triangular portion. To classify each action segment, we solve the l^1-minimization problem (6) to obtain a sparse linear representation and then use (7) and (8). Since individual segment decisions are expected to be somewhat noisy, we perform an additional step to filter out this decision noise. We fuse the decisions of all action segments in an action sequence using the majority rule to arrive at the final decision for the entire query video sequence. This improves the reliability by overcoming misclassifications in up to one-half of the test action segments.

6   Experimental Results

In this section, we report the results of performance evaluation of the proposed method on two publicly-available datasets: the Weizmann human action dataset[2] and the UT-tower human action dataset [3]. Although the KTH dataset[3] has been widely used to test the performance of action recognition methods, we omit it in our tests since it does not include silhouette sequences that are needed for a fair comparison.

6.1   Weizmann Human Action Dataset

This dataset consists of 90 low-resolution video sequences (180×144 pixels) that show 9 different people each performing 10 different actions. For each video sequence, a binary sequence of 2-D silhouettes is also available. As described in Sec. 5, we parse all silhouette sequences into overlapping 8-frame long silhouette segments with a 4-frame overlap. We refer to the resulting collection of segments as the silhouette *segment* dataset. Performance-evaluation is based on the leave-one-out cross validation (LOOCV) test. For each query silhouette segment from the segment dataset, we first remove all those segments which come from the same silhouette sequence as the query segment. Then, based on the remaining segments in the segment dataset, we determine the action label of the query segment using the proposed method. Details of the experimental setup can be found in [7]. The correct classification rate (CCR) is defined as the percentage of query segments that are correctly classified. Since the CCR is based on classifying individual segments, we call it SEG-CCR. In practice, however, we are usually interested in classifying a complete video sequence containing an action; not just one of its segments. Since segments provide time-localized action information, in order to obtain classification for the complete sequence, we apply the majority rule (dominant label wins) to the decisions obtained from individual segments of the video sequence

---

[2] http://www.wisdom.weizmann.ac.il/~vision/SpaceTimeActions.html
[3] http://www.nada.kth.se/cvap/actions/

Table 1. Action confusion matrix: Weizmann human action dataset, 8-frame segments with 4-frame overlap, SEG-CCR = 96.74%

	bend	jack	jump	sjump	run	side	skip	walk	wave1	wave2
bend	91.9	1.3	0	0.7	0	0	0	0	4.1	2.0
jack	0	99.4	0	0.6	0	0	0	0	0	0
jump	0	0	95.1	0	0	2.0	2.9	0	0	0
sjump	0	0.8	0	96.7	0	2.5	0	0	0	0.7
run	0	0	0	1.2	91.6	0	1.2	6.2	0	0
side	0	0	0	0	0	100	0	0	0	0
skip	0	0	1.0	0	4.2	0	92.7	2.1	0	0
walk	0	0	0	0	0	0	0	100	0	0
wave1	0	0	0	0.6	0	0	0	0	99.4	0
wave2	0	1.4	0	0	0	0	0	0	1.4	97.2

Table 2. LOOCV CCR comparison of the proposed method with state-of-the-art methods: Weizmann human action dataset, 8-frame segments with 4-frame overlap

Method	**Proposed**	Guo et al. [7]	Gorelick et al. [6]	Niebles et al. [9]	Ali et al. [1]	Seo et al. [11]
SEG-CCR	**96.74%**	97.05%	97.83%	-	95.75%	-
SEQ-CCR	**100%**	100%	-	90%	-	96%

as described in Sec. 5. In this case, we calculate a sequence-level CCR, that we call SEQ-CCR, defined as the percentage of query sequences that are correctly classified.

The proposed method attained a SEG-CCR of 96.74% and a SEQ-CCR of 100%. Table 1 shows the action "confusion" matrix based on SEG-CCR values. The element in row i and column j of the matrix indicates the percentage of action i segments which were classified as action j. The sum of all elements in every row is 100%. The confusion matrix indicates that while some actions, such as 'bend' and 'run', are more confusing, others, such as 'walk' and 'side', are easier to distinguish.

Table 2 compares the performance of the proposed method with some of the state-of-the-art action recognition methods, including our previous method [7] based on NN-classification on the feature-covariance manifold. It is clear that the proposed algorithm is very close in performance to our previous method and also approaches the performance of Gorelick et al.'s method [6].

6.2   UT-Tower Human Action Dataset

The UT-tower action dataset is used in the "Aerial View Activity Classification Challenge" at the ICPR 2010 Contest on Semantic Description of Human Activities (SDHA). The dataset consists of 108 video sequences with a frame resolution of 360×240 pixels and a frame rate of 10fps. The contest requires classifying video sequences into one of 9 categories of human actions. Each of the 9 actions is performed 2 times by 6 individuals for a total of 12 video sequences per action category. Ground truth action labels, bounding boxes, and foreground masks for each video sequence are provided. Only the acting person is included in the bounding box. In addition to the challenges associated with the low resolution of objects of interest in this dataset – the average height of human figures is about 20 pixels – there are additional challenges, such as camera jitter, shadows, and blurry visual cues (see [3] for details).

We conducted experiments using the same procedures as for the Weizmann dataset including LOOCV. The method proposed here attains a SEG-CCR of 96.15% and a SEQ-CCR of 97.22%. Table 3 shows the confusion matrices of SEG-CCR and SEQ-CCR values. Since the UT-Tower dataset is new and no action recognition results are publicly available for this dataset at the time of writing of this paper, in Table 4 we only compare the performance of the proposed method with our previous method [7].

Table 3. Action confusion matrices: UT-Tower human action dataset, 8-frame segments with 4-frame overlap

	SEG-CCR=96.15%									SEQ-CCR=97.22%								
	point	stand	dig	walk	carry	run	wave1	wave2	jump	point	stand	dig	walk	carry	run	wave1	wave2	jump
point	88.0	6.0	6.0	0	0	0	0	0	0	91.7	0	8.3	0	0	0	0	0	0
stand	4.4	94.2	1.4	0	0	0	0	0	0	16.7	83.3	0	0	0	0	0	0	0
dig	2.0	1.5	96.0	0	0.5	0	0	0	0	0	0	100	0	0	0	0	0	0
walk	1.4	0	0	98.6	0	0	0	0	0	0	0	0	100	0	0	0	0	0
carry	0	0	0	0	99.5	0.5	0	0	0	0	0	0	0	100	0	0	0	0
run	0	0	0	0	0	100	0	0	0	0	0	0	0	0	100	0	0	0
wave1	0	0	0.5	0	0	0	94.1	5.4	0	0	0	0	0	0	0	100	0	0
wave2	0	0	0	0	0	0	7.5	92.5	0	0	0	0	0	0	0	0	100	0
jump	0	0	0	0	0	0	0	0	100	0	0	0	0	0	0	0	0	100

All video sequences except those for pointing and standing are classified without error. Standing is sometimes confused with pointing whereas pointing is occasionally confused with standing. Both of these action categories are essentially static poses and are sufficiently similar to even cause confusion in human observers on account of the low resolution of the dataset.

Table 4. LOOCV CCR comparison of the proposed method with our previous method: UT-Tower human action dataset, 8-frame segments with 4-frame overlap

Method	**Proposed**	Guo *et al.* [7]
SEG-CCR	**96.15%**	93.53%
SEQ-CCR	**97.22%**	96.30%

The proposed method is also time-efficient and easy to implement. Our experimental platform was Intel Centrino (CPU: T7500 2.2GHz + Memory: 2GB) with Matlab 7.6. The computation of 13-dimensional feature vectors and the calculation of log-covariance matrices can be efficiently implemented on this platform, costing together about 4.3 seconds per silhouette sequence with spatial resolution of 111×81 and length of 89 frames. This method is also memory efficient since the training and query sets essentially store 13×13 log-covariance matrices instead of video data. Given a query sequence with 20 query segments and a training set with 1239 training segments, it takes about 4.5 seconds to classify all query segments (solving 20 l^1-norm minimization problems), i.e., about 0.22 seconds per query segment.

7   Concluding Remarks

In this paper, we proposed a new approach to action recognition in video based on sparse linear representations of log-covariance matrices of silhouette shape features. The proposed method is motivated by Wright et al.'s work [13] that has been successfully applied in the context of face recognition. The salient characteristic of our method is the fact that it uses log-covariance matrices to represent actions in a vector space. Our experimental results on the Weizmann dataset indicate that the classification performance of the proposed method is similar to that of recent successful methods, such as Gorelick's method [6] and our previous method [7]. At the same time, its computational complexity is relatively low in both feature extraction, on account of feature simplicity, and classification, owing to efficiencies in solving the l^1 minimization. On the challenging UT-Tower dataset, the proposed method outperforms our previous approach based on the same features and NN classification.

Acknowledgment

The authors would like to thank Prof. Pierre Moulin from the ECE Department at UIUC for suggesting the application of the recently-developed sparse linear representation framework for dictionary-based classification to log-covariance matrices as an alternative to NN-classification.

References

1. Ali, S., Shah, M.: Human action recognition in videos using kinematic features and multiple instance learning. IEEE Trans. Pattern Anal. Machine Intell. 32(2), 288–303 (2010)
2. Arsigny, V., Pennec, P., Ayache, X.: Log-euclidean metrics for fast and simple calculus on diffusion tensors. Magnetic resonance in medicine 56(2), 411–421 (2006)
3. Chen, C.C., Ryoo, M.S., Aggarwal, J.K.: UT-Tower Dataset: Aerial View Activity Classification Challenge (2010),
   http://cvrc.ece.utexas.edu/SDHA2010/Aerial_View_Activity.html
4. Dollar, P., Rabaud, V., Cottrell, G., Belongie, S.: Behavior recognition via sparse spatio-temporal features. In: IEEE Int'l Workshop VS-PETS (2005)
5. Donoho, D.L.: For most large underdetermined systems of linear equations the minimal l1-norm solution is also the sparsest solution. Comm. Pure Appl. Math. 59, 797–829 (2004)
6. Gorelick, L., Blank, M., Shechtman, E., Irani, M., Basri, R.: Actions as space-time shapes. IEEE Trans. Pattern Anal. Machine Intell. 29(12), 2247–2253 (2007)
7. Guo, K., Ishwar, P., Konrad, J.: Action recognition from video by covariance matching of silhouette tunnels. In: Proc. Brazilian Symp. on Computer Graphics and Image Proc. (October 2009)
8. Laptev, I., Marszalek, M., Schmid, C., Rozenfeld, B.: Learing realistic human actions from movies. In: Proc. IEEE Conf. Computer Vision Pattern Recognition (June 2008)

9. Niebles, J., Wang, H., Fei-Fei, L.: Unsupervised learning of human action categories using spatial-temporal words. In: Intern. J. Comput. Vis. (March 2008)
10. Schuldt, C., Laptev, I., Caputo, B.: Recognizing human actions: A local svm approach. In: Proc. Int. Conf. Pattern Recognition (June 2004)
11. Seo, H.J., Milanfar, P.: Action recognition from one example. IEEE Trans. Pattern Anal. Machine Intell. (submitted)
12. Starner, T., Pentland, A.: Visual recognition of american sign language using hidden markov models. In: IEEE Int. Conf. on Automatic Face and Gesture Recognition (1995)
13. Wright, J., Yang, A., Ganesh, A., Sastry, S., Ma, Y.: Robust face recognition via sparse representation. IEEE Trans. Pattern Anal. Machine Intell. 31(2), 210–227 (2009)
14. Yamato, J., Ohya, J., Ishii, K.: Recognizing human action in time sequential images using hidden markov model. In: Proc. IEEE Conf. Computer Vision Pattern Recognition (June 1992)

Variations of a Hough-Voting Action Recognition System

Daniel Waltisberg, Angela Yao, Juergen Gall, and Luc Van Gool

Computer Vision Laboratory, ETH Zurich, Switzerland

Abstract. This paper presents two variations of a Hough-voting framework used for action recognition and shows classification results for low-resolution video and videos depicting human interactions. For low-resolution videos, where people performing actions are around 30 pixels, we adopt low-level features such as gradients and optical flow. For group actions with human-human interactions, we take the probabilistic action labels from the Hough-voting framework for single individuals and combine them into group actions using decision profiles and classifier combination.

Keywords: human action recognition, Hough-voting, video analysis, low-resolution video, group action recognition, activity recognition.

1 Introduction

Recognizing human actions from video has received much attention in the computer vision community, though designing algorithms that can detect and classify actions from unconstrained videos and in realistic settings still remains a challenge. One difficulty is scene diversity, i.e. methods designed for sports analysis may not be well suited for surveillance. Furthermore, much of the work in action recognition has focused on single persons. In applications such as intelligent surveillance, where the goal is to detect unusual or dangerous events, however, the classification of group interactions becomes more critical as situations can only be understood by considering the relationship between persons.

We present here variations on a Hough-voting framework for action recognition, previously introduced in [8], as applied to two very different action recognition scenarios from the *ICPR 2010 Contest on Semantic Description of Human Activities*. In the first scenario, the Hough-voting framework is directly applied to classify actions on low-resolution videos, in which people performing actions are around 30 pixels high. In the second scenario, we classify group actions by combining the classification results of single individuals to strengthen the group action response.

The rest of the paper is organized as follows. In Section 2, we give a short summary of the Hough-voting framework described in [8]. In Section 3, we describe the combination of the classifier outputs of multiple people into group actions by using classifier combination rules and extending the model of decision profiles [6]. In Section 4, we show the classification results on low resolution

D. Ünay, Z. Çataltepe, and S. Aksoy (Eds.): ICPR 2010, LNCS 6388, pp. 306–312, 2010.

videos and on group action recognition. Finally, Section 5 summarizes the main results.

2   Hough-Voting Framework

The Hough-voting framework in [8] takes a two-staged approach. In an initial localization stage, the person performing the action is tracked. Then, in a secondary classification stage, 3D feature patches from the track are used to cast votes for the action center in a spatio-temporal action Hough space. In [8], a tracking-by-detection approach was used, though any other tracking method can be used as well since the tracking stage is disjoint from the classification stage. For classifying the action, random trees are trained to learn the mapping between the patches and the corresponding votes in the action Hough space.

2.1   Training

We train a random forest, which we term a "Hough forest", to learn the mapping between action tracks and a Hough space. Each tree is constructed from a set of patches $\{\mathcal{P}_i = (\mathcal{I}_i, c_i, d_i)\}$, where

$\mathcal{P}_i$ is a 3D patch (e.g. of $16 \times 16 \times 5$ pixels) randomly sampled from the track.
$\mathcal{I}_i$ are extracted features at a patch and can be multi-channeled to accommodate multiple features, i.e. $\mathcal{I}_i = \left(I_i^1, I_i^2, ..., I_i^F\right) \in \mathbb{R}^4$, where each I_i^f is feature channel f at patch i and F is the total number of feature channels.
c_i is the action label.
d_i is a 3D displacement vector from the patch center to the action track center.

From the set of patches, the tree is built from the root by selecting a binary test t, splitting the training patches according to the test results and iterating on the children nodes until either the maximum depth of the tree is reached or there are insufficient patches remaining at a node. Each leaf node stores p_c, the proportion of the patches per class label reaching that leaf, and $D_c = \{d_i\}_{c_i=c}$, the patches' respective displacement vectors.

The binary tests compare two pixels at locations $p \in \mathbb{R}^3$ and $q \in \mathbb{R}^3$ in feature channel f with some offset τ, i.e.

$$t_{f,p,q,\tau}(\mathcal{I}) = \begin{cases} 0 & \text{if } I^f(p) < I^f(q) + \tau \\ 1 & \text{otherwise} \end{cases} \tag{1}$$

First, a pool of binary tests with random values of f, p, q and τ are generated; the test which splits the patches with minimal class or offset uncertainty between the split is chosen. By switching randomly between the two uncertainty measures, the leaves tend to have low variation in both class label and center displacement.

2.2   Classifying and Localizing Actions

During test time, we extract densely sampled patches from the tracks and pass them through the trees in the Hough forest. Each patch arriving at a leaf votes into the action subspace proportional to p_c and into the space-temporal subspace of each class c according to a 3D Gaussian Parzen window estimate of the center offset vectors D_c. Votes from all patches, passed through each of the trees, are integrated into a 4D Hough accumulator. As the track has already been localized in space, we can marginalize the votes into a 2D accumulator in class label and time, with the maxima indicating the class label and temporal center of the track. For a formal description of the voting, we refer the reader to [8].

3   Combining Classifiers for Group Action Recognition

In our setting of group action recognition, we distinguish between symmetric or asymmetric interactions. Symmetric interactions are those in which all individuals perform the same movements, such as shaking hands. Asymmetric interactions, on the other hand, are those in the which the individuals behave differently. For example, when one person pushes another, there is an offender and a victim. We assume for simplicity that victims of all asymmetric actions behave in a similar way and add one generic victim class.

For each individual participating in an action, we get a single-person classification, and then combine them into group classifications using combination rules such as product rule, sum rule, min rule and max rule to strengthen the overall group response. A theoretical framework of these combination rules is given in [5]. A convenient and compact representation of multiple classifier outputs is the decision profile matrix [6] as the combination rules can be applied directly to the matrix. In the following, we review the model of decision profiles and extend them to handle both symmetric actions and asymmetric actions.

3.1   Decision Profiles

We define $c+1$ single action labels, corresponding to c group interactions and an additional victim label v. For each person l in a group interaction of L people, we have a single action classifier D_l, giving for each time instance t

$$D_l(t) = [d_{l,1}, \ldots, d_{l,c}, d_{l,v}],   \tag{2}$$

where each d corresponds to the support for a single action class. To combine the single action classifier outputs into group actions, we formulate a decision profile, DP, in matrix notation:

$$DP(t) = \begin{bmatrix} D_1(t) \\ \cdots \\ D_l(t) \\ \cdots \\ D_L(t) \end{bmatrix} = \begin{bmatrix} d_{1,1} & \cdots & d_{1,c}, & d_{1,v} \\ \cdots & & & \\ d_{l,1} & \cdots & d_{l,c}, & d_{l,v} \\ \cdots & & & \\ d_{L,1} & \cdots & d_{L,c}, & d_{L,v} \end{bmatrix}.   \tag{3}$$

For the combination of the single actions, the product, sum, min and max rule are directly applied to each column of the decision profile [6].

3.2   Extension for Asymmetric Group Actions

In our case, as we have added a victim class, we extend the above DP by dividing it into a symmetric and asymmetric block:

$$DP(t) = [DP_{sym}(t) \mid DP_{asym}(t)], \tag{4}$$

with $DP_{sym}(t)$ as defined in Equation (3), but for single action labels belonging to symmetric group interactions only. To handle the asymmetric group interactions, we consider each combination of single actions which could form the interaction. Equation (5) describes the combination for a two-person scenario, but can be easily adapted for more people. Assuming m asymmetric group actions with classifier outputs $d_{l,1}, \ldots, d_{l,m}$ and one victim class v with classifier output $d_{l,v}$, the asymmetric decision profile would be a $2 \times 2 \cdot m$ dimensional matrix defined as follows:

$$DP_{asym}(t) = \begin{bmatrix} d_{1,1} & d_{1,v} & d_{1,2} & d_{1,v} & \cdots & d_{1,m} & d_{1,v} \\ d_{2,v} & d_{2,1} & d_{2,v} & d_{2,2} & \cdots & d_{2,v} & d_{2,m} \end{bmatrix}. \tag{5}$$

While Equation (5) is a redundant representation of the single action classifications, we choose this formulation as the same classifier combination rules can be directly applied to the each column of the decision profile.

4   Experiments

4.1   Action Recognition in Low-Resolution Video

We apply the Hough-voting framework described in Section 2 to classify the actions in the UT Tower Dataset [2]. For building the tracks, we used the provided foreground masks and fit 40×40 pixel bounding boxes around the foreground blobs. To handle the low resolution of the video, we chose low-level features robust at lower resolutions [1,3], and chose greyscale intensity, absolute value of the gradients in x, y and time, and the absolute value of optical flow in x and y.

   We achieve an overall classification performance of 95.4%. The confusion matrix is shown in Figure 1. There is some confusion between similar actions, such as *standing* and *pointing*, or *wave1* and *wave2*, but all other actions are classified correctly.

4.2   Group Action Recognition

We demonstrate our approach of group action recognition on the UT-Interaction dataset [7], consisting of six classes of two-person interactions shown in profile view: *shake (hands)*, *hug*, *kick*, *point*, *punch* and *push*. We consider *shake* and *hug* as symmetric and the others as asymmetric interactions. For each class, there are two settings: *set 1* recorded from a parking lot with a stationary background and *set 2*, recorded on a lawn with some slight background movement and camera jitter.

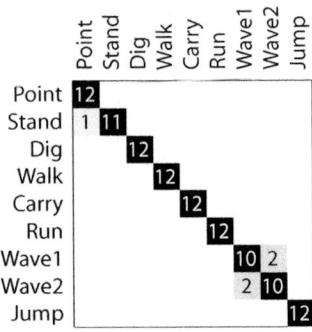

Fig. 1. Confusion matrix for classification on the UT-Tower Dataset

Single Person Actions. We use the described Hough-voting method to classify the single actions, using the same features as mentioned in Section 4.1. The tracks were built with a Hough forest trained for people detection [4] and a particle filter was used to assemble detections across time.

For simplification, only one classifier was trained for both the left and the right person; during testing, the classifier was applied to both the original and flipped version of the tracks and determined based on the higher response of the classifier if the person in the track stands on the left or right. Classification results for the seven single action classes are shown in Table 1.

Table 1. Classification performance of single actions according to track

	Set 1		Set 2	
	Left Track	Right Track	Left Track	Right Track
Shake	0.7	0.3	0.3	0.2
Hug	0.9	1.0	0.9	0.9
Kick	1.0	1.0	1.0	1.0
Point	0.8	0.63	0.6	0.6
Push	0.33	0.72	0.8	0.8
Punch	0.66	0.86	0.6	0.2
Victim	0.77	0.73	0.9	0.8
Average	0.74	0.75	0.73	0.64

Group Interactions. For evaluation of the group interactions, we use a leave-one-out cross validation for each set individually. Performance of the different combination rules are compared in Table 2. Confusion matrices of the min-rule for *set 1* and *set 2* are shown in Figures 2*(a)* and *(b)* respectively. Average performance of the best group classifier compared to the best single person classifier was higher by 13% in *set 1* and 7% in *set 2*. The min rule performs well for both sets. The product and sum rule have similar performance in both sets, but are more affected by a weaker individual classifier as is the case in *set 2* for right individual.

Table 2. Classification performance of group interactions for different fusion methods

	Set 1				Set 2			
	Min	Max	Product	Sum	Min	Max	Product	Sum
Shake	0.5	0.4	0.6	0.7	0.7	0.1	0.5	0.5
Hug	1.0	1.0	1.0	1.0	0.9	0.8	0.9	0.9
Kick	1.0	1.0	1.0	1.0	1.0	1.0	1.0	1.0
Point	1.0	0.6	1.0	1.0	1.0	0.5	1.0	1.0
Push	0.7	0.2	0.7	0.7	0.8	0.1	0.8	0.8
Punch	0.8	0.1	0.9	0.9	0.4	0.0	0.4	0.4
Average	0.83	0.55	0.87	0.88	0.8	0.42	0.77	0.77

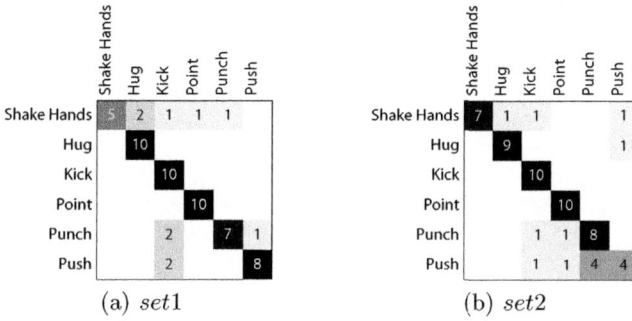

(a) *set1*   (b) *set2*

Fig. 2. Confusion matrix for classification of group actions for *(a) set 1* and *(b) set 2* using the min rule for classifier fusion

5   Discussion

The Hough-voting framework for action recognition, previously introduced in [8], was applied to two very different action recognition scenarios and showed flexibility and good results for both tasks. For classifying aerial video, we chose low-level features which were robust at low resolutions. For classifying group interactions, we presented a method for combining the classifiers of single-person actions. Overall performance was increased in comparison to single actions and the method can be easily adapted for scenarios with more than two people. A major advantage of this approach is that no additional training is needed for classifier combination.

Acknowledgements. This work has been supported by funding from the Swiss National Foundation NCCR project IM2 as well as the EC project IURO. Angela Yao was also supported by funding from NSERC Canada.

References

1. Chen, C.C., Aggarwal, J.K.: Recognizing human action from a far field of view. In: IEEE Workshop on Motion and Video Computing, WMVC (2009)
2. Chen, C.C., Ryoo, M.S., Aggarwal, J.K.: UT-Tower Dataset: Aerial View Activity Classification Challenge (2010), http://cvrc.ece.utexas.edu/SDHA2010/Aerial_View_Activity.html

3. Efros, A.A., Berg, A.C., Mori, G., Malik, J.: Recognizing action at a distance. In: ICCV (2003)
4. Gall, J., Lempitsky, V.: Class-specific hough forests for object detection. In: CVPR (2009)
5. Kittler, J., Society, I.C., Hatef, M., Duin, R.P.W., Matas, J.: On combining classifiers. IEEE Transactions on Pattern Analysis and Machine Intelligence 20, 226–239 (1998)
6. Kuncheva, L.I., Bezdek, J.C., Duin, R.P.W.: Decision templates for multiple classifier fusion: an experimental comparison. Pattern Recognition 34, 299–314 (2001)
7. Ryoo, M.S., Aggarwal, J.K.: UT-Interaction Dataset, ICPR contest on Semantic Description of Human Activities (SDHA) (2010), http://cvrc.ece.utexas.edu/SDHA2010/Human_Interaction.html
8. Yao, A., Gall, J., van Gool, L.: A hough transform-based voting framework for action recognition. In: CVPR (2010)

Author Index

GPSR Compliance

The European Union's (EU) General Product Safety Regulation (GPSR) is a set of rules that requires consumer products to be safe and our obligations to ensure this.

If you have any concerns about our products, you can contact us on ProductSafety@springernature.com

In case Publisher is established outside the EU, the EU authorized representative is:

Springer Nature Customer Service Center GmbH
Europaplatz 3
69115 Heidelberg, Germany

Batch number: 09490872

Printed by Printforce, the Netherlands